To Dr. John
my wonderful
friend who has
enriched my life
The Rab[illegible]
[illegible signature]

Also from me,
from the museum,
as Lester's side kick

[illegible signature]
1/15/2020

GEMS AND GEMSTONES }

Gems and Gemstones

TIMELESS NATURAL BEAUTY OF THE MINERAL WORLD

Lance Grande and Allison Augustyn

with photography by JOHN WEINSTEIN *and* LANCE GRANDE

{ THE UNIVERSITY OF CHICAGO PRESS }

Chicago and London

Lance Grande, Ph.D., is senior vice president and head of Collections and Research at The Field Museum, curator in the museum's Department of Geology, and author of over 100 scientific publications. He is also an adjunct professor of biology at the University of Illinois, lecturer at the University of Chicago, and board member of the Chicago Council of Science and Technology. He served as curator and general content specialist for the 2009 Grainger Hall of Gems exhibit team.

Allison Augustyn is a funding specialist at The Field Museum and was previously an exhibition developer and writer for exhibits, including the Grainger Hall of Gems, Maps, the Ancient Americas, George Washington Carver, and Nature Unleashed. She has a background in journalism.

The University of Chicago Press, Chicago 60637
The University of Chicago Press, Ltd., London

Printed in the United States of America

18 17 16 15 14 13 12 11 10 09 1 2 3 4 5

ISBN-13: 978-0-226-30511-0 (cloth)
ISBN-10: 0-226-30511-2 (cloth)

Library of Congress Cataloging-in-Publication Data

Grande, Lance.
Gems and gemstones: timeless natural beauty of the mineral world / Lance Grande and Allison Augustyn; with photography by John Weinstein and Lance Grande.
p. cm.
Includes bibliographical references and index.
ISBN-13: 978-0-226-30511-0 (cloth: alk. paper)
ISBN-10: 0-226-30511-2 (cloth: alk. paper) 1. Precious stones. 2. Gems.
I. Augustyn, Allison. II. Weinstein, John. III. Title.
QE392.G675 2009
553.8—dc22 2009015681

∞ The paper used in this publication meets the minimum requirements of the American National Standard for Information Sciences—Permanence of Paper for Printed Library Materials, ANSI Z39.48-1992.

To

JULI P. GRAINGER,

whose love for gems has led to beautiful things

CONTENTS }

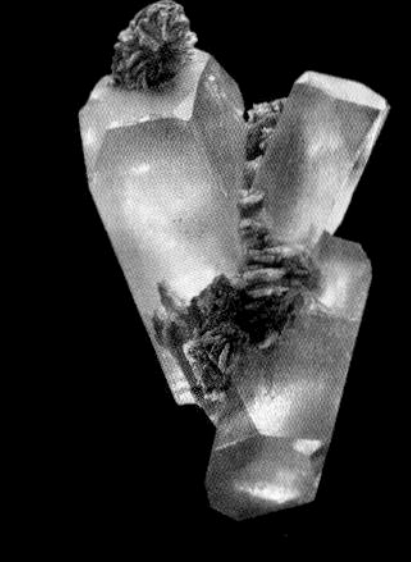

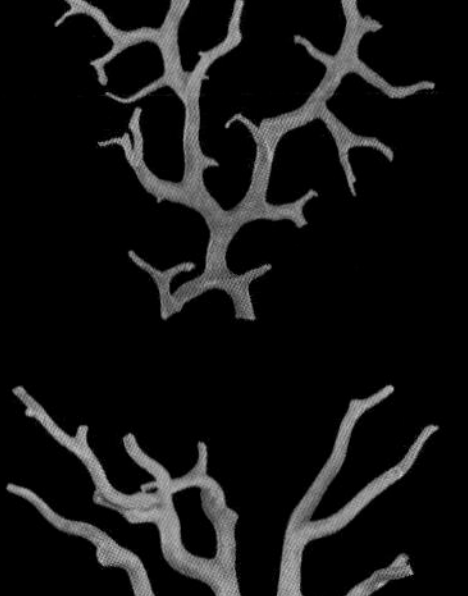

FOREWORD

Since its creation in 1893, The Field Museum of Natural History has had a hall of gems. As the museum has evolved over the past century, the gem hall has evolved with it, but with one constant over the years: the beauty and fascination of gems.

Now, in 2009, we are proud to present the new Grainger Hall of Gems, a stunning display of gems, gemstones, jewelry, and minerals from around the world and across the ages. We were able to do this thanks to generous support from The Grainger Foundation and David and Juli Grainger, longtime patrons and friends to the gem hall and the museum. As befitting a natural history museum, the Grainger Hall of Gems showcases not only scintillating cut gems and intriguing jewelry designs; it also features the natural, uncut crystals in all their amazing natural beauty. This juxtaposition of the natural form and the jeweler's art is the main thrust of the exhibition, and the inspiration for this companion volume written by Lance Grande and Allison Augustyn.

Visitors to the exhibition and readers of the book alike will be dazzled by one-of-a-kind pieces from the Hope Collection, the World Columbian Exposition, and the Royal Family of Ceylon; captivated by the "Chalmers Topaz" and the "Aztec Sun-god Opal" and the "Crane Aquamarine"; and entranced by jewelry crafted by Egyptian artisans circa 1400 BC as well as prominent designers of the twentieth and twenty-first centuries. They will also be amazed by the beauty of familiar and unfamiliar gems in their natural state, and learn how chemical processes deep in the earth can produce an aesthetic power unmatched by human craft.

There is fascination around every corner at The Field Museum, but gems have always had a special attraction for visitors to our institution. We invite you to immerse yourself in the timeless natural beauty of the mineral world within the dramatic space of the new Grainger Hall of Gems and the pages of this book.

John W. McCarter, Jr.
President, The Field Museum

PREFACE

After years of planning and development for the 2009 opening of the new Grainger Hall of Gems at The Field Museum in Chicago, we became so captivated by the project that we felt compelled to write this book. It is intended in part to be a companion piece to the exhibition, but also to be a stand-alone book about the natural beauty, diversity, classification, and significance of gems and natural gemstones. We have provided an extensive glossary at the back of this book for many of the technical terms. All of the words in this book that are set in a small ALL-CAPITALS FONT are defined in that glossary.

All of the beautiful pieces illustrated in this volume are from The Field Museum collections, and nearly all of them are on exhibit, mostly in the Grainger Hall of Gems. Catalog numbers for these pieces are provided in the figure captions as "FMNH" numbers. Most of the jewelry pieces from the exhibit illustrated in this book are uniquely designed artistic creations, and designers are identified in the figure captions wherever possible, although that information is unknown for many of the earlier pieces. Most of the pieces of jewelry designed for the Grainger Hall of Gems were made using previously unset gems from The Field Museum collection. A few pieces of jewelry in the exhibit are also anthropological treasures dating back hundreds or even thousands of years. The natural crystals, faceted gems, and jewelry illustrated in this book represent only part of what is in the exhibition hall. We therefore invite you to come to Chicago and experience the beauty and power of the gem exhibition in its entirety.

Lance Grande
Allison Augustyn
The Field Museum, Chicago
October 2009

INTRODUCTION *to* GEMS

Gem. Symbol of wealth, icon of beauty, and metaphor for the very best of anything. Gems are kept as a sign of prestige and power, and given as tokens of love and appreciation. They come in a kaleidoscope of colors and can be mineral or organic. Gems can command a person's gaze in the way they play with light and express rich color. They can evoke feelings of passion, greed, mystery, and warmth. Gems have been part of human culture for millennia; they have significant value, both economically and within folklore. But just what are gems, exactly? The U.S. Geological Survey (USGS) defines a GEM as "a gemstone that has been cut and polished," and a natural GEMSTONE as "a mineral, stone or organic matter that can be cut and polished or otherwise treated for use as jewelry or other ornament." Although some authors use the terms "gems" and "gemstones" interchangeably, we find the USGS distinction useful and use it here. In this volume we will use pieces in The Field Museum's Grainger Hall of Gems to illustrate the beauty of gems and natural gemstones, explore their natural origins, and explain the interrelationships between the different gem varieties.

The three main criteria that qualify a stone as a precious or semiprecious gemstone are BEAUTY, DURABILITY, and SCARCITY. BEAUTY is critical, because gems are forms of adornment, and their appeal is almost entirely a factor of their aesthetic qualities. People derive pleasure or even a sense of wonder from looking at true gems. DURABILITY is also important because the beauty of the stone (and what it expresses) is usually intended to be long lasting. There are many extremely beautiful MINERALS that are too soft to be practical as true gems. Abrasion from other stones or even windblown dust can quickly damage and dull the finish of a soft stone. Notable exceptions to the durability criterion are the organic gemstones (e.g., Pearls, Amber, and Coral). These items are relatively soft, yet often highly valued as gems. The third criterion to qualify as a true gemstone is SCARCITY. Of the more than 4,200 known mineral species, fewer than 40 qualify as true gemstones. Part of what gives most gems their value is the law of supply and demand. Many highly desirable gem varieties have always been in short supply, making them items of commerce or stored wealth over the centuries. The gem trade was thus born thousands of years ago and still thrives today.

Our presentation and CLASSIFICATION (fig. 12) of inorganic gems and gemstones will use two conventional categories used by mineralogists and gemologists: SPECIES (major types of gemstones based on unique chemical compositions, such as Corundum) and VARIETIES (subcategories of species identified as distinct types useful to gemologists). These two categories are often confused or used interchangeably in books on gems. We will clearly distinguish between the two here. Mineral names officially recognized as species such as "Beryl," as well as higher-level formal group names (e.g., CLASS, GROUP) are treated as proper nouns and are capitalized; variety names such as "emerald" are not. Back and Mandarino (2008: 1–256) also use capitals for mineral species names in their systemic listing section. Back and Mandarino's (2008) glossary of mineral species is used as the list of official (valid) species. Some varieties have individual names (such as "ruby" for the red variety of the species Corundum), while other varieties do not (such as the pink, blue, and yellow varieties of Corundum that are all referred to collectively as "sapphire" or "fancy sapphire").

The species are also systematically grouped into larger groups called CLASSES and SYSTEMS. The hierarchical organization of names used here creates natural groupings based on the chemical composition or origin of the gemstones. This organizational system can be illustrated as a tree and is discussed later in "The Classification of Inorganic Gems" (fig. 12). The different species of gems and natural gemstones in The Field Museum's Grainger Hall of Gems (including most known varieties of true gems) will be presented throughout this book, starting with Diamond, the hardest natural substance on earth.

Gems can be INORGANIC (of mineral origin found in rocks, or gravel derived from rocks) or ORGANIC (formed as a product or part of a living organism). We do not include inorganic minerals with a Mohs hardness of less than 6 (such as Fluorite, Rhodochrosite, or Apatite) in our list of gems because we consider them too soft to be considered true gemstones in the traditional sense. Some of these are very beautiful, but they are not generally recommended for jewelry due to their vulnerability to wear. Hardness, the Mohs scale, and several other issues of general interest will be discussed in sections below. We will also include discussions of a few organic gemstones (Pearls, Coral, and Amber) and one precious metal (Gold) because they are also featured in the Grainger Hall of Gems. All gems and gemstones illustrated in this book will include The Field

Museum catalog or lot numbers in the figure captions to facilitate interested researchers in seeking further information about individual pieces.

Our presentation of gems and natural gemstones in this book, and in the museum exhibition as well, has an overarching theme: the tie between finished gems and their natural origin. As a major natural history institution, part of The Field Museum's mission is to instill an appreciation for the beauty of nature. Gems, while beautiful as faceted and polished pieces of jewelry, are often just as beautiful or more beautiful as gemstones in their natural state. For example, an aquamarine or emerald crystal as originally found in a mine **VUG**, or pocket, with its natural crystal faces is often as beautiful as any cut-and-polished gem prepared for jewelry (fig. 110). As in so many cases, man can only attempt to mimic the intrinsic beauty of nature. So we illustrate many mineral varieties both as finished gems or pieces of jewelry and as natural gemstone specimens. The best **FACETING-GRADE** gemstones in their natural **CRYSTAL** form on pieces of natural **MATRIX** (the original piece of rock on which the crystal grew) are masterpieces of the mineral world and are sometimes more valuable than they would be as cut gems. This is because intact natural crystals are often much rarer on the world market than finished gems of the same species of mineral since most natural crystals of expensive gemstones are quickly cut into finished gems for jewelry. Attachment of the gemstone to original matrix is important for scientific reasons too: in addition to offering an aesthetic quality to a natural gemstone crystal, the matrix provides geologic clues to the environment in which the crystal developed.

BEAUTY

Among the qualities that grant a stone the title of "gem" or "gemstone," aesthetic allure trumps all. But what makes beauty? Clearly, this is a subjective characteristic that we cannot completely quantify; yet we know it when we see it. In the words of Ralph Waldo Emerson, "This love of beauty is Taste.... The creation of beauty is Art." There is beauty in nature, and there is beauty created within the scope of human artistry. When admiring raw natural gemstone crystals or cut-and-polished gems made from such crystals, we are attracted to the aesthetic and artistic features of form, color, clarity, and the reflection of light.

Form. Form is sensual, and true gems and crystals express this as no other nonliving substance. Inorganic gems and gemstones express geometric patterns of crystalline structure with flat, reflective surfaces intersecting at sharp corners. Their cold glassy beauty suggests the air of formality. In contrast, the

FIGURE 1.
Emerald as a natural gemstone (*top*) and as a cut-and-polished gem in a ring (*bottom*). The natural gemstone is from Colombia, and the largest crystal is approximately 18 carats and 17 mm high (FMNH H2345). The emerald in the ring is also from Colombia, and the faceted stone is 2.2 carats and 12 mm in height (FMNH G6752).

organic gems of Amber and Pearl show smoothly curved, free-form surfaces with a quality of soft, informal beauty.

The form of a gemstone can be beautiful both in its natural state and as a cut-and-polished gem. A perfect emerald crystal as originally excavated from a mine can be every bit as beautiful as a stone that was cut and polished by a gem cutter (fig. 1). Both the raw emerald gemstone crystal and the FACETED gem that was cut and polished from it are geometric shapes with a number of FACES (angled flat structural surfaces) that enhance the beauty of the stone. In the raw crystal, the faces are natural and their orientation is a product of Beryl's internal atomic structure that dictates the crystal's growth. In the faceted gem, the faces are carefully crafted to enhance the stone's natural beauty.

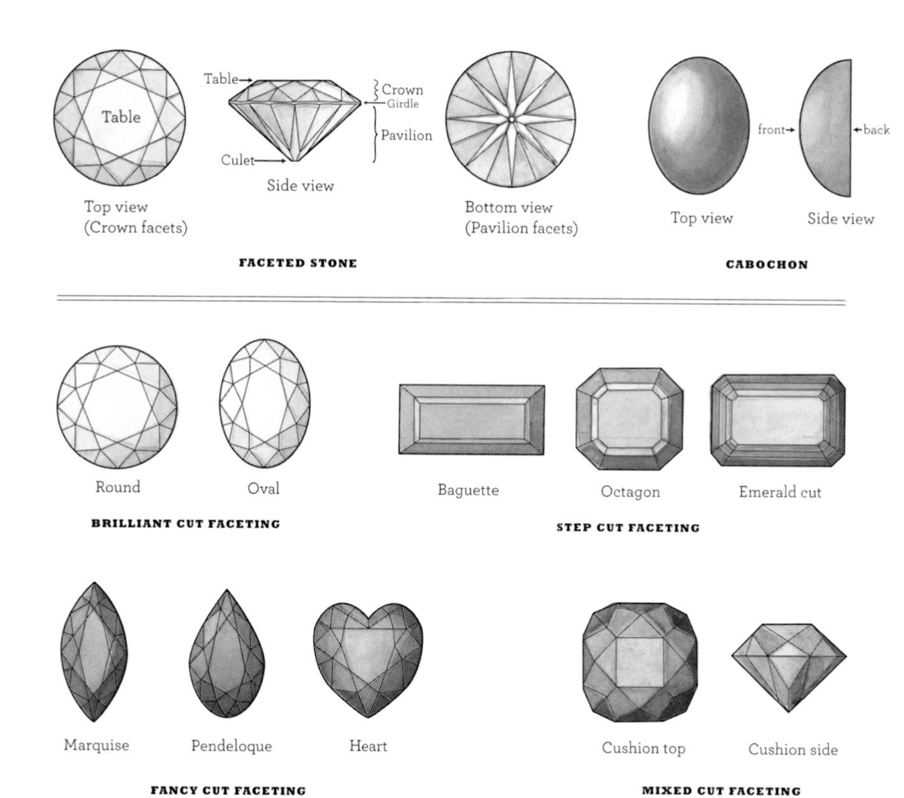

FIGURE 2. Anatomy of cut-and-polished gems, illustrating three views of a faceted stone (*left*) and two views of a cabochon (*right*).

FIGURE 3. Examples of different faceting patterns for gems. These illustrations represent but a few of the most popular faceting patterns.

To facet a gemstone for jewelry, new faces are carefully cut and strategically oriented. Generally, only TRANSPARENT gemstones (both colorless and colored) are faceted. While some would argue that it is often hard or impossible to improve on nature's beauty, naturally formed crystals are not usually the right size or shape to be set into pieces of jewelry. The facets are cut and oriented in a way to enhance a stone's best features and to facilitate the most effective use of light. To achieve this, the faceted stone is given more faces than are usually present in the natural crystal, and the facets are carefully oriented to achieve the maximum impact with play of light and reflection. The orientation of the facets will vary depending on the type of cut, and BRILLIANCE is achieved by orienting the inside surface of each FACET to reflect the maximum amount of light back through the front of the stone. The major types of faceted cuts (fig. 3) are BRILLIANT CUTS (highlighting brightness and FIRE by maximizing the amount of light reflected through the front of the stone with 58 precisely positioned facets), step cuts (highlighting the rich hues of color more than brilliance by providing a larger front window or TABLE view into the stone), fancy cuts (sometimes to make the most of an irregularly shaped or flawed gem), and mixed cuts (highlighting a combination of color and brilliance). Modern faceted cuts, such as the brilliant cut, are the culmination of hundreds of years of experimentation and development. The skill, precision, type, and quality of the faceting job are often major considerations in determining a gem's commercial value. Gems that are well cut will optimize the color or brilliance of the stone. Gems that are poorly cut will lose potential value and aesthetic appeal.

OPAQUE and TRANSLUCENT gemstones such as Amber, precious Opal, Turquoise, and star sapphires are usually made into cabochons rather than faceted stones. A CABOCHON is a stone with the surface polished into a dome (fig. 2). The back sides of cabochons are generally flat to facilitate mounting into rings or other pieces of jewelry. Where faceted cuts highlight the inner light and transparency of a gemstone, cabochon cuts emphasize the stone's surface or features just below the surface, such as PLAY OF COLOR, iridescence, INCLUSIONS, starring effects, or cat's-eye banding. In times past, before FACETING became the refined art it is today, many transparent gemstones were also cut as cabochons.

Another way gemstones are occasionally shaped is through artistic sculpturing. Nephrite jade, for example, is commonly used for carving because of its toughness, which makes small carved details resistant to breakage (fig. 4). In rare instances, PRECIOUS GEMS are also carved. Any gem can be carved with the use of fine tools and Diamond abrasive, which is harder than any other mineral. The most difficult gem to carve is Diamond, because there is nothing harder with which to carve it. Yet even Diamond can be carved using great skill and patience. This is done using fine burrs and dental drill-like tools with tips of hardened steel charged with Diamond dust in oil, and can take years to complete. Carved Diamonds are quite rare, given the time and risk involved in sculpting a Diamond with anything other than flat facets. Sculpting a complex image on a Diamond must be done with patience and skill. One such example is a sculpted Diamond in a stickpin with a portrait of King William III of Holland. This piece (figs. 5, 6), which was given to The Field Museum in 1894, reportedly took the artist five years to carve.

Color. Color gives both distinctiveness and character to a gem. Some gems are the very essence of color, and gemstones that most richly display certain colors are among the most coveted. What could be redder than a fine ruby or greener than a fine emerald? For that reason, color is the single most important factor in grading and determining the value of colored gems. The color of a gem is evaluated based on its HUE, TONE, and SATURATION. HUE is the particular color shade or shades, TONE is the relative lightness or darkness of color, and SATURATION is the intensity of color. Gray and brown MODIFIERS or overtones often decrease the value of a gem.

The color of a gem results from the way it absorbs and reflects white light.

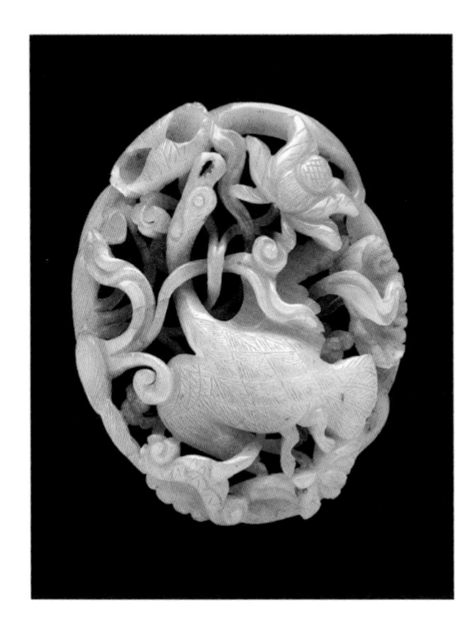

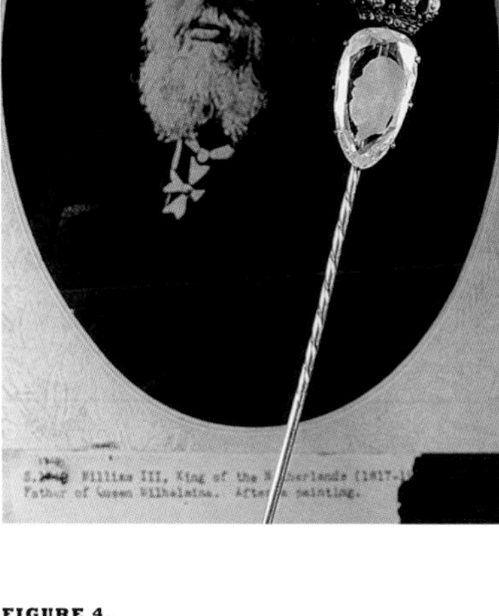

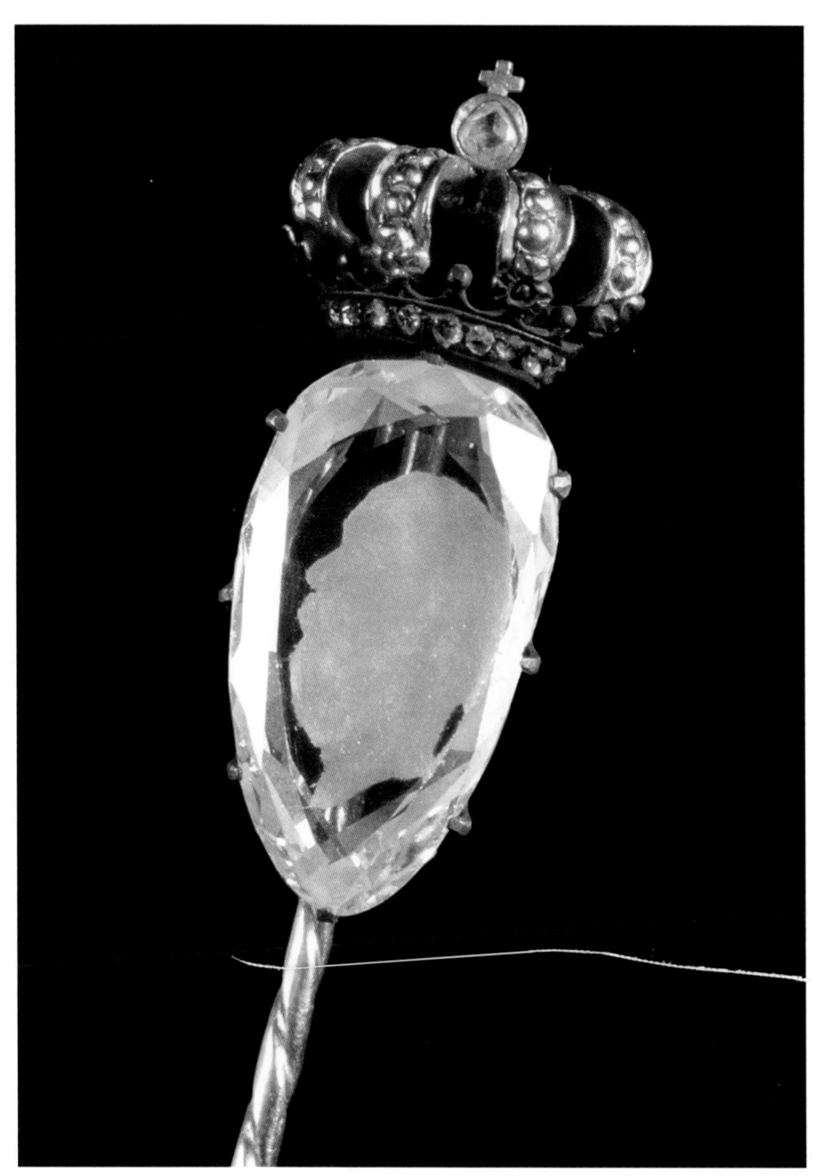

FIGURE 4.
Top left: Nephrite carving showing the sturdy fine detail made possible by the stone's natural toughness. From the Yuan dynasty, fourteenth-century China, and measuring 100 mm from top to bottom (FMNH A116646).

FIGURE 5.
Bottom left: A 2.1-carat colorless Diamond mounted on a Gold stickpin. The shallow-cut pear-shaped Diamond, measuring 15 mm in height, has the portrait of William III of the Netherlands (*pictured behind*) engraved on the table facet. It was made in the late nineteenth century by Diamond cutter M. C. M. de Vries of Amsterdam, who spent five years doing the engraving (FMNH H15).

FIGURE 6.
Right: Close-up of the engraved Diamond from the stickpin in fig. 5, showing the details of the engraving and the delicate Gold and Diamond crown.

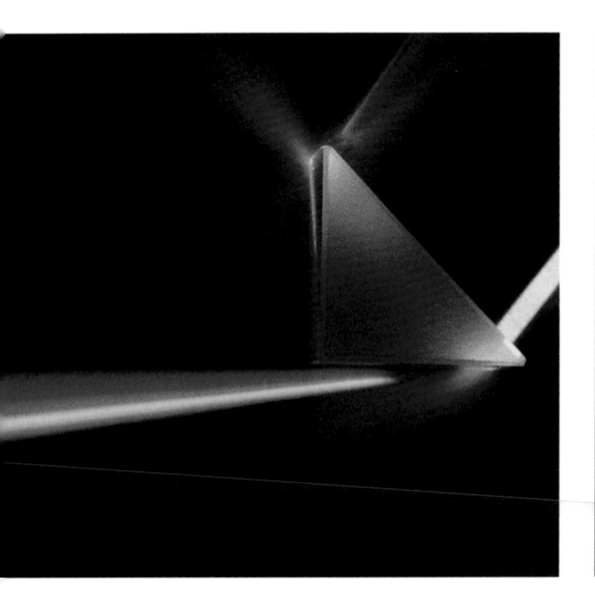

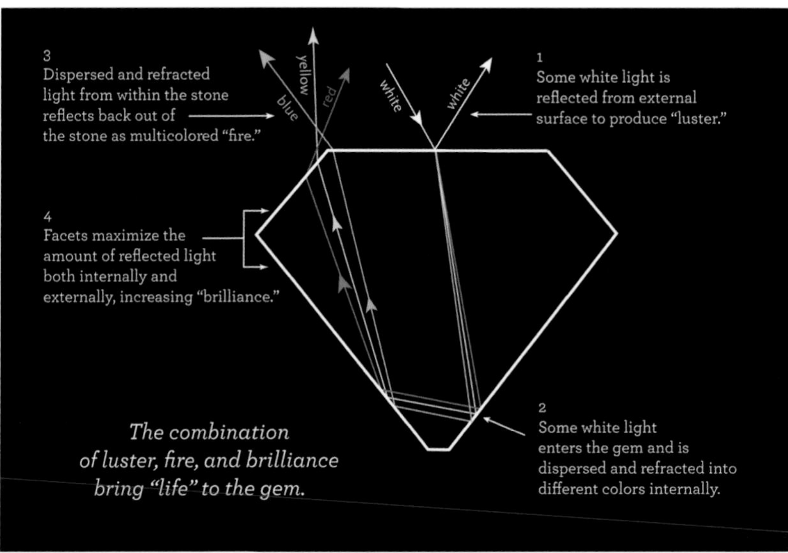

FIGURE 7.
Left: White light broken into a spectrum of colors after passing through a crystal prism. Photo by Don Farrall / Getty Images. *Right:* Diagrammatic illustration of luster, fire, and brilliance—three properties that bring life to a gem.

White light is a combination of all visible colors of light. Shine a beam of white light through a prism or sunlight through a rainstorm and the white light is broken down into its components as a rainbow (fig. 7). When light hits a colorless or white stone, no individual colors are absorbed by it, so we see the complete spectrum of white light reflected back to our eyes. When light strikes a COLORED STONE, the gem absorbs some colors of the spectrum while the rest are reflected back. The colors that are reflected back are the colors that we see. When light hits a black stone, all colors of visible light are absorbed by the stone, and we see only black.

The differences in color usually result from the different elemental compositions of the gemstone varieties. For example, many gemstone species such as Corundum and Beryl are colorless if they are absolutely pure; but if certain minute trace elements or impurities are present when the crystal forms (perhaps as little as 100 parts per million), those elements can create different colored varieties of gemstones entirely. These trace elements usually consist of metals, such as Iron (Fe), Chromium (Cr), Cobalt (Co), Copper (Cu), Manganese (Mn), Titanium (Ti), Nickel (Ni), or Vanadium (V). If a crystal of Corundum contains trace amounts of Chromium, then it is generally red in color and called a ruby. If a crystal of Corundum contains trace amounts of Titanium and Iron, then it is blue in color and called a sapphire. With gemstone varieties such as ruby and sapphire where the color is dependent only on trace elements and impurities, the color is called ALLOCHROMATIC. In contrast, gemstone varieties such as peridot and Almandine in which color is derived primarily from elements that are major parts of their essential chemical composition are said to have IDIOCHROMATIC color. Normally allochromatic species (such as Corundum,

Beryl, Elbaite, and Quartz) have more color variation and varieties than idiochromatic species (Forsterite). Beryl, depending on which trace elements are present, can be rich green (emerald), blue (aquamarine), pink (morganite), red (bixbite), yellow (heliodor), or colorless (goshenite). Forsterite gem varieties (Magnesium Iron Silicate), such as peridot and chrysolite, are always green or greenish-yellow, because the color is a product of the Iron in its basic chemical structure. Individual stones of peridot, Almandine, and other idiochromatic varieties tend to be uniformly colored throughout the body of the stone. Allochromatic varieties, such as amethyst and blue sapphire, are often unevenly colored, with light and dark banding or zones of colorlessness.

Some gems have desirable ZONATION OF COLOR, such as bicolor and watermelon Tourmalines; some gems have mixed-color hues, such as the purple-blue in tanzanite or the pinkish-orange in padparadscha Corundum; and some gemstones even change color in different types of light, such as alexandrite, which is greenish in sunlight and reddish-brown in incandescent light. Some gems, like precious Opals (also called noble Opals), have what is referred to as PLAY OF COLOR, which appears as a display of blazing colors. Gems with bright, intense color tend to be more highly valued than those that are extremely dark or light, but when it comes to color, the standard of beauty is also partly a matter of personal taste.

Clarity. Clarity speaks to the purity of substance. For a faceted gem to approach the grade of "flawless," it must also approach absolute clarity at 10× magnification. The clarity of gems is a critical aspect of beauty for transparent stones (although ranking second to color in importance for most colored stones). Clarity is assessed by the quantity of visible inclusions, fractures, or other internal defects that affect the transparency of the stone. INCLUSIONS are sometimes smaller crystals or foreign bodies that in some cases formed before the host gemstone did, generally with a composition different from the host. The host stone grew around the inclusions and eventually enclosed them entirely. In other instances inclusions, such as fine Rutile in sapphire, form slowly over GEOLOGIC TIME as the host mineral cools. Some flaws in gemstones are not inclusions at all but are instead internal cleavages, cracks, or fractures that formed after the host finished growing. Glass-clear, without visible inclusions or other distracting flaws, is considered to be the ideal for most transparent gem types. The standard terms for high grades of clarity are EYE-CLEAN, meaning there are no inclusions visible to the naked eye (without magnification), or CLEAN, meaning there are no inclusions visible even under 10× magnification (the highest grade of clarity). For some colored varieties of gemstone, perfect clarity is rarely, if ever, seen. For example, some varieties of Beryl such as aquamarine and morganite are often found as eye-clean gems,

CLARITY GRADE	CHARACTERISTICS
Flawless (FL)	*No inclusions or surface blemishes are visible to a skilled grader using 10× magnification.*
Internally Flawless (IF)	*No inclusions and only surface blemishes are visible to a skilled grader using 10× magnification.*
Very, Very Slightly Included (VVS1 and VVS2)	*Inclusions are difficult for a skilled grader to see under 10× magnification.*
Very Slightly Included (VS1 and VS2)	*Inclusions are clearly visible under 10× magnification but can be characterized as minor.*
Slightly Included (SI1 and SI2)	*Inclusions are noticeable to a skilled grader using 10× magnification.*
Included (I1, I2, and I3)	*Inclusions are obvious under 10× magnification and may affect transparency and brilliance.*

{ TABLE 1 } *The GIA Clarity Scale for Colorless Diamonds.* The Clarity Grading Scale developed by the Gemological Institute of America (GIA) is the most widely accepted grading system for Diamonds in the world. Because the differences between some of the categories are somewhat subjective, it takes a specialized and experienced gemologist to grade a Diamond precisely for clarity. The GIA Clarity Scale contains 11 grades, with most Diamonds falling into the VS (very slightly included) or SI (slightly included) categories. In determining a clarity grade, the GIA system considers the size, nature, position, color or relief, and quantity of clarity characteristics visible under 10× magnification.

but other Beryl varieties such as emerald and bixbite almost always contain substantial inclusions. For this reason, a clean emerald with ideal color is today more valuable than colorless Diamond, and one of the rarest, most valuable gemstones on earth.

There are different grading systems for Diamonds and colored gems. In 1953 the Gemological Institute of America (GIA) developed a clarity grading scale for Diamonds that is now the most widely accepted grading system in the world (table 1). This grading scale was developed particularly for colorless Diamonds, the most common of all precious gems and by far the largest component of today's international gem trade. This scale is not applicable to colored gems because many colored varieties of gems tend to have more natural inclusions than do high-quality Diamonds. Even the highest-quality grades available for some colored varieties will have at least some inclusions. Consequently, the GIA also developed a separate grading system for colored gems (table 2). Grading colored gems is further complicated because each variety must be graded within its own qualitative context. Because some varieties such as aquamarine and Topaz are commonly found free of inclusions, while others

CLARITY CATEGORIES	TYPE I	TYPE II	TYPE III
Eye-clean	*The stone appears clean to the unaided eye.*	*The stone appears clean to the unaided eye.*	*The stone appears clean to the unaided eye.*
Slightly included	*Minute inclusions difficult to see with the unaided eye.*	*Minor inclusions somewhat easy to see with the unaided eye.*	*Noticeable inclusions apparent to the unaided eye.*
Moderately included	*Minor inclusions somewhat easy to see with the unaided eye.*	*Noticeable inclusions apparent to the unaided eye.*	*Obvious inclusions very apparent to the unaided eye.*
Heavily included	*Inclusions are prominent and have a negative effect on appearance or durability.*	*Inclusions are prominent and have a negative effect on appearance or durability.*	*Inclusions are prominent and have a negative effect on appearance or durability.*
Severely included	*Inclusions are prominent and have a severe effect on appearance, durability, or both.*	*Inclusions are prominent and have a severe effect on appearance, durability, or both.*	*Inclusions are prominent and have a severe effect on appearance, durability, or both.*

{ TABLE 2 } ***The GIA Colored Stone Grading System.*** Because clarity is less of a factor for many colored stones, the GIA has a separate grading system for them. As with colorless Diamonds, grading clarity of colored stones requires a qualified and experienced gemologist. See table 3 for definitions of clarity type.

such as emerald and red Tourmaline are almost never free of inclusions, the GIA created a "type" classification for colored gem varieties indicating the relative scarcity or RARITY of inclusion-free stones for each variety (table 3). Grading standards of clarity are much more severe for Type 1 gem varieties (varieties that are often inclusion-free) than for Type 3 varieties (varieties that almost always contain very noticeable inclusions). "Superior clarity" in aquamarine or Topaz is not the same as "superior clarity" in emerald.

Inclusions are not always undesirable in gems. The best example of this is Corundum. When a sapphire is heavily included with extremely thin fibrous inclusions of the mineral Rutile, three intersecting bands of light cross the surface of the stone creating a six-rayed STAR effect. This gem variety is called star sapphire, and it is both desirable and valuable (see fig. 34). Rubies with similar inclusions are star rubies (fig. 28). The starring effect is called ASTERISM. Four-rayed, six-rayed, and, rarely, twelve-rayed, or even twenty-four-rayed stars also occur. Similarly, gems such as cat's-eye Chrysoberyl have a single band of light crossing the stone, creating an "eye" effect (see fig. 55). The CAT'S-EYE EFFECT is called CHATOYANCY. Organic gemstones also sometimes have desirable inclusions, most notably insect inclusions in Amber (see chapter on Amber and fig. 267).

TYPE	RELATIVE CLARITY	EXAMPLES (BY VARIETY)
Type I	*Transparent stones are usually eye-clean with no inclusions visible to the unaided eye. The stones in this type are usually of such high clarity that even minor inclusions can detract from their desirability for use in jewelry.*	*blue Topaz, kunzite, triphane, tanzanite, aquamarine, pale green Beryl, heliodor, morganite, goshenite, all colors of Chrysoberyl except alexandrite, blue or colorless Zircon, green Tourmaline*
Type II	*Transparent stones typically show some eye-visible inclusions that don't detract from the gem's overall beauty. Many stones with inclusions visible to the unaided eye are faceted for use in jewelry.*	*amethyst, citrine, alexandrite, peridot, Quartz, Garnet, ruby, all colors of sapphire, iolite, Spinel, all colors of Tourmaline except green or watermelon, Zircon other than blue or colorless, Topaz other than blue, fire Opal, transparent Feldspar*
Type III	*Transparent stones almost always show eye-visible inclusions, but even specimens with obvious or prominent inclusions are often faceted for use in jewelry.*	*emerald, bixbite (red Beryl), and watermelon Tourmaline*

{ TABLE 3 } ***GIA Clarity Type Classification for colored gem varieties, based on GIA criteria.*** Classification for colored gem varieties, based on GIA criteria. This applies only to transparent inorganic minerals that normally appear in the gem trade as faceted stones.

Reflection of light. Beauty is greatly enhanced by the way a gem interacts with light, both externally and internally (fig. 7, *right*). When light strikes a gem, it can be transformed into sparkle, fire, and brilliance, bringing the gem to life. In transparent gems, light is reflected from both the external surface of the gem and from internal surfaces within the gem. Reflected light from the external surface includes LUSTER. Luster can range from hard and highly polished or ADAMANTINE (as in Diamond) to glassy or VITREOUS (as in most other transparent faceting-grade gemstones) to WAXY (as in Turquoise) to METALLIC (as in Gold). The amount of light reflected as luster varies depending on the variety, transparency, and polish of the gem. In polished, colorless gem-grade Diamond, for example, luster plays only a small part of the total optical effect, with about 17 percent of the light reflected from the surface of the gem and about 83 percent from within the gem. In contrast, an opaque black Diamond or Turquoise has about 100 percent of the reflected optical effect coming from the surface of the stone, and little or no optical effect reflected from within the gem.

Some reflected light in transparent gems emanates from inside the gem. The facets of a gem are placed in such a way that light entering the gem's upper surface reflects off the inside surfaces of the gem's facets. This light is eventually projected back out through the upper surface of the gem toward the eye of the viewer, adding to the gem's brilliance. Light reflected from within a gem can be specially transformed and aesthetically enhanced by REFRACTION and DISPERSION. REFRACTION is the bending of light once it enters the gem. The strength of refraction is quantified by a numerical scale called a REFRACTIVE INDEX. The refractive index ranges from near 0 in air, to 1.3 in water; among common gemstones, it is highest in Diamond, where it is 2.4. Because refraction is a function of the wavelength of light, it bends each color of the light spectrum at a slightly different angle; blue, for example, is bent more than yellow, and yellow more than red. The result of this is called DISPERSION of light, which is the splitting of a single white beam into a rainbow of colors (fig. 7). In effect, the facets of a white Diamond disperse and reflect light at many different angles, creating a myriad of rainbows within the interior of the stone. The resulting flashes of multicolored light we see within the stone are what we call FIRE. The intensity and frequency of flashes of white and colored light are called sparkle, or BRILLIANCE. Diamonds and other highly valued colorless stones are faceted in such a way as to internally reflect the maximum amount of light back through the front of the stone to project the most sparkle and fire. This ultimately gives the gem its brilliance and LIFE.

* * *

With all of these measurements and terms, can we now better define the beauty of gemstones? In some ways, yes. The physical traits of form, color, clarity, and reflection of light have been basic categories of evaluation for centuries. But the details of exactly what is most desirable and what is least desirable have changed over the course of human history. Tens of thousands of years ago, necklaces and bracelets were made from shells, bones, and teeth, possibly indicating a sense of aesthetics now foreign to Western culture. Black Tourmalines were highly prized gems used for mourning jewelry during the Victorian era in Britain, but today they have little or no value as gemstones. Gem industry standards aside, elements of taste and value evolve from generation to generation through a process of ever-changing social customs and commercial marketing. And after all is said and done, beauty is more than simply a calculation of physical traits; it is also emotional. In order to appreciate or express beauty, you have to feel it. No one can completely quantify what beauty is, although form, color, clarity, and reflection of light are clearly important. In the end, the complex beauty of a gem or natural gemstone is truly in the eye of the beholder.

DURABILITY

In order for a beautiful stone to qualify as a true gem, it must be durable. Durability is what allows the beauty of a gem to be long lasting, and long-lasting beauty gives a gem the license to be a symbol of enduring power, commitment, or commemoration. So how do we measure the durability of a gemstone? To answer this, we must examine two different properties of durability: HARDNESS and TOUGHNESS.

HARDNESS is a measure of durability based on a mineral's ability to resist scratching and abrasion. This is important in terms of preserving clarity, luster, sparkle, and other desirable properties involving the reflection of light from a gemstone. Hardness also preserves a smooth clear TABLE (the top window-like surface of a faceted stone), which helps reveal the inner richness of colored stones with good clarity. Gems should be harder than materials that they will come into contact with in order to maintain their smooth lustrous surfaces over a long period. The MOHS SCALE OF HARDNESS, devised by German mineralogist Friedrich Mohs in 1822, is the scale by which hardness is measured in minerals (table 4). This relative scale uses ten species of minerals as standards of increasing hardness. These minerals in order of increasing Mohs hardness are Talc (1), Gypsum (2), Calcite (3), Fluorite (4), Apatite (5), Orthoclase (6), Quartz (7), Topaz (8), Corundum (9), and Diamond (10). This means that Gypsum can scratch Talc, Calcite can scratch Gypsum and Talc, and so on up the scale to Diamond, which can scratch all other known minerals.

There are intervals between the values, known as half values, although they are not absolute values. For example, a Mohs value of 5½ is not necessarily *exactly* halfway between Mohs 5 and Mohs 6; it is simply *somewhere* between 5 and 6. Some simple spot tests for hardness include fingernails at 2½ (fingernails will scratch Mohs 1 and 2), Copper at 3½ (Copper will scratch Mohs 1–3), and window glass at 5½ (window glass can be scratched by Mohs 6–10).

At the bottom of the Mohs scale is Talc (Mohs 1). It is the softest mineral and is well known for its use as a skin powder (talcum powder). On the upper end of the scale is Diamond, the hardest known natural mineral in the universe. Only another Diamond can scratch a Diamond. This is why Diamonds need less repolishing than any other gem: they hold their polished surface and will not be scratched during normal wear. The Mohs scale is a relative scale rather

MOHS STANDARDS	GEMSTONE SPECIES AND GROUPS DISCUSSED IN THIS BOOK
Talc (1)	
(Between 1 and 2)	*Graphite (1-2)*
Gypsum (2)	
(Between 2 and 3)	*Amber (2½), Gold (2½)*
Calcite (3)	*Pearls (3), Coral (3)*
(Between 3 and 4)	
Fluorite (4)	
(Between 4 and 5)	
Apatite (5)	
(Between 5 and 6)	*Turquoise (5-6)*
Orthoclase (6)	*Opal (5½-6½), Orthoclase (6)*
(Between 6 and 7)	*Zoisite (6-6½), Forsterite (Olivine) (6½-7), Actinolite (nephrite) (6-6½), Andradite Garnet (6½-7), Spodumene (6½-7), Benitoite (6½), Jadeite (6½-7), Albite (6-6½), Albite-Anorthite (6-6½)*
Quartz (7)	*Quartz (7)*
(Between 7 and 8)	*Beryl (7½-8), Tourmaline (7-7½), Pyrope Garnet (7-7½) Almandine Garnet (7-7½), Spessartine Garnet (7-7½), Grossular Garnet (7-7½), Uvavorite Garnet (7-7½), Phenakite (7½-8), Zircon (7½), Cordierite (7-7½)*
Topaz (8)	*Topaz (8), Spinel (8)*
(Between 8 and 9)	*Chrysoberyl (8½)*
Corundum (9)	*Corundum (9)*
(Between 9 and 10)	
Diamond (10)	*Diamond (10)*

{ TABLE 4 } *Mohs scale of hardness, based on one mineral's ability to scratch another.*

than an absolute one, and it emphasizes convenience of application rather than precise measurements of the intervals between the 10 Mohs values. The intervals between values at the top of the scale are much larger than the intervals between values at the bottom of the scale. For example, Mohs 9 (Corundum) is twice as hard as Mohs 8 (Topaz), but Mohs 10 (Diamond) is four times as hard as Mohs 9 (Corundum). In fact, the absolute difference in hardness between Corundum (Mohs 9) and Diamond (Mohs 10) is about three times as much as the difference between Talc (Mohs 1) and Corundum (Mohs 9).

There is variation in Mohs hardness for many mineral species. This variation is due to a number of factors, including (1) different amounts of impurities in different samples of the mineral species being tested; (2) different hardness along different crystal faces of the crystal being tested; and (3) possibly even variations in the scratching tools being used to measure the hardness. Also, some minerals have additional factors contributing to variation in hardness, like some Zircon gemstones, where deteriorating radioactive impurities within the crystal structure can soften the stone over the centuries. Consequently, Mohs hardness values are frequently given by some authors as ranges for species (emerald as Mohs 7½–8, for example).

Mohs hardness is also a consideration in choosing an abrasive powder for polishing gems. Corundum grit can be used to polish gems that are softer than Mohs 9, such as Tourmaline or Beryl, but Corundum cannot scratch or abrade Diamond in any way and is useless for polishing Diamond. Because of Diamond's extreme hardness, it has many important industrial uses. DIAMOND GRIT is the best polishing agent for finishing faceted gems, and Diamond granules and chips are key elements of blades and bits for cutting and drilling metals, rocks, glass, masonry, concrete, and other hard materials. Of all Diamonds mined, only about 20 percent are used for gems. The rest are lower-grade materials used as cutting and polishing abrasive for industry.

What is a suitable hardness for a gem? A hardness of 7 or greater is desirable, because Quartz (Mohs 7) is one of the most common minerals in nature. Quartz is all around us: fine Quartz sand can be a component of windblown particles and dust that can rapidly wear away and dull the polished surface of softer minerals worn as jewels. This is one reason why minerals like Fluorite (Mohs 4), Rhodochrosite (Mohs 4), and Apatite (Mohs 5), although beautiful as faceted stones, do not make good gems for jewelry. Faceted gems generally rely on maintaining their highly polished glassy faces to maintain their beauty. Organic gems are an exception to this convention. Because organic gems are not generally faceted, usually have curved or irregular surfaces, and are not always transparent, the wear and abrasion on these gemstones are usually far less visible. Thus, even though these gems are relatively soft, their beauty can be long lasting. Popular examples of organic gemstones include

Pearls (Mohs 3) and Amber (Mohs 2½). Organic gems and softer inorganic gems such as precious Opal (Mohs 5½–6½) and Turquoise (Mohs 5–6) require extra care while being worn to ensure their preservation and lasting beauty. They are particularly vulnerable to damage when worn in rings, where there is danger of immersion in hot water, a bump against a hard surface, or even regular exposure to wind and outdoor elements.

The hardness of a gemstone or crystal is determined by its chemical bonds (the attraction of its atoms to each other). The stronger the binding force between the atoms of the gemstone, the harder the stone is. Different types of chemical bonds can result in profoundly different minerals, even for the same elemental composition. A prime example is Diamond (Mohs 10) and Graphite, or "pencil lead" (Mohs 1 or 2). These two minerals are of the same exact elemental composition, pure Carbon; but the Carbon atoms are bonded together very differently. The Carbon atoms in Graphite are bonded together in one-atom-thick planes or sheets that are stacked on top of one another. Although the lateral bonds within the sheets are extremely strong, the forces between these sheets are very weak. The sheets slip over one another and come apart easily when Graphite is drawn on a surface, leaving a gray streak such as a pencil line on paper. The slippery relationship between the planes of atoms is why Graphite feels "soft" and why it is used as a lubricant in motor oil. In contrast, the Carbon atoms of a Diamond are all bonded to each other as a single tightly bound structure, rather than in individual sheets. A Diamond is, in fact, a huge single molecule of Carbon.

As the hardest natural substance in the universe, Diamonds are not easy to cut. Even with Diamond dust on the saw blade, the cutting, shaping, and faceting of a Diamond are time-intensive processes. Fortunately, Diamond has perfect cleavage. CLEAVAGE is the property of a mineral to break cleanly along a flat plane of structural weakness that results in a smooth, flat surface. Cleavage is related to the mineral's atomic structure and the differential forces of the bonds holding the atoms together. Diamond has cleavage in four different directions. Carefully using a sharp chisel and hammer to deliver a sharp blow in the direction of cleavage will produce a clean break that can save much time in the process of cutting and polishing. CLEAVING Diamonds is a skilled job that is entrusted only to highly trained specialists. The resulting cleaved surface must still be polished, and many additional facets must still be ground into the stone to produce the finished gem. When setting a Diamond in a piece of jewelry, a good jeweler or Diamond setter will orient a Diamond in such a way that the cleavage planes are not in vulnerable directions that would allow accidental cleaving from a bump or knock against a hard surface. Homogenous minerals without cleavage break along curved surfaces and are said to have CONCOIDAL FRACTURE; these include glass and Quartz.

FRACTURE TOUGHNESS	ROCK TYPES (WITH MOHS HARDNESS STANDARDS)
Poor	*Topaz (8), Talc (1), Gypsum (2), Fluorite (4), Spodumene (6½–7), Orthoclase (6–6½), Albite (6–6½), Albite-Anorthite (6–6½)*
Poor to fair	*Zoisite (6–6½), Zircon (7½), Opal (5½–6½), some Calcite (3), Benitoite (6½)*
Fair	*Tourmaline (7–7½), Apatite (5), Pearls (3)*
Fair to good	*Forsterite (Olivine) (6½–7), Garnet (6½–7½), some Calcite (3), Turquoise (5–6)*
Good	*Quartz (7), Beryl (7½–8), Spinel (8), Cordierite (7–7½)*
Good to excellent	*Diamond (10)*
Excellent	*Corundum (9), Chrysoberyl (8½)*
Excellent to exceptional	*Jadeite (6½–7)*
Exceptional	*Actinolite (nephrite) (6–6½)*

{ **TABLE 5** } ***Scale of toughness based on resistance to fracture.*** From Solenhofen (2003).

TOUGHNESS, also referred to as fracture toughness, is the other measure of durability besides hardness. Toughness is the ability to resist shattering, chipping, or fracture upon impact. It is the opposite of brittleness. Although Diamond is the *hardest* natural substance, it is still brittle. Drop a Diamond on a hard floor or hit it with a hammer, and it can shatter into many pieces. Knock it against a hard surface by accident, and it can chip or even crack. Thus, the most common damage to a Diamond is chipping or cracking, rather than scratching or wear.

The toughest natural substance is nephrite jade. Nephrite is so tough that historically it has been used to make ornate carvings, ax heads, and hammer heads. Yet nephrite jade is relatively soft (Mohs hardness of 6–6½). A Diamond will easily scratch a piece of nephrite, but smash a piece of nephrite with a Diamond, and it is the Diamond that will shatter, leaving the nephrite intact. The toughness of nephrite is due to its internal microstructure of fiber-like or needle-like crystals arranged in interlocking, randomly oriented bundles. Unlike the Mohs scale, the scale of toughness is expressed in qualitative rather than quantitative terms, ranging from poor to exceptional. Table 5 displays a general summary of mineral toughness for some of the minerals discussed here.

Of the two durability criteria, hardness is generally considered the most important for precious gems. Although brittle gemstones like Topaz and even Diamond can be difficult to cut, facet, and polish, once finished for a piece of jewelry, it is their hardness that best helps gems maintain their beauty over time.

SCARCITY

In order for a stone that is beautiful and durable to qualify as a precious gem, it must also be scarce. Scarcity itself does not guarantee that a stone will qualify as a precious gem, but if a stone is both scarce and popular, it eventually becomes precious and valuable. It is largely an issue of supply and demand; the more that the demand outstrips the supply, the scarcer an item becomes. There are two primary ways that gems can be scarce: TRUE RARITY and MARKET MANIPULATION.

Stones with TRUE RARITY fall into two main categories: those that are OVERALL RARITIES and those that are CONDITION RARITIES. Gems with true overall rarity include bixbite (red Beryl, sometimes also called red emerald), one of the rarest gems on earth. Bixbite comes primarily from one small, recently discovered area in the Wah Wah Mountains of Utah and is thousands of times more rare than Diamond. Ironically, bixbite is so rare that its own scarcity prevents it from being more expensive than it is already. Because there has never been a large enough supply of fine bixbite, the gem cannot support a major marketing campaign to enhance the stone's popularity. Even so, in 2008 fine faceted stones of 1 carat or greater were selling for $5,000 or more per carat. Though roughly the same price as emeralds, this is a relative bargain, because gem-quality red Beryl is hundreds of times more rare than emerald. On the other hand, some truly rare minerals do not have significant value because of a lack of popularity. Phenakite, for example (p. 140), is a hard (Mohs 7½–8), very scarce, and bright, colorless stone that facets well. Unfortunately, it resembles clear Quartz too closely to be popular as a rare gem.

The overall rarity of particular gem species can change markedly over time. Colorless Diamonds, for example, were very rare in the thirteenth century, when they were reserved mainly for royalty; but now they are relatively common and their high value is sustained mainly by controls on distribution and massive advertising campaigns. In 1652 fine amethyst was rare in Europe and was as valuable as Diamond. Prices for amethyst remained very high until Civil War times, when Brazilian amethyst started to be imported in large quantities, causing it to eventually become much more common and practically worthless. Today amethyst remains one of the most beautiful colored stones of the gem world, but it is not of significant value because of its extreme abundance and availability. Another example is tiger's eye, which was very popular and scarce in the United States from 1880 to 1890, when it sold for $6 per carat. Then a huge deposit was found in South Africa, and within a few years, the supply sent the value from $11,200 per pound to 25 cents per pound.

So what does increased abundance over time do to a gem like amethyst or tiger's eye? It shifts it from the category of PRECIOUS GEM to SEMIPRECIOUS GEM or to COMMON GEM. We can subjectively classify gems in three categories based on scarcity and value (whether due to true rarity or to market manipulation). PRECIOUS GEMS traditionally have included varieties such as white and colored Diamonds, ruby, emerald, and sapphires, although there are more gems that belong in this category today because of their extreme rarity and market popularity. SEMIPRECIOUS GEMS traditionally have included varieties such as Tourmaline, Garnet, aquamarine, and Topaz, although some varieties of Garnet such as fine demantoid or Grossular and varieties of Tourmaline such as flawless rubellite, indicolite, or cuprian Elbaite should now be in the precious category because of rarity and modern valuation. COMMON GEMS have included commonly available varieties such as amethyst, citrine, agate, tiger's eye, and jasper. These three categorizations are very subjective and not particularly useful today because they do not always adequately reflect relative scarcity or market value, particularly considering the issue of condition rarity.

CONDITION RARITIES are gemstones that are exceptional examples of their variety. For example, most of the highest-quality emeralds on the market have significant inclusions because that is normal for emeralds. Color is the key for grading normal emeralds, and a moderate number of inclusions are normally acceptable. A flawless, inclusion-free emerald of ideal color and carat weight of over 2 carats is so rare that it is often completely unavailable. Such an emerald is an extremely valuable condition rarity and worth many times that of a typical gem-quality emerald. Other gems that are condition rarities when found as sizable inclusion-free pieces with good color include rubies, red Beryl, and watermelon Tourmalines.

Another way gems can be scarce is through MARKET MANIPULATION. The best example of this is Diamond, the world's best known and most successfully marketed gem. Small (fewer than 2 carats) white gem-quality Diamonds are today very common. How many people do you know personally who have a Diamond in an engagement ring or other piece of jewelry? Extrapolate that to the over 150 million women of 18 years of age or older in the United States. Even if only half of that number have Diamonds, that is a lot of Diamonds. If you extrapolate further to include the entire population of the world, the number of existing Diamonds in jewelry today is staggering. In addition, there are huge deposits of Diamonds around the world yet to be mined, providing an ample supply for new stones well into the future. In contrast, gems like tanzanite (known only from a few mines in Tanzania), bixbite (known primarily from a small region of Utah), and Benitoite (known only from a small mine in San Benito County, California) are far less common and appear to have very limited minable deposits. In spite of the abundance of gem-quality Diamonds, De Beers has done a masterful job of promoting and advertising them while at the same time tightly controlling their distribution for over 100 years. Demand for Diamonds has been maintained by De Beers' highly successful marketing campaign, and by a history of carefully limiting the release of Diamonds into the marketplace and selling only to strategically selected suppliers. A global campaign to promote Diamond wedding engagement rings has also been extremely successful and is now so entrenched in the culture of industrialized nations that there is a steady demand for millions of small Diamonds each year. The advertising slogan "Diamonds are forever" is not intended so much to promote Diamonds as a symbol of long-lasting love and commitment as it is a way to keep purchased Diamonds from reentering the market. That is, if Diamonds are passed down from generation to generation or taken to the grave, the market for new Diamonds to new customers will be less diminished by a "pre-owned" Diamond aftermarket. The volume of Diamond sales globally exceeds the sale of all other gems combined, and for much of the twentieth century, the De Beers cartel sold 85 to 90 percent of the Diamonds mined worldwide (although they control far less of the market today). De Beers went so far as to purchase stones from potential competitors at high prices to keep the supply limited and the price high. In the process of restricting the flow of Diamonds into the market, De Beers has hoarded huge quantities of Diamonds over the years. In 2001 De Beers was holding $5 billion (in 2001 dollars) worth of uncut stones in its London vaults alone, according to a 2001 article by Nicholas Stein. Other mining companies have since tried to emulate the De Beers strategy of market manipulation with other gem varieties, such as the company TanzaniteOne with tanzanite. It works—the prices of tanzanite have subsequently increased significantly over the last several years.

Diamonds once were true rarities. In the thirteenth century, an act of Louis IX of France established a sumptuary law reserving Diamonds for the king. The later discovery of huge Diamond-rich deposits and the eventual development of large-scale mining practices in the twentieth century eventually resulted in abundant supplies of these stones. Near-flawless white Diamonds are much more common today than many colored gemstone varieties of similar quality, such as emerald, ruby, bixbite, alexandrite, tanzanite, red Spinel, red Topaz, several sapphire and Garnet varieties, most Tourmaline varieties, and many others. That being said, some special varieties of Diamonds are still true rarities. Natural colored Diamonds—particularly red, green, pink, and blue—rank with the rarest of gem varieties. Fine colored Diamonds are very popular today, with quality blue and red stones selling for up to $1 million per carat or more.

Scarcity will always remain a variable factor for many gemstones because new mines are discovered frequently, sometimes making formerly rare varieties much more common. Changes in distribution monopolies can also change. Since major Diamond-mining operations have opened in Canada, Australia, and other regions not controlled by De Beers, the market share formerly held by De Beers alone has fallen to about half of what it was in the 1980s.

Scarcity has a significant effect on the *value* of a gem, but it is not the only factor. As we mentioned earlier, some varieties such as Phenakite are true rarities but are simply not popular as gems. Alternatively, some familiar varieties such as amethyst are extremely popular, but not true rarities. The popularity and scarcity of gem varieties continually change over time, and it is the interplay of these two properties that determines the complex forces of market values for the gem trade.

There is also a significant market for fine, complete natural crystals of gemstones still in their original rock matrix (on the rock base on which they originally grew). Such crystals are more properly called mineral specimens. Mineral specimens of true faceting-grade gemstones are very rare and valuable, particularly specimens of emerald, blue sapphire, ruby, bixbite, and several varieties of Tourmaline and Garnet. Such specimens can command prices of tens of thousands to hundreds of thousands of dollars today, and many faceting-grade gemstone specimens in matrix, such as blue sapphire, are often completely unavailable on the world market. Rarity of such material is due to several factors. Historically, when fine gemstone crystals were discovered in a mine or vug, the gem-quality crystals were broken off and collected to be processed into faceted stones for jewelry. Little, if any, of the gem-grade material was saved as mineral specimens. Only in the last several decades has the mineral specimen market developed to the point where there is great economic incentive to collect gem-quality stones as mineral specimens in matrix. These pieces remain scarce or extremely rare for many gemstone varieties. International laws also continue to impede the collecting of some gemstones as mineral specimens. It is today illegal to export unfinished faceting-grade gemstones from certain countries, for example, unfinished faceting-grade blue sapphire in matrix from Madagascar. This restriction is meant as a protection of Madagascar's own gem-cutting industry (previously most raw stones were being exported and faceted abroad, contributing less to the Malagasy economy).

The collector market for gem-quality mineral specimens continues to grow and expand. The result of this is that in addition to the international market in finished gems and jewelry, there is now also an international market in gem-grade natural gemstone specimens, particularly those attractively attached to a natural matrix base. This is most evident in huge international trade shows, such as the annual Tucson Gem and Mineral Show, which has been held every year since 1955. In 2008 the Tucson show had thousands of vendors exhibiting and selling from all over the world. This show takes over the entire town of Tucson, Arizona, for the early part of February each year, where gem dealers, mineral dealers, jewelers, collectors, wholesalers, retailers, and plain tourists all converge, seeking rare, as well as common, natural gemstone specimens and finished gems.

WEIGHTS USED BY THE GEM TRADE

Another factor in determining the value of a gem is its weight. The basic standard of weight for a gem is the **CARAT**. Prior to the twentieth century, the weight of a carat varied from country to country and from time to time, ranging from just under 1/5th of a gram to somewhat over 1/5th of a gram. Since 1907 Europe and the United States, later followed by most other countries, settled on the metric carat, with a standardized value at 1/5th of a gram per carat, or 5 carats per gram. The use of the term "carat" today implies the metric carat. Very small valuable gemstones are sometimes weighed in **POINTS**. Each point is 1/100th of a carat. The carat weight unit of a gem should not be confused with the term **KARAT**, which is a measure of purity for precious metal (e.g., 24-karat Gold is pure Gold; 12-karat Gold is half Gold/half **ALLOY**).

The term "carat" is ancient, originally derived from the weight of a particular seed, supposedly either from the *kuara* tree of Africa, or possibly the locust tree. There is some controversy over the identity of the exact tree from which the seed standard was established, which may explain the fact that the actual weight varied from country to country prior to the twentieth century.

The standard unit of weight for Gold and other precious metals is the **TROY OUNCE**. The troy ounce is heavier that the **AVOIRDUPOIS OUNCE**, which is the standard used in grocery stores for weighing meat and produce. The troy ounce is equivalent to about 31.1 grams, or about 10 percent more than the avoirdupois ounce, which is 28.35 grams. The troy ounce dates back to the Middle Ages to a system used in Troyes, France, and today is used only for precious metals.

Another very small weight unit from the Middle Ages that is occasionally used for precious metals is the **PENNYWEIGHT** (abbreviated dwt). This unit equals 1/20th of a troy ounce, or approximately 1.555 grams, and is used by jewelers in calculating the necessary amount of precious metal for casting settings for gems.

the FORMATION *of* GEMS

How are gemstones formed in nature? We can attempt to answer this question through a combination of direct observations of nature, synthesis of gemstones in the laboratory, and reconstruction of the natural process through scientific theory. We can also distinguish between two very different ways in which gemstones form in nature: organically—from biological processes; and inorganically—from geological and chemical processes. The formation of organically derived gems is discussed further on page 259.

INORGANIC GEMS make up the vast majority of gemstone varieties. Inorganic gems are created by physical chemistry, geological processes, and extreme forces of nature. Ultimately, the growth of an inorganic gemstone's external shape is a product of its internal arrangement of atoms, chemical composition, and other factors of physical chemistry. Our understanding of the natural growth process of gemstones is largely theoretical and based partly on the study of synthetic gemstones created in laboratories, and partly on our knowledge of geology and physical chemistry. We can observe the complete formation of synthetic gemstones because they are created in a lab over a period of days or weeks. We cannot observe the complete formation of natural gemstones in nature because they may develop over longer periods of time (although we are not sure how long) and usually under enormous heat and pressure. Additionally, the process generally occurs deep underground where we cannot observe it.

What we do know is that inorganic gemstones originally form from a liquid source and their formation is influenced by three factors: temperature, pressure, and chemical ingredients. The principal crystal growth processes that lead to gemstone formation are (1) freezing from molten rock, and (2) precipitation from mineral-rich HYDROTHERMAL SOLUTIONS. We can use some simple analogies to help us visualize the processes.

Crystal growth through freezing from molten rock (MAGMA), in its simplest form, can be compared to the freezing of ice in water as the temperature cools, although the freezing point of gemstone crystals is at a much higher temperature than that of ice. Freezing is simply the transformation of a substance from liquid to solid as the temperature falls. The freezing point at which solid ice forms from water is generally 32°F (at normal surface pressure). In contrast, the freezing point at which most gemstones form from magma is hundreds or thousands of degrees Fahrenheit. Deep in the earth's crust and mantle (fig. 8), temperatures and pressures are much greater than on the surface of the earth. For example, at the crust/upper-mantle boundary, which is thought to average about 30 kilometers deep, the temperature is estimated to be about 930° to 1,650°F. At the mantle/outer-core boundary, which is thought to be about 2,930 kilometers deep, the temperature is estimated to be about 7,200°F. At 7,200°F and enormous pressure, most gemstone minerals would still be in a molten liquid state. As the magma moves upward through the crust, it begins to cool, and various minerals crystallize from the molten rock as they reach their respective freezing points. Many gemstones, such as Diamonds, were formed billions of years ago, when the upper subsurface temperatures of the earth were much greater than they are today, and the mantle was closer to the surface. The overall temperature of the earth has cooled and the crust has thickened considerably since the earth's early formation.

Crystal growth through precipitation from solution, in its simplest form, can be observed through a simple experiment to grow crystals of sugar, or rock candy. Take a quart-size container of water at room temperature and stir in sugar by the tablespoon to the point at which no more sugar will dissolve even with vigorous stirring. This makes a transparent solution of water that is saturated with sugar. Next put the clear solution into a large saucepan and bring it to a boil. Once the solution is boiling, remove it from the fire and continue to dissolve tablespoons of sugar until it is saturated once again at the higher temperature. The heated water holds much more dissolved sugar than the room-temperature solution does, because the saturation capacity of water increase as temperature rises. (Increasing the pressure on the solution will also increase the saturation capacity of water, although this is not so easy to achieve in a simple kitchen experiment.) As the solution cools, the saturation point for dissolved solids will fall, and it will become SUPERSATURATED with sugar. This will cause solid sugar crystals to form out of the solution. If you hang a string in the solution for 20 to 30 minutes while it is cooling, you will eventually see crystals form (it helps to keep a weight on the end of the string to keep it straight).

Gemstone crystals can form in much the same way, but with much higher temperatures and pressures. It might be hard to imagine Quartz crystals like

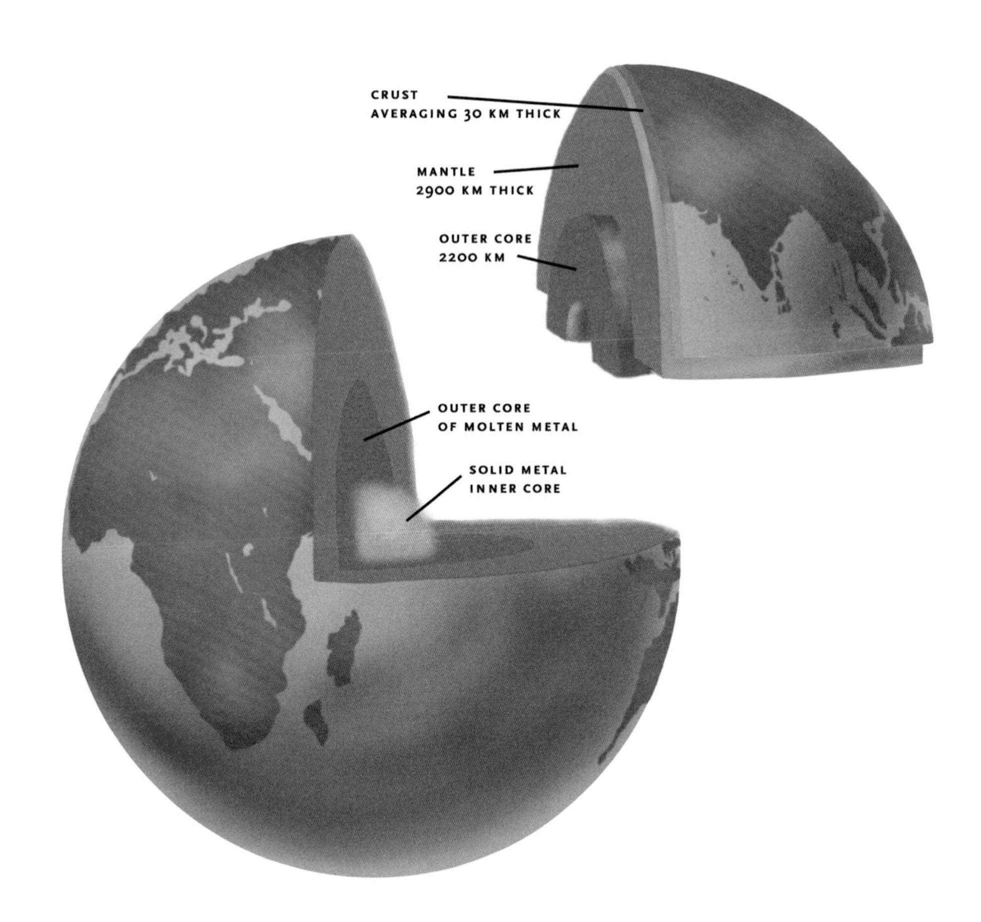

FIGURE 8.
Section through a model of the earth showing the various layers from the outer surface (crust) to the center (inner core).

amethyst or citrine resulting from a process similar to the one that produces sugar crystals, because the Silicates that we observe from day to day, such as Quartz or glass, do not noticeably dissolve in water at normal pressure (i.e., the water in your drinking glass does not dissolve the glass, nor do the waves of the surf dissolve the beach sand). But even Silicates such as Quartz will dissolve into solution at very high pressures and temperatures of the sort found deep underground and will crystallize back out of solution as it cools. We know this because we can approximate similar conditions of pressure and heat in a laboratory pressure chamber. In fact, this is part of the process of growing industrial Quartz crystals, which are used to make everything from optical Quartz for microscopes and cameras to semiconductors for computers and cell phones. The conditions in laboratories where Quartz crystals are made duplicate the pressures that exist deep under the earth's surface. As you go deeper and deeper under the surface of the earth into the lower crust, the pressure and heat continue to increase. Very deep beneath the surface, water flowing through rock under great heat and pressure becomes rich in dissolved minerals. This super-hot, mineral-rich water is called a HYDROTHERMAL SOLUTION,

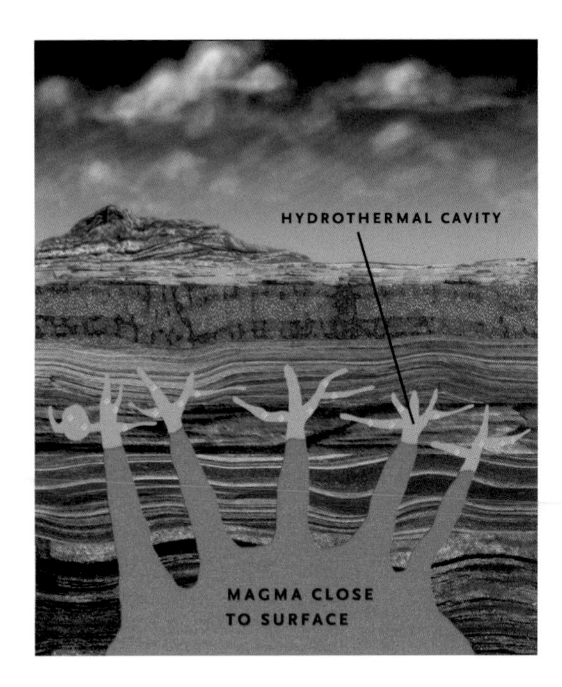

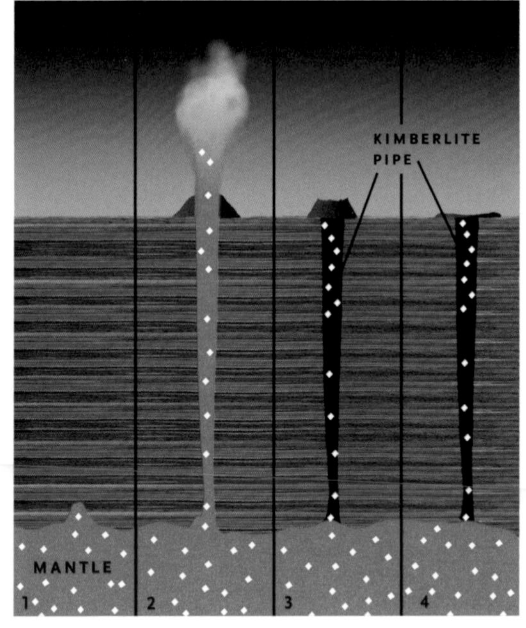

FIGURE 9.
Diagrammatic representation of hydrothermal vents underground where superheated water, supersaturated with dissolved minerals, forms crystals (*left*), and representation of magma from the mantle bearing Diamonds and other deep minerals coming to the surface and eventually solidifying to form kimberlite pipes of the sort mined for Diamonds (*right*).

and it collects in rock cavities called hydrothermal vents (fig. 9). Just as dissolved sugar crystallizes as water cools and evaporates from the pan of sugar-saturated water, Quartz or other minerals crystallize out of hydrothermal solutions underground within rock cavities as the temperature and pressure decrease. It is believed that many of the finest crystals in nature have grown from these solutions.

To better understand crystal growth and where inorganic gemstones form, we need to understand some general geology. There are three basic rock types produced on our planet—IGNEOUS, SEDIMENTARY, and METAMORPHIC—and gemstones can be found in all three.

IGNEOUS ROCKS ("fire-formed" rocks) form from molten rock (MAGMA or LAVA) either aboveground as EXTRUSIVE IGNEOUS ROCKS, or underground as INTRUSIVE IGNEOUS ROCKS. Extrusive igneous rocks form from volcanic eruptions and typically consist of dark-colored, opaque basalts and ash. In general, extrusive igneous rocks are not usually a good source of quality gemstones. Exposure to surface temperature and pressure during solidification of the rock causes rapid chilling and does not generally allow for segregation and growth of large crystals. One exception is in basaltic rocks containing empty cavities from gas that was trapped in the magma before it solidified. Sometimes these cavities later fill with solutions rich in silica that create linings of amethyst, as well as other varieties of Quartz and other mineral crystals. A common example of such a crystal-lined cavity is called a GEODE or DRUSY CAVITY (see fig. 10). Drusy cavities are also occasionally found within sedimentary rocks

FIGURE 10.
Above: A vug or geode with an inner lining of amethyst crystals. Length of specimen is 3,800 mm; mined from San Eugenio, Uruguay (FMNH M14072).

FIGURE 11.
Right: Pegmatite specimen with natural crystals of Almandine Garnet, Muscovite, and aquamarine imbedded in white Albite Feldspar. Front side (*top*) and close-up of back side (*bottom*). Aquamarine crystal on back is 30 mm in length, and the entire pegmatite piece is 150 mm long. From Skardu, Pakistan (FMNH H2325).

where fossils have been dissolved away, leaving cavities that are later filled with crystals.

Intrusive igneous rocks are the product of much slower cooling and solidification processes than extrusive rocks and are formed within large reservoirs of molten rock deep beneath the earth's surface. These rocks typically consist of lighter-colored granites and are the source of most species of inorganic gems. Although often formed deep underground, over time erosion carves away overlying rock, bringing the intrusive rock to the surface or close enough to the surface to mine. The type of granite that is the most productive for a variety of large colored gemstones is called a PEGMATITE (fig. 11). Gems form in pegmatites through a combination of solidifying from molten rock and, more importantly, from precipitation from mineral-rich hydrothermal solutions in cavities and open seams and wide cracks within the pegmatite. Different chemical combinations result in different gemstone varieties. When the pegmatite and hydrothermal solution is rich in the element Boron, Tourmaline crystals can result. If it is rich in the element Beryllium, varieties of Beryl crystals can form. When the gemstone crystals form completely within the rock matrix of the pegmatite through solidification of magma, they are commonly highly fractured and filled with inclusions. But when the gemstones, or at least the growing ends of gemstones, form as free crystals within the hydrothermal cavities, they can form beautiful gem-quality crystals with smooth geometrically constructed faces. Hydrothermal solution-filled cavities allow gemstones to develop undisturbed, sometimes growing into large, transparent, spectacular crystals. Gemstone miners blast and excavate through rock in search of such pockets, or VUGS, which are sometimes also referred to as "nature's jewel boxes."

Sometimes pipes of magma moving to the surface transport gems formed deep beneath the earth's surface, as long as the gems have high-enough melting points not to be melted or otherwise destroyed by the great heat of the magma. One prime example of such a transported gem is Diamond. Diamonds that are mined today are thought to have been formed about 3 billion years ago around 100 to 150 miles below the surface of the earth in the upper mantle. These gems were transported by deep pipes of magma extending from the upper mantle to near the earth's surface. These carrot-shaped igneous pipes of now-solidified rock are called KIMBERLITE pipes. Kimberlite is the main type of rock mined for Diamonds. Only about 1 of every 200 kimberlite pipes is minable for Diamonds, and within the minable kimberlite pipes, it can take as much as 100 tons of ore to recover 2 carats of gem-quality Diamond. Other gems that form deep in the earth and are transported for long distances to the earth's surface include peridot and Pyrope Garnet.

SEDIMENTARY ROCKS form from eroded fragments or chemicals leached from other rocks. Most sedimentary rocks start out as layers of sediments that accumulate within a body of water. As these layers accumulate to great thicknesses, the overlying sediments create huge pressure and heat on the lower layers. Eventually the heat and pressure cause the lower layers of sediments to consolidate to form rock through a number of processes, starting with compaction. The compaction of the sediments eventually squeezes all water out of the sediments in a process called DEWATERING and chemically binds the particles of sediment together in a process called CEMENTATION. The overall process of sediments becoming sedimentary rock is called LITHIFICATION.

It is believed that this process can sometimes take thousands of years. Few gemstones form from sedimentary rocks, but one exception is Opal. Opal forms through circulation of silica-rich water through sedimentary rocks, cavities in rocks, or hard CARBONATE organic materials such as fossilized bone or shells. Opal is consequently found in veins, fissures, and as opalized fossils. Although the source of silica-rich water that forms Opal is often from surface groundwater, it can also come from hydrothermal solutions rising from below the surface of the earth.

Gemstones may also be discovered in sedimentary deposits weathered from pegmatites and kimberlites. Over millions of years, erosion can reduce even pegmatites or other rocks to piles of rubble or gravel in a streambed. Because gemstones are generally very durable, they often survive the processes of weathering and erosion intact and are found as water-worn pieces of gravel. If you screen-wash stream sediment of eroded igneous and metamorphic rocks, you can occasionally find gemstones among the Quartz pebbles and other wear-resistant rock (fig. 285). Such gem-bearing gravel deposits are called PLACER DEPOSITS. Quick note: Because the sedimentary process did not actually create the gemstone itself, gemstones eroded from igneous or metamorphic rocks cannot really be considered gemstones of sedimentary rock origin. It is simply nature's way of mining gems from mostly non-sedimentary sources.

METAMORPHIC ROCKS, or "changed" rocks, form when igneous, sedimentary, or previously formed metamorphic rocks are altered by heat and pressure, turning them into a different type of rock. Most metamorphic rocks are derived from sedimentary rocks and fall into one of two categories: (1) those derived from sedimentary rocks rich in silica, and (2) those derived from Carbonates such as limestone. Metamorphic rocks derived from silica-rich sources may contain such gemstones as almandite Garnet, Tourmaline, Corundum, and spessartite Garnet. Garnets, for example, are sometimes found in mica schists, which are metamorphosed from mudstone and clay. Metamorphic rocks derived from Carbonates may contain Corundum, Grossular Garnet, Quartz, Spinel, Tourmaline, Zircon, or Zoisite. Rubies, for example, are sometimes formed from metamorphosed clay within a limestone that has been metamorphosed into marble.

Whether organic or inorganic, there are many ways in which nature produces a gemstone. The process of gemstone formation is an artifact of the evolution of the earth itself, and it has been in progress for billions of years. Next we will look at the CLASSIFICATION and diversity of gemstones. Just as there are shared special characteristics between species of organisms that help us to understand their organization and interrelationships, there are also shared special characteristics between varieties of gemstones that can help us better understand their organization and interrelationships.

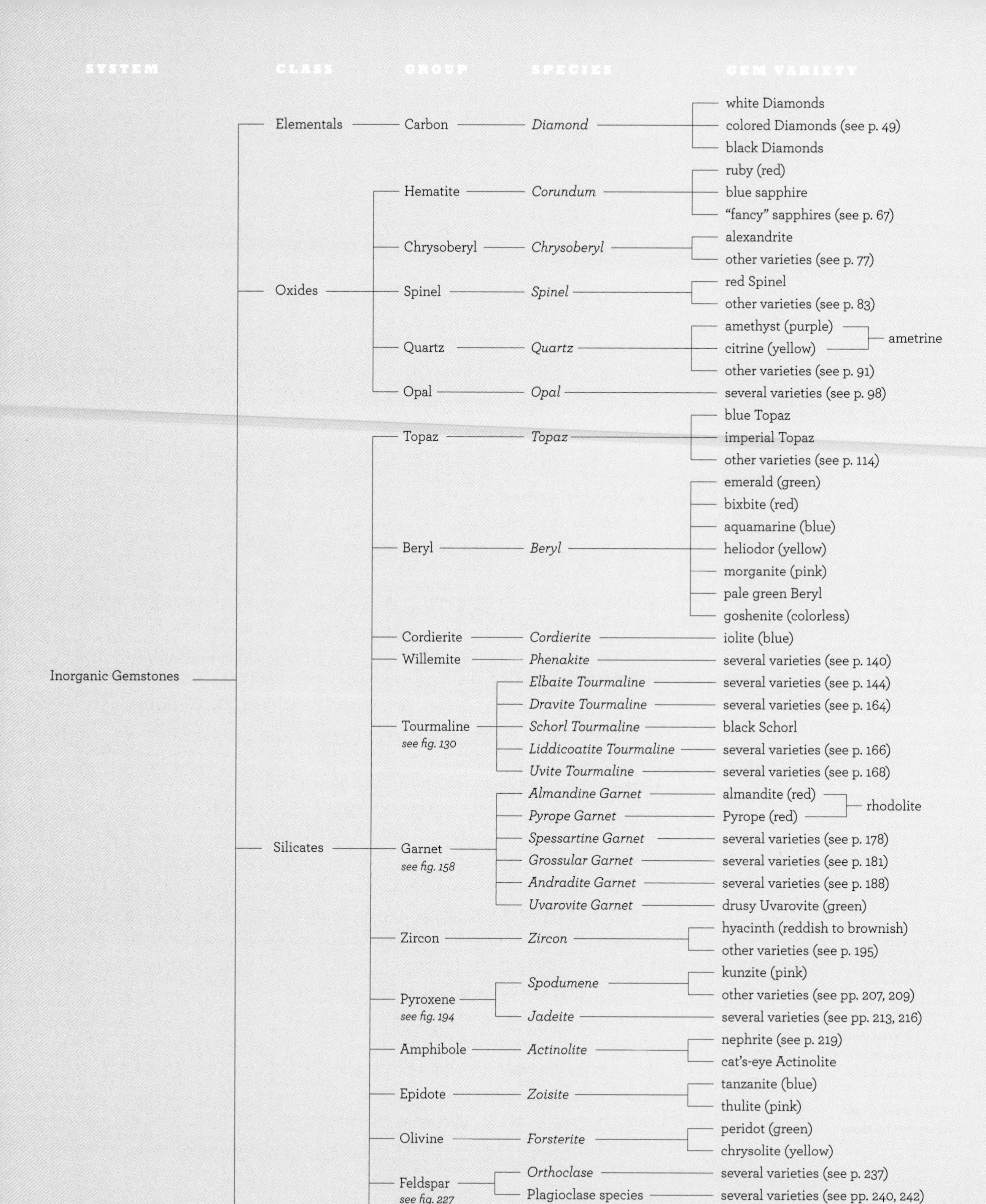

SYSTEM
CLASS
GROUP
SPECIES
GEM VARIETY
Inorganic Gemstones
Elementals
Carbon
Diamond
white Diamonds
colored Diamonds (see p. 49)
black Diamonds
Oxides
Hematite
Corundum
ruby (red)
blue sapphire
"fancy" sapphires (see p. 67)
Chrysoberyl
Chrysoberyl
alexandrite
other varieties (see p. 77)
Spinel
Spinel
red Spinel
other varieties (see p. 83)
Quartz
Quartz
amethyst (purple)
citrine (yellow)
ametrine
other varieties (see p. 91)
Opal
Opal
several varieties (see p. 98)
Silicates
Topaz
Topaz
blue Topaz
imperial Topaz
other varieties (see p. 114)
Beryl
Beryl
emerald (green)
bixbite (red)
aquamarine (blue)
heliodor (yellow)
morganite (pink)
pale green Beryl
goshenite (colorless)
Cordierite
Cordierite
iolite (blue)
Willemite
Phenakite
several varieties (see p. 140)
Tourmaline
see fig. 130
Elbaite Tourmaline
several varieties (see p. 144)
Dravite Tourmaline
several varieties (see p. 164)
Schorl Tourmaline
black Schorl
Liddicoatite Tourmaline
several varieties (see p. 166)
Uvite Tourmaline
several varieties (see p. 168)
Garnet
see fig. 158
Almandine Garnet
almandite (red)
Pyrope Garnet
Pyrope (red)
rhodolite
Spessartine Garnet
several varieties (see p. 178)
Grossular Garnet
several varieties (see p. 181)
Andradite Garnet
several varieties (see p. 188)
Uvarovite Garnet
drusy Uvarovite (green)
Zircon
Zircon
hyacinth (reddish to brownish)
other varieties (see p. 195)
Pyroxene
see fig. 194
Spodumene
kunzite (pink)
other varieties (see pp. 207, 209)
Jadeite
several varieties (see pp. 213, 216)
Amphibole
Actinolite
nephrite (see p. 219)
cat's-eye Actinolite
Epidote
Zoisite
tanzanite (blue)
thulite (pink)
Olivine
Forsterite
peridot (green)
chrysolite (yellow)
Feldspar
see fig. 227
Orthoclase
several varieties (see p. 237)
Plagioclase species
several varieties (see pp. 240, 242)
Benitoite
Benitoite
two varieties (see p. 250)
Phosphates
Turquoise
Turquoise
several varieties (see p. 252)

the CLASSIFICATION *of* INORGANIC GEMS

You ask what is the use of classification, arrangement, systematization? I answer you: order and simplification are the first steps toward mastery of a subject . . .
THOMAS MANN, *The Magic Mountain (1927)*

In order to better understand the diversity of gemstone varieties, we must start with some sort of classification system. When we organize objects created by nature, science strives to develop logical, natural classifications. The differences and special similarities between gemstone varieties are numerous, wide-ranging, and often confusing. Why do some varieties have similar names? Why are all rubies Corundum, but not all Corundum rubies? Why do we group gemstones the way we do? How does it all fit together? To help answer these questions, we will explain the rationale and makeup of our organizational system and illustrate the interrelationships of gem varieties with a series of tree diagrams like the one in figure 12.

With natural history objects, whether living or inorganic, we classify and organize varieties of objects based on their relationships and natural characteristics. Whether we are classifying living species, whose structure and characteristics are derived from their DNA, or inorganic mineral species, whose structure and characteristics are derived from physical chemistry, we group them according to unique features. Just as we group all animal varieties with hair and mammary glands together in a class called Mammalia, we group all gemstone varieties whose basic chemical composition is $Be_3Al_2(SiO_3)_6$ (also known as Beryllium Aluminum Silicate) in a group called Beryl. The gray wolf and the domestic dog are classified as subgroups of the biological species *Canis lupus* much like emerald and aquamarine are classified as subgroups of the mineral species Beryl. Species and other unambiguous groups are defined by special

FIGURE 12.
Facing page: Classification of inorganic gemstone varieties discussed in this book. References to other figures on the branches of this diagram are for expanded details and variety lists for those branches.

similarities shared by all of their members. And because we group these varieties based on what appear to be uniquely derived natural features (hair, chemistry, etc.), we consider these groups to be "natural" groups that express natural interrelationships between varieties. Our natural gemstone groups imply that an emerald is more closely related to an aquamarine than it is to a Pearl, just as a mouse is more closely related to a dog than it is to a plant. Thus, in addition to providing an organized classification of gemstone variety names, our classification system is an organized presentation of many properties associated with particular gemstone varieties. For example, all Beryl varieties—such as emerald, aquamarine, morganite, and heliodor—have (1) the same basic chemical composition, (2) a hexagonal crystal structure, and (3) a Mohs hardness of 7½–8. If an emerald is a Beryl and an aquamarine is also a Beryl, what makes them different? The Beryl varieties are distinguished from each other primarily by trace elements whose concentrations are generally too small to be included in the basic chemical formula. These trace elements are nevertheless responsible for the color differences between emerald, aquamarine, and other Beryl varieties.

A ranking system of groups allows us to express relationships and characteristics in an efficient hierarchical manner (fig. 12). Inorganic mineral types are defined and classified primarily on the basis of specific chemical properties, crystal structure, and various physical properties. In this book we rank gemstones into several formal categories. In order of highest (most inclusive) to lowest (least inclusive) these are SYSTEM, CLASS, GROUP, SPECIES, and VARIETY. In the sections on Tourmalines, Garnets, and Feldspars, we add a sixth category called SUBGROUP between group and species because of the complexity of these groups (figs. 130, 158, 227). We also use the category of INTERMEDIATE between species and variety. Intermediates are usually a BLEND or "hybrid" between two mineral species (e.g., figs. 158, 227, 235).

We define SYSTEM here as being either organic or inorganic. The plants, coral polyps, and MOLLUSKS that make organic gems originated only tens to perhaps hundreds of millions of years ago. Inorganic gemstones, on the other hand, originated billions of years ago and represent a different and older line of physical history. Most varieties of gemstones fall within the inorganic system.

Within the "systems" of gemstone varieties, there are a number of subdivisions that we refer to as CLASSES. Gemstones within the inorganic system fall primarily into one of four classes: ELEMENTALS, OXIDES, SILICATES, and PHOSPHATES. Classes are defined by shared chemical characteristics of the highest level. Minerals in the ELEMENTAL class are made of a single type of atom, such as Diamond (pure Carbon, chemical formula C) or Gold (pure Gold atoms, chemical formula Au). In general, there are not many naturally occurring Elemental minerals, because most native elements combine with others elements to form stable compounds. All other classes of gemstones consist of compounds of different types of atoms. OXIDES, for example, are minerals in which Oxygen is the principal ANION, often bonding with a metal element. In minerals of this class, the Oxygen symbol O or the hydroxyl symbol OH appears at the right end of the chemical formula, as in Corundum (Al_2O_3). SILICATES are minerals in which silica combines with other elements to form the anions, and all Silicates have a Silicon-Oxygen combination SiO as part of their chemical makeup, as in Beryl [$Be_3Al_2(SiO_3)_6$]. Silicates are the most abundant minerals in the earth's crust, although most Silicate gemstone varieties are uncommon to rare. PHOSPHATES are minerals in which the anions are principally Phosphates, with PO_4 as part of their chemical makeup, as in Turquoise [$CuAl_6(PO_4)_4(OH)_8 \cdot {}_4H_2O$]. Phosphates, such as Turquoise, can be intensely colored. Phosphate gemstone varieties are few; nearly all true gemstones are either Oxides or Silicates.

Within the classes of gemstone varieties, there are a number of subdivisions we refer to as GROUPS. Groups contain species, and species can be subdivided into varieties. As we use them here, groups can contain one to several species, although Back and Mandarino (2008) recommend that a group should consist of at least three species. Each SPECIES of inorganic gemstone has a different basic chemical structure. Each VARIETY of gemstone within a single species has the same basic chemical structure and differs only by trace elements, or "impurities" that give each variety a different color. Many publications use the terms "species" and "variety" interchangeably for gems, but we prefer a more rigorous use of the terms in order to better express the chemical relationships among gemstones. For valid species names of minerals, we follow the standard guide *Fleischer's Glossary of Mineral Species* (Back and Mandarino 2008). All valid species-level names, such as "Beryl," and higher category names such as group-level and class-level names are treated as proper nouns and capitalized. Variety names, such as "emerald," are not, except in the cases where the variety name is the same as the species name, such as "Pyrope." This is done primarily to help the reader identify the more vernacular variety names from the more formal mineralogical names. Variety names are often more controversial and subject to interpretation than species, group, and class names.

In MINERALOGY, the mineral SPECIES are the most significant terminal groups. The species-level groups categorize mineral types according to primary chemical composition and crystal structure, which is most useful for scientific work. In GEMOLOGY, the mineral VARIETIES are often the most significant terminal groups. The varieties reflect aesthetic variations within mineral species. Within a single species, one variety may be extremely valuable while another may be nearly worthless. Thus the mineralogist and the gemologist generally have different priorities in their classification schemes. Our basic classification outlined here is intended to serve both interests as much as possible and to illustrate the relationships among the species and varieties of gemstones.

Figure 12 is a good intuitive illustration of the interrelationships among all of the categories of inorganic gemstones discussed in this book. Some of the terminal branches of figure 12 are expanded elsewhere in this book, and the page numbers for those expansions are given on that figure. A similar illustration for the categories of organic gemstones discussed in this book is figure 249, except that there are no formal species names for groups of organic gems. Consequently, we use the term SUPERVARIETY instead of species for the primary variety groups of organic gems.

We use the preceding classification as a map to present the following gem and gemstone varieties from the Grainger Hall of Gems. At the end of many of the gem sections below, we cite author and publication dates that key into the reference section at the end of the book.

INORGANIC GEM TYPES

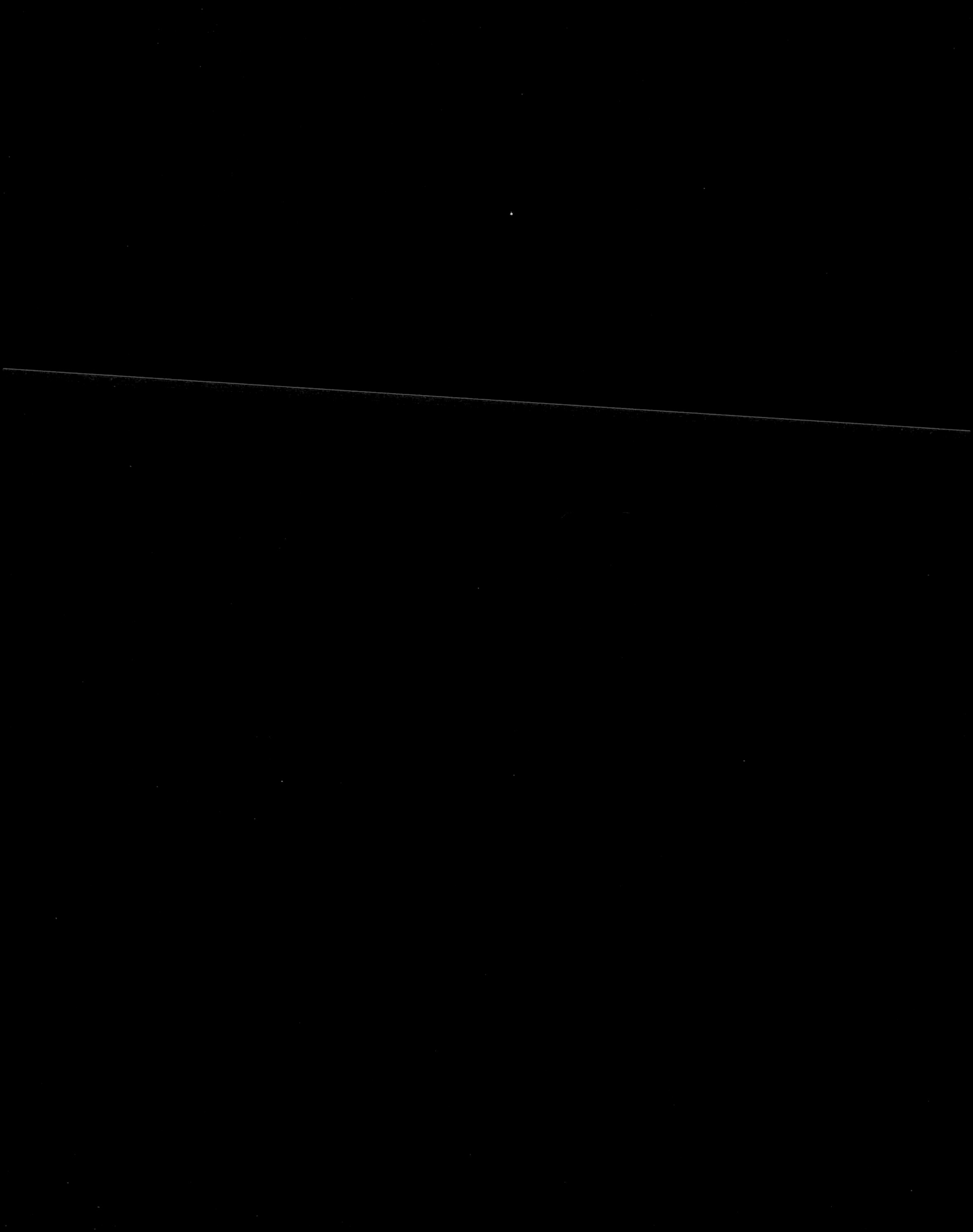

Diamond

SYSTEM Inorganic

CLASS Elemental

GROUP Carbon

SPECIES *Diamond*

GEM VARIETIES colorless Diamond; colored Diamonds; black Diamond (carbonado)

What name is more synonymous with precious gems than Diamond? A gemstone of great beauty and the hardest known natural mineral in the universe, Diamond has long been a precious adornment of the highest order. Diamonds were known in India at least 2,500 years ago, but the practice of cutting them did not come about until the thirteenth or fourteenth century AD. The delayed appearance of Diamond cutting was partly due to an early perception that cutting a Diamond would destroy its magical powers. It was also due to the fact that the technological knowledge and skill for cutting Diamond did not become entrenched until the thirteenth or fourteenth century. Records of commerce in Diamonds date back at least to the twelfth century, and Diamonds have had a long and colorful history in nearly every civilized society over the last several centuries. To the jeweler, Diamond is regarded as the king of gemstones, both as a centerpiece stone and as an accent stone for other types of gems. To the mineralogist, Diamond is intriguing because of its many important physical properties, including its unsurpassed hardness, its structure as a single giant molecule of Carbon, and its heat conductivity. To industry, Diamond is the most important of all gemstone minerals because of its many uses as the ultimate abrasive and for its use in making the world's best cutting tools.

Although colorless Diamond is the standard variety for its species, Diamonds occur in several colors. Gem-quality, vividly colored blue, green, red, pink, and orange Diamonds are exceedingly rare, much more valuable than other Diamonds, and in today's market can sell for $1 million or more per carat. Fancy vivid yellow and absolutely colorless stones of exceptional clarity are also very valuable, especially if they are 4 carats or more in weight. Diamonds with tinges of pale yellow, brown, or black are significantly less valuable than fine colorless stones.

Colorless Diamond is far from the rarest of gemstones and is not even the most valuable, but it is unquestionably the most important. It constitutes about 90 percent of the modern precious gem trade. In 2004 more than 25 million carats were mined for use as gems, and another 100 million carats of non-gem-grade Diamonds (called BORT) were mined for use in industry. In addition to the 100 million carats of industrial-grade Diamonds that are mined, another 3 billion carats of synthetic industrial-grade Diamonds are produced annually for use in industry (also based on 2004 figures). The popularity of Diamonds as gems increased significantly during the twentieth century due to many factors, including successful marketing campaigns, improved cutting and polishing techniques, increased mining productivity, and growth in the world economy. About half of all Diamonds originate from central and southern Africa, while the rest come mostly from Canada, India, Brazil, Russia, and Australia. Diamond mining is a frequent subject of ethical controversy in some areas due to working conditions for miners, fair-market concerns, or their use in transferring great wealth for criminal activities, as portrayed in the recent movie *Blood Diamond*. "Blood Diamonds" or "conflict Diamonds," for example, are those sold by revolutionary or insurgent groups to fund terrorist activities. Strong efforts have been made by the Diamond industry to curtail the number of conflict Diamonds entering the market. Ethical aspects of Diamond mining are discussed in more detail on page 315.

The places where most Diamonds are cut and polished are remarkably concentrated. In 2003, about 90 percent of gem-quality Diamonds were cut in India. Other important centers of wholesale trade for Diamonds include Antwerp, New York, London, Tel Aviv, and Amsterdam. The company that is most famous, by far, for successful marketing and control of the Diamond trade is the De Beers company. De Beers is a South African–based company that developed the standard for effective control and promotion of the Diamond trade (see pp. 20–22).

The hardness of Diamond has been known for many centuries. Even the name Diamond, taken from the ancient Greek *adamas*, means "invincible." Nothing can scratch the surface of a Diamond except another Diamond, which is one reason that a Diamond can hold its unparalleled polish for such exceptional lengths of time. From time to time, Diamonds must be cleaned to remove dirt, oils, or other substances that have settled on the stone's surface, but Diamonds rarely need repolishing within a lifetime. The hard, highly polished LUSTER of Diamond is so distinct that it sets a standard for the rich submetallic luster called ADAMANTINE. The hardness (or resistance to scratching) of Diamond should not be confused with its relative toughness, or resistance to shattering on impact. Diamond is the hardest natural substance, but it is far from the toughest, and in fact it is brittle. If you drop a Diamond on a hard surface or hit it with something hard, it can crack or shatter.

Diamond has been the major force of the gem trade for well over a century (see discussion on p. 311). Diamonds (other than certain black Diamonds) are formed deep (50–250 miles) underground at extremely high temperatures and pressures, and the Diamonds mined today were formed billions of years ago and were later brought to the surface through volcanic activity. Early Diamond mining prior to the mid-1700s was primarily from ALLUVIAL DEPOSITS in India. By the mid- to late 1700s, Diamond mining in Brazil began to overtake the Diamond production of India, partly because the Indian deposits were becoming commercially exhausted and commercial development of the Brazilian mines was expanding. Finally, massive mining operations of Diamond from kimberlite rock began with the discovery of Diamonds in the Kimberley region of South Africa in the mid- to late 1800s. South Africa has remained a major producer of stones, but over the last century major deposits have also been opened up in a number of other regions, including New South Wales, Australia (Inverell); Russia (Yakutia, Udachnaya); Ghana (Birim River); Sierra Leone (Yengema); Congo (Tshikapa); Botswana (Jwaneng); Brazil (Minas Gerais); Namibia (Oranjemund); Angola; and Canada (Northwest Territories). Several of the mines in Africa have become the subject of ethical scrutiny over the last decade (see "Ethics" chapter).

For further reading about Diamonds, we recommend O'Donoghue (2006), Bauer (1968), and Bruton (1978).

COLORLESS DIAMOND

SYSTEM Inorganic
CLASS Elemental
GROUP Carbon
SPECIES *Diamond*
VARIETY colorless Diamond
COMPOSITION Carbon (C)
TRACE ELEMENT FOR COLOR none
HARDNESS ON MOHS SCALE 10

Few gems can match the superb FIRE and BRILLIANCE of Diamond, particularly in flawless, completely colorless or blue-white stones. This is due to Diamond's many superior optical properties, luster, transparency, and resistance to wear. Diamonds occasionally bear the nickname "ice," especially in old crime novels. The most obvious explanation for this tag is their colorless clarity. But the analogy also expresses two other physical properties. First, Diamond is "slippery," even at an extremely small (nano) scale. Synthetic Diamond coatings are being researched as possible low-friction coating material for high-tech moving parts. Second, Diamond is often cold to the touch. It conducts heat better than any other known element, so when you touch a Diamond that is at a temperature below body temperature, it conducts heat away from your skin, making it feel cold.

A colorless Diamond has the ability to split white light into all the colors of the visible spectrum and reflect a multicolored fire back to the eye of the viewer. It is simultaneously suggestive of pure ice and fiery brilliance in a way that adds cool refinement and vibrant life to any piece of jewelry. Diamonds will also become fluorescent under UV light. This is a highly useful property for sorting out genuine Diamonds from imitations. The art of cutting and polishing Diamonds to make the most of their superb optical qualities has been perfected over recent centuries. Early Diamond cutting was very minimal and focused on minimizing waste in the cutting process rather than bringing out the brilliance of the stone. The craft took advantage of the natural direction of CLEAVAGE and the natural crystal faces of the stone. Prior to the fourteenth century, Diamonds were not faceted. In the fourteenth century, the first known guild of Diamond cutters and polishers, located in Nürnberg, Germany, created the "point cut." This cut followed the natural octahedral shape of the raw Diamond (see figs. 13 and 14), eliminating nearly all of the waste in the finishing process, and left a pyramidal point on the top center of the stone. During the fifteenth century, the finishing process evolved to highlight Diamond's optical properties by cutting the point of the octahedron off and polishing the surface into a window-like TABLE facet, and adding a few other polished facets to

FIGURE 13.
Left: Large gem-quality natural Diamond crystal on bort matrix. The gemstone part (not including the dark bort matrix) is a high-quality, colorless octahedral crystal weighing 20 carats and measuring 18 × 17 mm. Specimen was originally discovered in Kimberley, South Africa (FMNH H19-1).

FIGURE 14.
Right: Series of raw, natural Diamond crystals of various shapes and colors from South Africa and Brazil. Crystals range from 0.8 to 5.9 carats, and from smallest to largest are FMNH H10-8, H14-3, H14-5, H2363-3, H14-8, and H10-13.

the stone. Through the following centuries, more and more facets were added to the Diamond, focusing on bringing out the brilliance of the stone rather than simply maximizing final carat weight. In the eighteenth century, early versions of the BRILLIANT-CUT Diamond started to develop, with its primitive forerunner, the OLD MINE CUT, containing about 50 facets at varying angles. Through centuries of experimentation and improvement in cutting design, the art of Diamond faceting finally reached a pinnacle with the so-called modern brilliant cut used today. This is currently the standard cut for Diamonds, which maximizes the amount of light reflected back through the table of the stone, and highlights a Diamond's special fire and brilliance (see pp. 8, 13). Today's modern ROUND BRILLIANT CUT, or ideal cut, has 57 to 58 precisely angled facets and was designed by an engineer named Marcel Tolkowsky in the early twentieth century. His design uses mathematical principles to reflect the maximum amount of brilliant light entering the stone back through the table of the stone to the eye of the viewer. The skillful cut of a stone is today a prime factor in determining its value. Part of what eventually made it practical to incorporate so many facets into Diamond cuts was the development of high-speed cutting and polishing machines. Because of the extreme hardness of Diamond, hand polishing and cutting brilliant-cut stones would be overly time-consuming and impractical. The hardness of Diamond also explains why it is exceedingly rare to find true carved Diamonds (Diamonds with images hand-carved into them). The small carved Diamond in figure 5 took the nineteenth-century engraver five years to complete.

The unsurpassed hardness of Diamonds is also part of what earns it its superior reputation. There is no harder natural substance. As hard as it is, it is not an exceptionally tough stone and can shatter if given a hard blow. Hardness and toughness are two different factors of durability (see pp. 14, 18), and confusing the two could be potentially disastrous. PLINY THE ELDER (AD 23–79) called Diamonds "unconquerable" and wrote: "The best way to test [a Diamond] is upon the anvil; strike even upon the point of [the Diamond] with a hammer as hard as you can, it defies all blows." No one knows how many Diamonds were turned to dust because of this advice. Pliny also advised that the only way to soften Diamond was to soak it in goat's blood. Since Pliny's time, there have been significant advances in both mineralogy and gemology!

The history of Diamond mining goes back several thousand years. Mining took place mainly in southern India until the eighteenth and nineteenth centuries, when Diamonds were discovered in Brazil and later in Africa. Nearly all gem-quality Diamonds are of the colorless category, although many of these actually have very slight tinges of yellow or brown.

Colorless Diamond is the BIRTHSTONE of people born in April and is the gem associated with the 30th and 60th or 75th wedding anniversaries in

FIGURE 15.
Diamond necklace from the 1850s, in English style manufactured in India. Several hundred old mine-cut Diamonds of various sizes totaling about 20 carats, set in 9-karat Gold (FMNH H2237-1).

FIGURE 16.
Diamond necklace from 1904, with colorless Diamonds set in Platinum. Large pear-shaped stone is a shallow-cut 4-carat near-flawless colorless Diamond measuring 11 mm in height (FMNH H2307).

FIGURE 17.
Top: Diamond pendant, "Cumulus," designed by Lester Lampert in 2008 for the Grainger Hall of Gems. Total of 21.73 carats of brilliant-cut colorless Diamonds set in 18-karat white Gold, the largest of which is an 8-carat round brilliant-cut stone. Pendant is 70 mm in height (FMNH H2525).

FIGURE 18.
Bottom: Diamonds set in Platinum dress clips with floral design, from the 1940s. Designed by Mauboussin for Trabert and Hoeffer. Mixture of baguette and cushion-cut Diamonds totaling 25 carats in weight. The larger clip is about 70 mm in length (FMNH H1546.1, H1546.2).

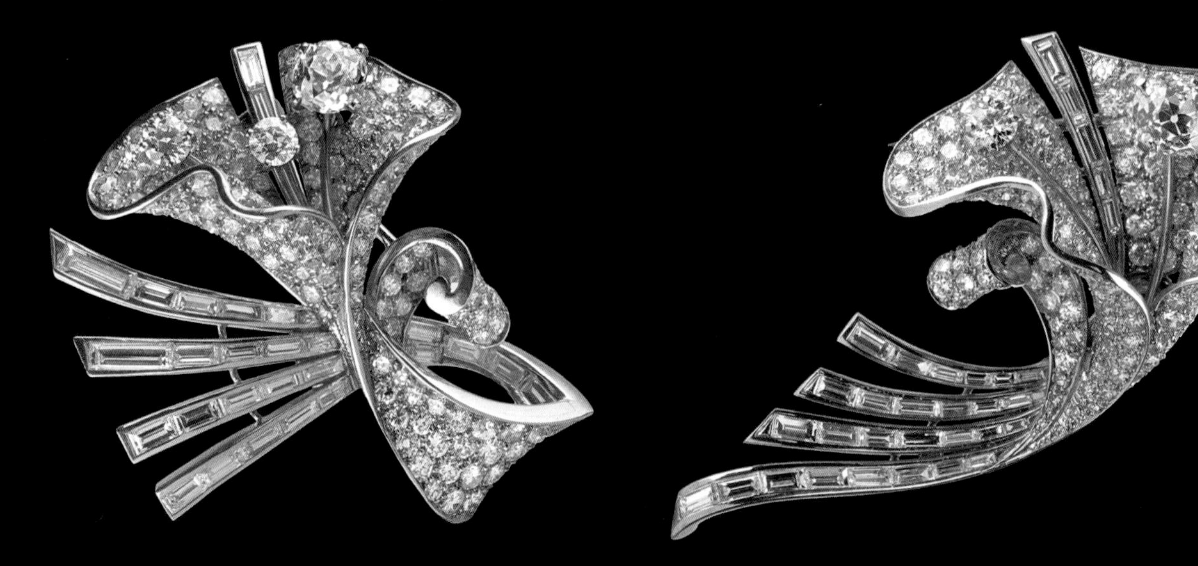

FIGURE 19.
Left: Brooch designed by Henry Dunay, mid-twentieth century, with round brilliant-cut Diamonds set in 18-karat Gold, 70 mm in height (FMNH H2452). *Right:* Earrings designed by Tiffany & Co., mid-twentieth century, also with round brilliant-cut Diamonds set in 18-karat yellow Gold (FMNH H2458).

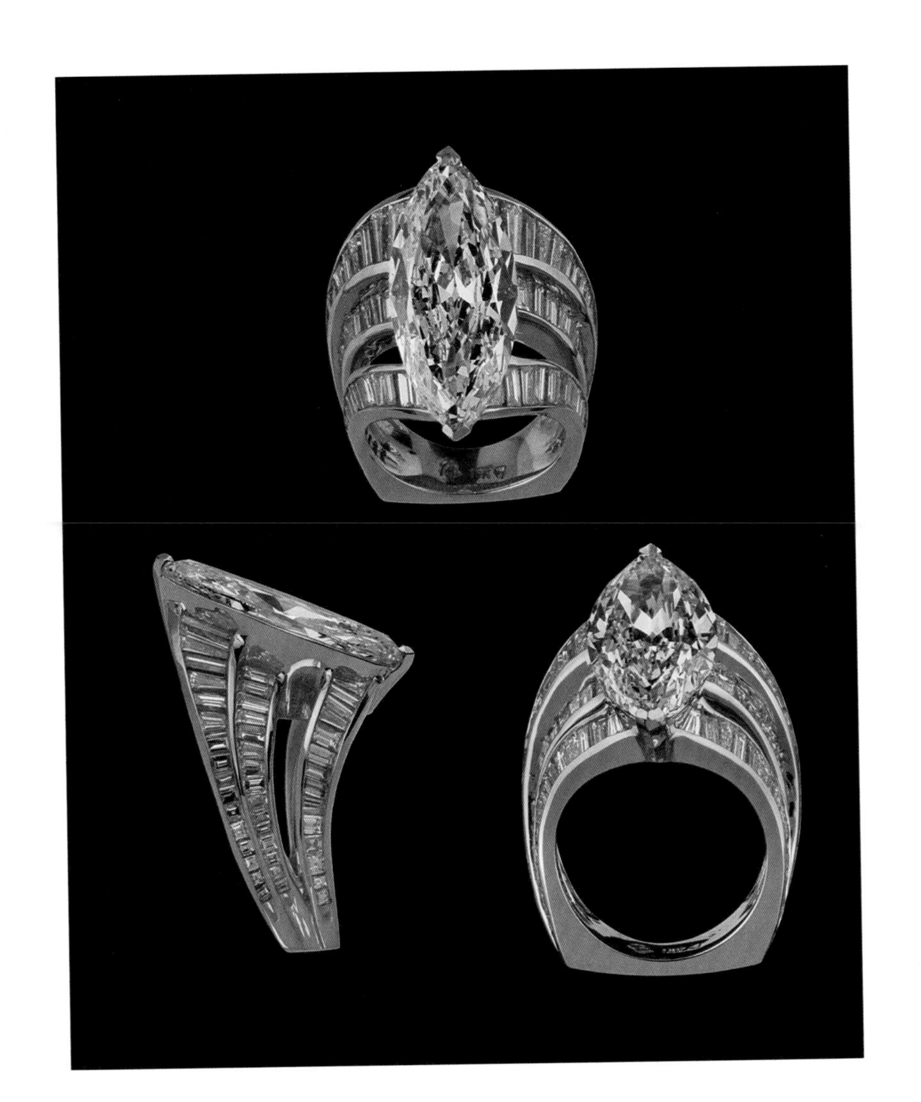

FIGURE 20.
A 7.42-carat marquis-cut Diamond mounted in an 18-karat yellow Gold setting with 92 baguette Diamonds, making a total of 11.24 carats of colorless Diamonds. The big Diamond is 22 mm in height. This ring, "Ascension," was designed by Lester Lampert in 2008 for the Grainger Hall of Gems. Slightly tilted top, side, and front views (FMNH H2526).

Western culture. To the ancient Greeks, they were tears of the gods; to the Romans, splinters from falling stars (crystallized lightning). Diamonds have had many symbolic meanings in various cultures, ranging from fearlessness, power, and invincibility, to sex, reproductive power, and good fortune. In Western culture, it is commonly associated with long-lasting affection. It is also the ultimate "glamour" gem of celebrities. Zsa Zsa Gabor, oft-divorced Hungarian American actress and socialite, summed up this sentiment in her much-quoted statement, "I never hated a man enough to give his Diamonds back."

In 1905 the largest gem-quality colorless Diamond discovered to date was found in South Africa. It was called the Cullinan Diamond and weighed 3,106 carats. This was cut into several gems, the largest of which is the 530-carat Great Star of Africa that forms the head of the Royal Sceptre in the Crown Jewels of the United Kingdom. One of the most spectacular pieces in the Grainger

Hall of Gems is a Diamond brooch and pendant called "Cumulus," designed by Lester Lampert of Lester Lampert, Inc. This piece is set in 18-karat white Gold and contains 21.73 carats of white Diamonds, the largest of which is an 8-carat brilliant-cut stone. The piece uses a cluster design with overlapping Diamonds set on a multi-level "cloud-like" shape in partial bezels that accent the extraordinary brilliance of the gems (fig. 17).

COLORED DIAMONDS

SYSTEM Inorganic

CLASS Elemental

GROUP Carbon

SPECIES *Diamond*

VARIETIES red Diamond; pink Diamond; purple Diamond; blue Diamond; green Diamond; orange Diamond; yellow Diamond; brown ("champagne" or "cognac") Diamond

COMPOSITION Carbon (C)

TRACE ELEMENT FOR COLOR Nitrogen (N) or Boron (B)

HARDNESS ON MOHS SCALE 10

Normally, absolutely colorless Diamonds are more valuable than stones with very slight color tints. But when a Diamond's color is intense or vivid and of a rare color, the colored stones are worth many times that of the colorless stones. Most vividly colored Diamonds (red, blue, green, pink, and orange) are exceedingly rare. Although not as rare as the five colors mentioned above, vivid or "fancy" yellow is also uncommon and can often be valued at a premium over colorless stones. These should not be confused with pale yellow stones, which are not as valuable as fine colorless stones. Brown Diamonds, or the so-called champagne or cognac Diamonds, are abundant and not as valuable as other color varieties. The champagne and cognac designations are part of a recent marketing effort to increase the popularity of these previously unpopular stones. Their value is currently well below that of fine colorless stones.

Because of the scarcity of some colored Diamond varieties, there is less known about the precise factors causing the coloration. Vivid red Diamonds are perhaps the most rare and valuable of all Diamonds and known by only a few stones over a carat in weight. They are thought to derive their red color from defective crystal growth patterns during the formation of the stone. The largest known vivid red Diamond gem is the Moussaieff Red, a 5.11-carat stone cut from a 13.9-carat rough stone. This stone came from Brazil, although today Australia is the main source of red and pink Diamonds. One of the finest vivid red Diamonds known is a small .95-carat stone from Brazil that in 1987 sold for $880,000 at auction. Today such fine reds might bring millions of dollars per carat.

Pink Diamonds are also rare, though not as rare as their vivid red cousins. The largest known pink Diamond is the Steinmetz Pink, a 59.6-carat stone cut from a 100-carat rough stone found in southern Africa. This stone took twenty months to cut. Some hues of pink Diamonds range more toward purple in color, and true purple Diamond is extremely rare. The Diamond deposits in the Kimberley district of Western Australia discovered in the 1970s have produced a number of pink and red Diamonds, and today these deposits are the main source of pink Diamonds in the world.

Blue Diamonds are probably the most famous of all colored Diamonds because of the famous Hope Diamond, now housed in the Smithsonian Institu-

FIGURE 21.
Yellow Diamond, natural octahedral crystal in kimberlite matrix. Diamond is about 1 carat in weight. From Kimberley, South Africa (FMNH E4942-1).

FIGURE 22.
Facing page: Colored Diamonds in jewelry. *Top:* One-carat, brilliant-cut, fancy yellow Diamond set in 18-karat yellow Gold ring with small round brilliant-cut colorless accent Diamonds, in side, top, and front views (FMNH H2281). *Bottom:* Brooch with pink, yellow, orange, brown, and colorless Diamonds set in 18-karat white and yellow Gold. Width of brooch is 52 mm (FMNH H2601).

tion in Washington, DC. The blue color is usually derived from traces of the element Boron. There are many shades of blue Diamonds, ranging from light grayish-blue to deep navy blue. Intense blue stones are exceedingly rare. In May 2008, a 3.73-carat vivid blue Diamond sold for $4.93 million at a Sotheby's auction in Geneva, Switzerland, setting a record price per carat for a gemstone purchased at auction ($1.33 million per carat). The largest intense blue Diamond is, in fact, the Hope Diamond, weighing 45.52 carats. The stone from which the Hope Diamond was cut was originally discovered in Golconda, India, in the seventeenth century. The original stone was a crudely cut 115-carat triangular shape. Over the years it changed hands and was recut several times; from the original 115-carat triangular stone in India, to the 69-carat "French Blue" in the Crown Jewels of France, to a 45.52-carat stone in a number of private collections (including that of the Hope family, after which the stone in

its current form was named). In 1958 the Hope Diamond was donated to the Smithsonian Institution in Washington, DC, by Harry Winston, a New York Diamond merchant. He reportedly sent it through the U.S. mail wrapped in plain brown paper (registered and insured for $1,000,000, at the cost of $145.29). High-quality blue Diamond rarely is offered for sale, but when it is, it is extremely expensive.

Natural green Diamonds are another very rare type of gem. They can be yellowish-green to apple-green to blue-green in color. The largest known intense green natural Diamond is the 41-carat Dresden Green, which is apple-green in color. The Dresden originally came from India, although little is known of its history prior to its purchase by Frederick Augustus II of Saxony in 1743. It is named after the capital of Saxony, Germany, where it has been displayed for much of the last two centuries. The green color is due to natural irradiation. It is sometimes extremely difficult to tell natural green Diamonds from lab-irradiated specimens.

Orange Diamonds are another extremely rare color variety of Diamond. The largest known vivid orange Diamond is the Pumpkin Diamond, a 5.54-carat stone mined from the Central African Republic in the late 1990s. It is currently valued at $3 million.

Fancy yellow Diamonds, sometimes referred to as canary-yellow Diamonds, get their color from traces of Nitrogen impurities. One of the largest yellows, and the largest known vivid yellow, is the 128.54-carat Tiffany Diamond, found in Kimberley, South Africa, in 1878. It was cut from a rough stone weighing 287.42 carats. The largest fancy brownish-yellow Diamond is "The Incomparable," a 407.48-carat stone from the Democratic Republic of the Congo. There is a fancy yellow Diamond ring in the Grainger Hall of Gems (fig. 22).

Brown is the most common variety of colored Diamonds, and the largest producer of brown Diamonds is Australia. Some brown shades can be irradiated to produce red, blue, and green shades, but these treated stones can be readily distinguished from their natural counterparts and are worth only a fraction of the natural colored stones. At this time, brown Diamonds are worth less than colorless stones, although light-colored brown stones with pink overtones can be extremely beautiful as accent stones. One example is a fine Persian Turquoise and champagne Diamond necklace set in 22-karat Gold in the Grainger Hall of Gems (fig. 247). The largest known faceted brown Diamond is the Golden Jubilee, a 545.67-carat stone cut from a 755-carat rough stone, discovered in South Africa. At the time this book was written, it was also the largest faceted Diamond of any kind.

BLACK DIAMOND (CARBONADO)

SYSTEM Inorganic

CLASS Elemental

GROUP Carbon

SPECIES *Diamond*

VARIETIES carbonado black Diamond; bort black Diamond

COMPOSITION Carbon (C)

TRACE ELEMENT FOR COLOR none

HARDNESS ON MOHS SCALE 10

The black Diamond is technically not a colored Diamond, because black is the absence of all color. There are two types of black Diamond; carbonado and bort. Carbonado is opaque, and bort can be opaque or slightly translucent. Both types have been used as gemstones.

One reason for carbonado's opacity is the fact that, unlike other Diamonds, which are made of a single crystal, carbonado is usually polycrystalline—that is, it is made of a mass of microcrystals that are stuck together. This makes carbonado more porous than other Diamonds. It also makes carbonado the only variety of Diamond without cleavage, given that it is not a single crystal. Carbonado does not occur in most of the world's conventional mining fields. They are known from only Brazil and the Central African Republic. Because of their very limited geographic occurrence, and because of the significant amounts of Hydrogen and Nitrogen in the stones, they are thought to have come from space, perhaps formed in stellar supernovae explosions billions of years ago and transported to early earth by asteroids. This theory is hotly debated among scientists today.

Some faceted black Diamond material on the market today also consists of opaque bort. Bort is Diamond that is heavily included with Graphite, and its major application today is as an industrial-grade stone that is pulverized for use as an abrasive material. Bort, like carbonado, can be opaque, but in some cases it is slightly translucent. Bort that has even opaque black throughout the

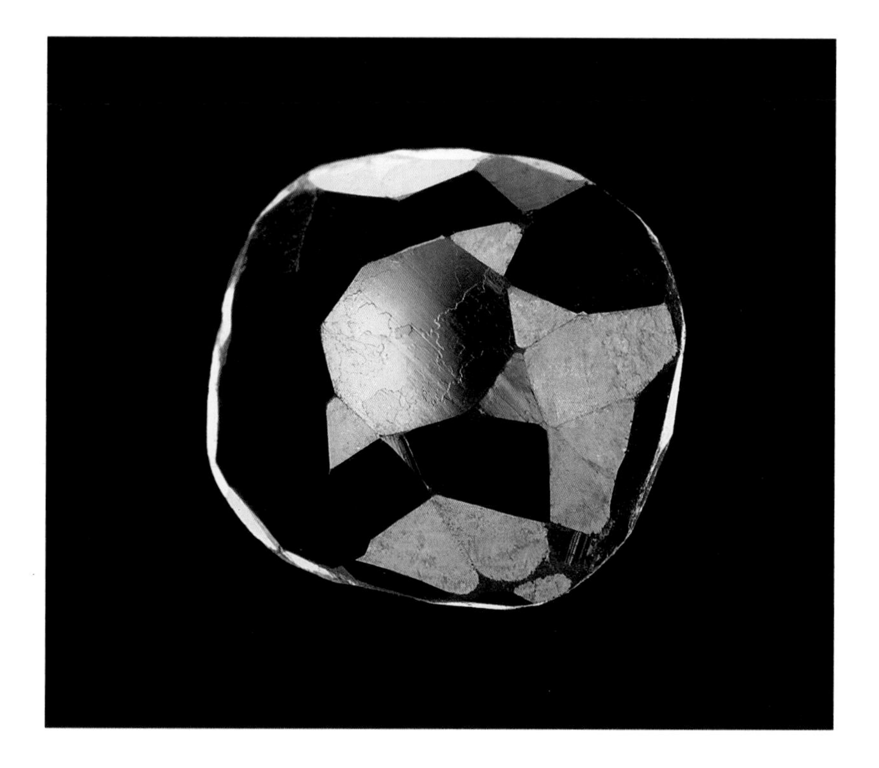

FIGURE 23. Large faceted black Diamond, from Bahia, Brazil. Gem weighs 8.1 carats and measures 10 × 10 mm (FMNH H8).

FIGURE 24.
Round-cut 2.65-carat black Diamond set in 18-karat white Gold. Designed by Ashish Jhalani of MySolitaire.com of New York. Setting © MySolitaire .com. Top, side, and front views (FMNH H2364).

stone, like carbonado, is sometimes faceted today for jewelry. Faceted black Diamonds of over 4 carats are uncommon, but the popularity of black Diamond lags behind all other varieties, and therefore small stones are inexpensive.

The largest known faceted black Diamond is the "Spirit of de Grisogono," a 312.24-carat stone cut from a rough stone of 587 carats that was found in west central Africa in the late twentieth century. The stone is currently set in a somewhat impractical white Gold ring together with 36.69 carats of colorless Diamonds.

Corundum

SYSTEM Inorganic

CLASS Oxide

GROUP Hematite

SPECIES *Corundum*

GEM VARIETIES red Corundum (ruby); blue Corundum (blue sapphire); other colored Corundums ("fancy" sapphires); colorless and black Corundum (leuco and black sapphire)

The species Corundum includes several of the world's most precious gemstone varieties, ruby and sapphire, as well as material vital to industry. Among natural substances, it is second only to Diamond in hardness, making it an excellent abrasive, used for machining metals, plastics, and wood, and in the production of emery. EMERY is a mix of Corundum and other substances, used to make sandpaper, grinding wheels, fingernail files, and other abrasive and polishing tools.

Pure Corundum is colorless. Only trace elements too sparse to be included in the chemical formula make the difference between colorless Corundum, ruby, and sapphire. The chemical formula of pure Corundum is Al_2O_3, and tiny traces of Chromium in the mix make ruby, while tiny traces of Titanium and Iron make sapphire. In the gem trade, natural Corundum is king among all colored gemstone species except certain colored Diamonds. Corundum can be colorless, black, or a variety of other colors. Red-colored Corundum gemstones are called ruby, and all other gem Corundums are called sapphire. Near-flawless vividly colored red ruby is by far the most valuable of all the Corundum gems.

Some Corundum gemstones have good ASTERISM ("starring" effect). If such a stone is red in color, it is called a star ruby (fig. 28) and if the stone is blue, white, yellow, orange, green, purple, gray, or black in color, it is called a star sapphire (fig. 34). The asterism in Corundum is caused by a dense network of extremely fine, hair-like inclusions of the mineral Rutile. While these stones are valuable gems in their own right, some can be heat-treated to dissolve the Rutile to produce transparent rubies and sapphires of good color. Cat's-eye Corundum varieties also occur if the inclusions are appropriately distributed, although these are extremely rare.

Gem-quality Corundum is very rare in nature, and the vast majority of all Corundum mined is dull opaque rock suitable only for industrial use. Synthetic Corundum is now being manufactured in enormous quantities for industrial and gem use. Synthetic Corundum was first produced in 1837 and has been produced on a commercial scale for more than 100 years. It is widely used for such purposes as mechanical parts requiring extreme resistance to wear, scratch-resistant watch crystals, windows for spacecraft, and lasers. Synthetic Corundum gems can be distinguished from natural Corundum gems by the lack of mineral and fluid inclusions indicative of natural crystals, and qualified gemologists can easily determine whether a Corundum gem is natural or synthetic. Natural Corundum gems (ruby and sapphire) have SILK inclusions, consisting of microscopic hair-like crystals of the mineral Rutile arranged in a particular pattern. These inclusions are sparse enough to allow transparency but observable under a jeweler's loupe or a microscope. If they are extremely dense, the stone exhibits asterism, as discussed above.

Natural Corundum usually forms in METAMORPHIC ROCKS like marble that are rich in Aluminum, or in INTRUSIVE IGNEOUS rocks like PEGMATITE. Corundum gemstones are most often mined from PLACER DEPOSITS (loose gravel) because rubies and sapphires are dense and resistant to the weathering processes that erode their host rock (often marble or pegmatite Feldspar). Natural Corundum is found in many places around the world. Although several color varieties are named after countries, such as Thai ruby (dark red), Burma ruby (medium to light rich glowing red), Ceylon ruby (light red to pinkish-red), or Kashmir sapphire (bright rich blue), today these names often have little to do with the country of origin. Because of the widespread use of color enhancement through HEAT TREATMENT and other techniques, and because of color variation within some mining deposits, these vernacular names often refer only to the color shade and appearance rather than the actual locality of the source mine.

Heat treatment of Corundum gems to enhance color and clarity is common practice in the gem trade today. It is widely accepted because it is only a slight manipulation of natural processes, as long as no attempt is made to add dyes, other trace elements, or fillers to the stone. It is often difficult or impossible to distinguish lab heat-treated stones from purely natural ones. Nevertheless, most stones that have been heat-treated in a lab are indicated as such by reputable gem dealers. IRRADIATION enhancement of color by exposure of the gemstone to a radioactive source is also occasionally used with Corundum gems (although to a lesser extent than heat treatment).

Presently, the most valuable variety of natural Corundum is ruby, with the finest stones of over 8 carats sometimes selling for $400,000 per carat or more. Among the varied colors of sapphire (blue, pinkish-orange, pink, orange, yellow, golden, purple, green, and transparent), the most valuable are the cornflower-blue Kashmir sapphire and the pinkish-orange padparadscha. These varieties are rare, highly sought-after gems and can be very expensive in sizes of 5 carats or more. Most Corundum gems are dense with inclusions, and they sell for substantially less than the near-flawless stones.

Non-gem, industrial-grade Corundum crystals can be quite large. The largest that we know of is a 335-pound (760,000-carat) crystal in the Geological Survey Museum of Pretoria, South Africa. Gem-grade Corundum is found in many different regions around the world and is discussed in the following sections on Corundum varieties. For further reading on Corundum gems, we recommend Hughes (1990, 1997), O'Donoghue (2006), and Bauer (1968).

RED CORUNDUM (RUBY)

SYSTEM Inorganic

CLASS Oxide

GROUP Hematite

SPECIES *Corundum*

VARIETIES ruby; star ruby

COMPOSITION

Aluminum Oxide (Al_2O_3)

TRACE ELEMENTS FOR COLOR

Chromium (Cr) and occasionally Iron (Fe)

HARDNESS ON MOHS SCALE 9

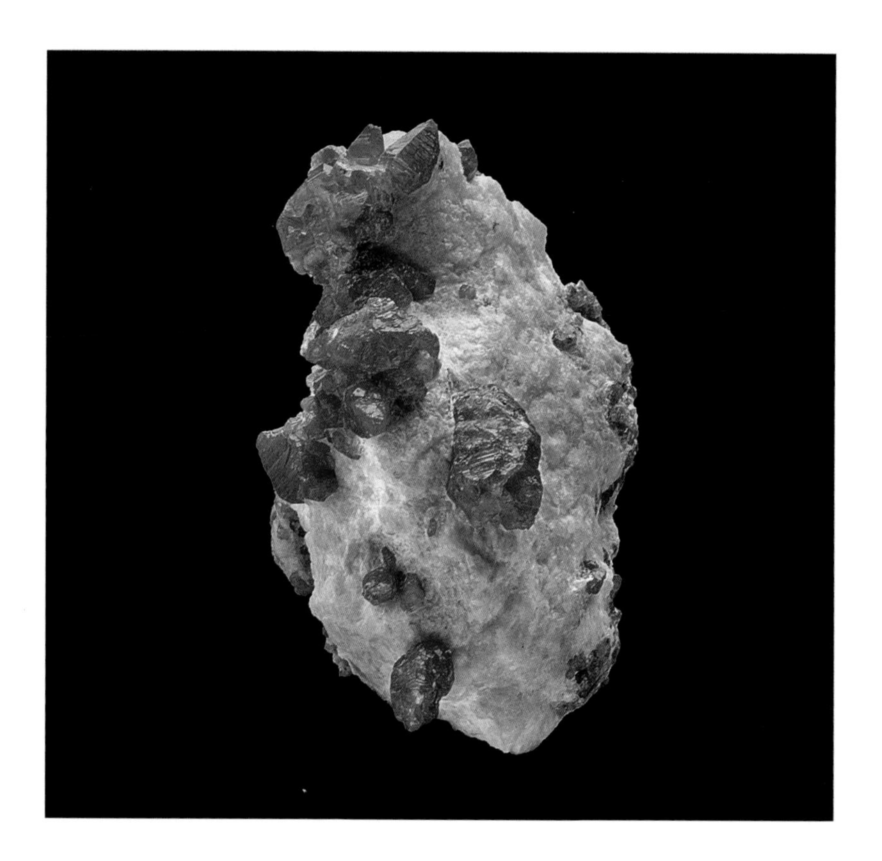

FIGURE 25. Natural crystals of ruby on white marble matrix. Specimen measures 120 mm high and is from Yen Bai Province, Vietnam (FMNH H2326-1).

Fine ruby is one of the most valuable gemstones on earth. Its magnificent red color is so intense that the very name ruby is considered to be a synonym for the color red. Many times rarer than colorless Diamonds, fine rubies have all the qualities that make a great gemstone: spectacular color, sparkling brilliance, very limited availability, and a hardness second only to Diamond.

Different combinations and amounts of the trace elements result in slightly different shades of red. The deeper red gemstones have historically come primarily from places like Mogok, Myanmar (formerly Burma), and Thailand; while the pinker-hued gemstones usually come from localities like Jagdalek, Afghanistan; Northern Pakistan; Palin, Cambodia; Southwestern Sri Lanka (formerly Ceylon); and Yen Bai, Vietnam (see the Vietnamese piece in fig. 25). As mentioned in the previous section, skillful heat treatment of rubies is common practice today, altering the color of stones from certain localities to match those from other localities. For example, people often want their rubies to match those from Myanmar. Most of the world's most valuable rubies have come from the Mogok mines of Myanmar, and the most valuable color in the world is an intense crimson color commonly called pigeon blood red. Clarity of a ruby is also important, but no natural rubies are perfectly flawless. In fact, the peculiar inclusions are part of what distinguishes natural rubies from

FIGURE 26.
Ruby necklace, with 13 faceted rubies totaling 90 carats in weight, set in Platinum with several hundred small baguette and round accent Diamonds (FMNH H2327-1).

synthetic rubies. One of the world's most valuable faceted fine rubies is the "Carmen Lucia" ruby, a 23.1-carat Burmese gem with extraordinary color and clarity that is part of the Smithsonian Institution's collection. Large Burmese gems of this quality are extremely rare and expensive. In 2006 a faceted 8.62-carat Burmese ruby of similarly fine quality known as the "Graff Ruby" sold at auction for $3,637,480 ($421,981 per carat). Larger-faceted rubies exist, such as the 48-carat "Mandalay Ruby" faceted from rough stone found in Mogok, Myanmar, but this stone had not been sold for decades and has no current sales information as of this writing. There are several pieces of ruby jewelry in the Grainger Hall of Gems, including a Platinum necklace with 90 carats of rubies and several hundred accent Diamonds (fig. 26).

Some rubies show a six-point asterism or **STAR** (see fig. 28) like their Corundum cousin the star sapphire. The star is the result of reflections from a dense network of hair-like inclusions of the mineral Rutile regularly oriented in accordance with the symmetry of the crystal's face, usually rendering it opaque, or slightly translucent. The Rutile needles formed out of liquids that entered the Corundum crystal through cracks and fissures as it was cooling. To properly bring out the star, the stone is cut into a **CABOCHON**, with careful orientation of the optical axis perpendicular to the back of the cabochon.

There are a number of vernacular "ruby" names used for red transparent stones that are not actually rubies, or even Corundum. "Balas ruby" and "spinel ruby" are actually Spinels. After close examination in the nineteenth century, some famous large "rubies" in the British Crown Jewels turned out not to be rubies at all but instead to be Spinels, such as the 352.5-carat "Timur Ruby" and the 170-carat "Black Prince Ruby." "Alabandine ruby" and "Garnet ruby" are Garnets, "copper ruby" is cuprite, "Siberian ruby" is Tourmaline, and "Brazilian ruby" is pink Topaz. Pyrope Garnet often strongly resembles ruby and has adopted similarly misleading names such as "American ruby," "Australian ruby," "bohemian ruby," "Rocky Mountain ruby," and a number of other names too numerous to list completely here. Also, "pink ruby" is sapphire by definition (red Corundum is ruby; pink Corundum is sapphire). "Geneva ruby," "Verneuil ruby," and "Chatham ruby" are all names for synthetic ruby. Ruby is one of the earliest precious gems to be made synthetically. In 1837 Marc Gaudin was able to make synthetic rubies by fusing Aluminum Oxide and trace quantities of Chromium at high temperature. Later, methods were greatly improved until large quantities of ruby could be synthesized particularly for industrial use. See pages 297–302 for further discussion of synthetic processes.

The earliest records for the mining of rubies go back over 3,500 years in Myanmar (Burma) and over 2,500 years in Sri Lanka. The ancient Hindus considered rubies to be "Ratnaraj," or the "King of Precious Stones." Thirteenth-century medical literature from India claims that rubies could cure digestive

FIGURE 27.
Ruby and Diamond pendant, with 23-carat ruby cabochon set in 18-karat yellow Gold with 6 carats of brilliant baguette-cut Diamonds. Pendant is 40 mm in width (FMNH H2384-1).

FIGURE 28.
Star ruby cabochon, from Sri Lanka, weighing 8.5 carats. Stone is 11 mm in height (FMNH H24-1).

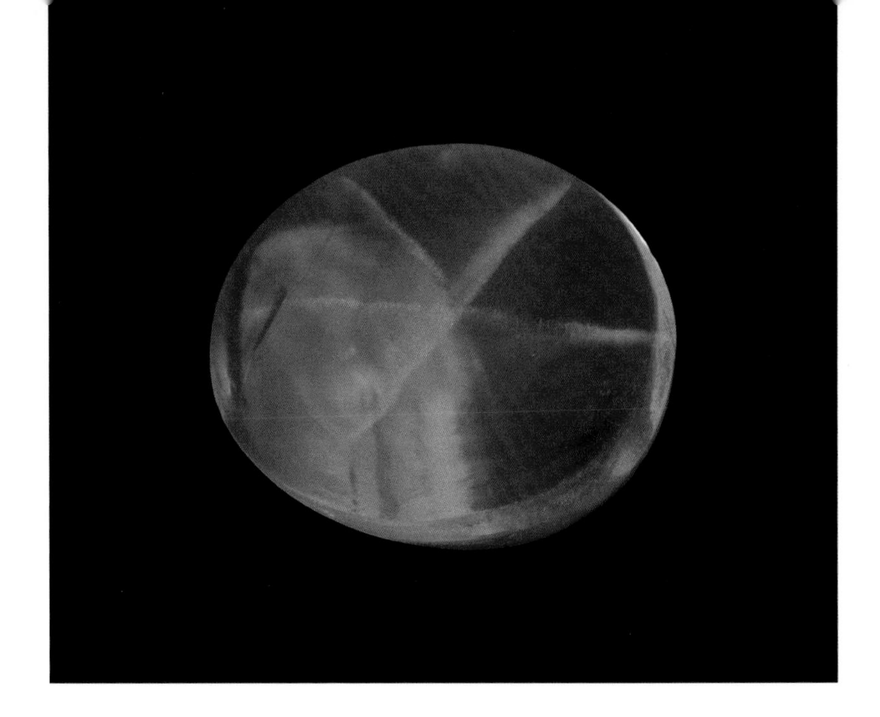

disorders. Burmese warriors inserted rubies under their skin to protect them in battle. Through the years, various cultures have considered rubies to be a symbol of wealth, power, and passion. They have been considered TALISMANS of spirituality, wisdom, protection from misfortune and illness, and triumph in love. The ruby is the modern birthstone of people born in July and is the gem associated with a 40th wedding anniversary in Western culture.

BLUE CORUNDUM (BLUE SAPPHIRE)
SYSTEM Inorganic
CLASS Oxide
GROUP Hematite
SPECIES *Corundum*
VARIETIES blue sapphire; color-change sapphire; blue star sapphire
COMPOSITION
Aluminum Oxide (Al_2O_3)
TRACE ELEMENTS FOR COLOR
Titanium (Ti) and Iron (Fe)
HARDNESS ON MOHS SCALE 9

Blue sapphire is to blue what ruby is to red: the very essence of the color blue. It can be the color of a cloudless sky, leading some ancient cultures to believe that fine sapphire reflected and lent its color to the heavens above. Technically, all colors of Corundum other than red are sapphires (see next section), but the first color that "sapphire" suggests to most people is blue. The various shades of blue in sapphire are the result of different trace combinations of Titanium and Iron. The most valued blue sapphires are neither very dark nor pale. They are a soft, velvety, vivid blue shade called "cornflower blue" or "Kashmir blue" or colors of similar hue, but because darker stones have become a more affordable standard today, most consumers today would not recognize Kashmir or cornflower blue as the top grade that they are. The major mining localities for sapphire are similar to those for ruby and include Myanmar (formerly Burma), Thailand, Sri Lanka (formerly Ceylon), and Kashmir, India. Important other mining localities for blue sapphire are Madagascar, Australia, Nigeria, Zimbabwe, Malawi, Tanzania, Pakistan, Cambodia, Colombia, Brazil, and China. There are also important deposits in the United States, particularly Montana.

FIGURE 29.
Below: Natural crystal of blue sapphire. Specimen measures 60 mm high and weighs 263.4 carats, and is from Kashmir, India (FMNH H1804-1).

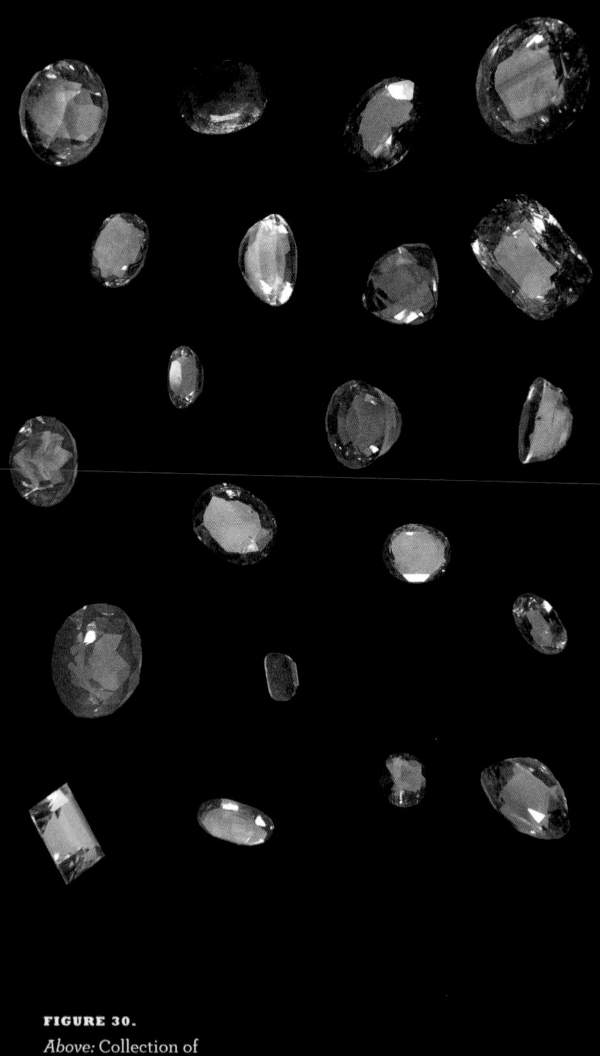

FIGURE 30.
Above: Collection of faceted blue sapphires from Sri Lanka showing the variation of color hue, and ranging in weight from 1.1 carats to 19.1 carats (FMNH H43-2, H43-3, H50, H51, H54, H58, H60, H61, H72, H1289, H1749–H1751, H1773, H1774, H1793, H1795, H1798, H1799). *Right*: Large blue star sapphire of 131.8 carats from Sri Lanka measuring 27 × 27 × 18 mm (FMNH H66-1).

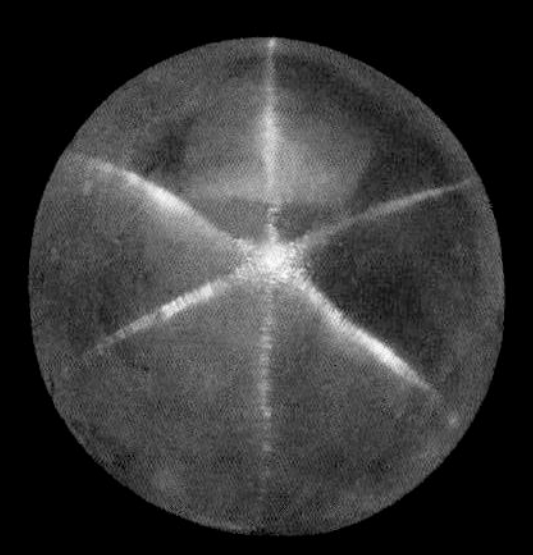

A number of authors have commented that the sapphires of Yogo, Montana, have been seriously undervalued.

Most blue sapphires show inclusions or uneven color distribution. Sapphires that are of ideal color and nearly flawless, even under a 10-power jeweler's loupe, are extremely rare. Like rubies, high-quality natural blue sapphires show fine, hair-like inclusions of Rutile called "silk" that do not interfere with the transparency of faceted gems. The finest sapphires, though not as valuable as the finest rubies, are still extremely valuable in sizes over 10 carats. Of all the colors of sapphire, the top-grade blue sapphires are the most valuable, followed closely by the orange-pink padparadscha. In 2007, a 22-carat Kashmir blue sapphire sold for $3,064,000 (over $139,000 per carat) at auction in New York. In contrast, small 1- to 2-carat commercial-grade dark blue sapphires sell for only $100 to $500 per carat. The 423-carat Logan sapphire, mounted in a brooch, is the largest faceted, near-flawless medium blue on public exhibit, part of the collection of the Smithsonian Institution. One truly exceptional blue sapphire in the Grainger Hall of Gems is a fine, transparent 60.2-carat gem carved into the form of a graceful face and set in Platinum with Diamonds (fig. 31).

Another type of rare and extremely valuable blue sapphire is the color-change sapphire, which shows a marked **ALEXANDRITE EFFECT**. The stone is blue to violet in natural daylight but changes to purplish blue or purplish red when lit by incandescent light. Color-change also occurs in some Garnet, Spinel, Tourmaline, and, most famously, alexandrite Chrysoberyl. Color-change sapphires over 10 carats are very rare. There is a near-flawless 56-carat blue sapphire in the Grainger Hall of Gems that shows color change. This gem is set in a white Gold pendant designed by Marc Scherer (fig. 33).

In some stones where the network of fine, hair-like Rutile inclusions is dense enough to render the stone translucent or opaque, asterism results. These stones are called star sapphires. Most stars are six-rayed but twelve-rayed stars also exist. Although the value of star sapphire is significant for high-quality stones with sharp asterism and good color, the value is not generally as high as it is for nearly flawless transparent stones. The most valuable star sapphires are natural stones that are intense pure blue (rather than grayish-blue), translucent, and with sharply defined, vivid stars. The largest blue star sapphire known to us is the 536-carat Star of India, which is a light gray-blue gem that was found in Sri Lanka and is now in the collection of the American Museum of Natural History in New York. There is an extremely fine 38.8-carat dark blue star sapphire set in a necklace designed by Lester Lampert in the Grainger Hall of Gems (fig. 34).

Heat treatment of blue sapphires is a widespread practice. The blue color of sapphire can be enhanced with properly applied heat, which rearranges the atoms of the gemstone. Very dark blue color can be lightened somewhat by

FIGURE 31.
Beautiful 60.2-carat blue sapphire carved into the form of a face, set in an Edwardian-style 18-karat Platinum pendant from the early twentieth century, with more than 300 small Diamonds. Chain is also Platinum with Diamonds. Carved face is 27 mm high, 21 mm wide, and 17 mm deep (FMNH H1580).

FIGURE 32.
Top: Blue sapphire and Diamond necklace in 14-karat rose Gold (FMNH H2278-1).

FIGURE 33.
Bottom left: Large faceted blue to purplish color-change sapphire from Sri Lanka weighing 56 carats set in 18-karat white Gold pendant with round brilliant-cut Diamonds. Designed by Marc Scherer of Marc and Co. in 2008 for the Grainger Hall of Gems. Pendant measures 50 mm from top to bottom and was named "Serendipity" by the designer (FMNH H2545). *Bottom, middle and right:* The sapphire before it was set, in daylight (*middle*) and incandescent light (*right*).

FIGURE 34.
Star sapphire pendant, "Etoilles," designed by Lester Lampert in 2008 for the Grainger Hall of Gems. The sapphire, from Sri Lanka, is 38.8 carats and is set in 18-karat white Gold (FMNH H2540).

a few minutes at 1200°C. Pale blue stones that are semi-opaque can often be made transparent by heating to 1200°C in open air. Patchy colored stones can be heated to eliminate the patches of color and then recolored more evenly with the application of heat and other enhancement techniques. Natural blue sapphire is generally much more valuable than heat-treated sapphire. Less stable enhancement treatments involve using chemicals to alter the color, but this method only changes a thin surface layer, and if the stone is chipped, the unaltered inside of the stone can become visible. Expertly applied heat treatment and other enhancements can be difficult to detect in sapphires.

Synthetic blue sapphire is made by the same process that produces synthetic ruby: by the VERNEUIL FLAME-FUSION process (see pp. 298–300). In 1947 a synthetic process for producing star sapphires and star rubies was introduced by the Linde Company, which resulted in mass quantities of high-quality synthetic star Corundums in the market. Synthetic star sapphires can be detected from natural stones because the Rutile needles are much smaller than those in natural stones, but usually 50-power magnification is needed to see this.

The earliest records of mining sapphires are similar to those of rubies: about 3,500 years ago in Burma (now Myanmar) and over 2,500 years ago in Sri Lanka. Early sapphire mining, like today, focused on searching gem gravels or placer deposits eroded from marbles, intrusive igneous rocks, or other sources. The Greek word *sapir*, or *sapphiros*, used in Exodus 28:18, is often translated as "sapphire." The actual meaning of the Greek word *sapir* is "blue."

Currently, it is thought that in the time of the Bible the terms *sapir* or *sapphiros* referred to lapis lazuli, a relatively common blue opaque mineral traded among Middle Eastern nations in biblical times. Blue sapphire was not really known in the Middle East region until the time of the Roman Empire. Blue sapphires have been considered as talismans of truth, authority, and protection, as well as healing. The blue sapphire is the birthstone for people born in September, and it is the gem associated with the 45th and 70th wedding anniversaries in Western culture.

OTHER COLORED CORUNDUMS ("FANCY" SAPPHIRES)

SYSTEM Inorganic
CLASS Oxide
GROUP Hematite
SPECIES *Corundum*
VARIETIES pink sapphire; yellow or golden sapphire; bicolor sapphire; orange sapphire; padparadscha; green sapphire; purple sapphire; colored star sapphires
COMPOSITION Aluminum Oxide (Al_2O_3)
TRACE ELEMENTS FOR COLOR dependent on color (see text)
HARDNESS ON MOHS SCALE 9

Although blue is the first color that comes to mind when most people think of sapphire, there are several other colors of great beauty and popularity, sometimes grouped under the name "fancy sapphires." Colored sapphires include pink sapphire, green sapphire, yellow or golden sapphire, purple sapphire, and, perhaps the most exotic, a pinkish-orange variety that has the distinction of having its own name, padparadscha. Rare bicolor forms also exist, such as yellow and blue stones (see fig. 41). Trace elements responsible for color in each variety are primarily mixtures of Chromium (Cr), Iron (Fe), and Vanadium (V) for the pinkish-orange padparadscha, various traces of Iron for yellow and green sapphires, Chromium for pink sapphires (as for rubies), and Vanadium for purple. Many of the finest colored fancy sapphires come from Sri Lanka, in the alluvial plains at the base of the Saffragam Mountains, and most of the colored sapphires in the Grainger Hall of Gems are from that region.

Padparadscha is mined almost exclusively in Sri Lanka, although it has also been found in Vietnam, Tanzania, and Madagascar. The largest faceted padparadscha known to us is a 100.18-carat stone from Sri Lanka in the collection of the American Museum of Natural History in New York. Yellow sapphires (aka "pushparaga" and "Oriental Topaz") come from most of the same deposits that produce blue sapphires and rubies, although many of the largest are either from Sri Lanka or Queensland, Australia. The largest faceted yellow sapphire we know of is a 400.06-carat stone called the "Ceylon Sinflower," discovered in Sri Lanka. This stone is currently owned by a private individual. Natural green sapphire, sometimes called "Oriental emerald," is rare. Green stones are found primarily in Queensland and New South Wales (Australia), Thailand, and Sri Lanka. We know of no extremely large faceted green stones, but a 195-carat rough stone called the "Stonebridge Green," found in Queensland, Australia, is currently in a private collection. A 33-carat faceted purple (violet) sapphire is in the collection of the American Museum of Natural History in New York. That stone was originally found in Thailand, but deep purple sapphires are also found in Tanzania.

FIGURE 35.
Facing page, right: Collection of faceted pink sapphires from Sri Lanka showing the variation in color hue, with gem weights ranging from 1.5 carats to 10.0 carats (FMNH H29, H41, H50, H79, H80, H1292, H1743–H1746, H1794, H1796, H1780, H2235, H2236, H2395).

FIGURE 36.
Facing page, left: Pink star sapphire cabochon from Sri Lanka weighing 4.8 carats and measuring 9 × 9 × 6 mm (FMNH H62).

FIGURE 37.
Facing page, center: Modern pendant with 4 colorless and 29 pink sapphires from Sri Lanka, showing a wide range of pink hues. Sapphires range in size from .3 to 5.2 carats each and are set in 14-karat white Gold. Designed by Thomas Guth of Oak Park Jewelers in 2008 for the Grainger Hall of Gems. Named the "Danielle Rose" by the designer (FMNH H2524).

Among the various colors of sapphire, only the padparadscha is as valuable as highest-quality blue sapphire. Padparadscha is also rarer than blue, and stones larger than a few carats are rarely seen for sale. Even small stones of 1 or 2 carats, if particularly fine, can command thousands of dollars per carat.

Pink sapphire should not be confused with the pinkish-orange padparadscha, a more valuable stone. Pink or "hot pink" sapphire, as it is sometimes known, has no trace of orange in it. The relationship of pink sapphire to ruby is confusing and controversial due to changing attitudes and perceptions of the gem trade and market. They are both more or less red-colored varieties of Corundum, with a complete gradation of shades ranging from faint red (pink "sapphire") to deep red (ruby) (figs. 35 and 37). So where do we draw the line between "sapphire" and "ruby"? Or should we even draw a line? Until the late nineteenth century, any form of red Corundum, no matter how light or "pink," was considered to be a ruby. Today what we call pink sapphire was previously thought of as light ruby. In the early nineteenth century or thereabouts, pink Corundums were termed "female" rubies as opposed to the deeper red "male" rubies. It wasn't until the early twentieth century that the term "pink sapphire" caught on for those "female rubies." This certainly differs from the practice with blue sapphire, where both light blue and dark blue hues occur, but both bear the name blue sapphire. The subdivision of the red Corundums into two differently named categories with an uncertain dividing line between them seems ambiguous, and therefore a problematic classification. In 1989 the International Colored Gemstone Association (ICA) agreed that this was an unnecessarily confusing classification and tried to fix the problem by recommending that all red Corundum be labeled ruby, regardless of how faint the red color is. The color pink is simply a very faint red color after all. This should have fixed the problem, but the term "pink sapphire" was already too etched into the mind of the powerful American gem market by then. Today the term "pink sapphire" continues to be widely used by both the consumer and the gem industry.

Purple sapphire, like the pink, is basically another shade of red that could be considered a type of ruby for many of the arguments put forth in the discussion about pink sapphires above. Deep purple sapphires are scarce and currently similar in value to orange sapphires. Good-quality purple or orange sapphires can sell for thousands of dollars per carat, and stones over 10 carats are rare. Tanzania is an excellent source of fine purple and mauve sapphires, often of a pastel shade.

For yellow sapphire, golden yellow or orange-yellow is worth more than pure yellow, but the distinction has lessened in the last century because the color of pure yellow can be enhanced to resemble other yellows with heat treatment. While not as rare as padparadscha or as valuable as most other colors of gem-quality Corundum, yellow sapphire is still a prized gemstone. Good

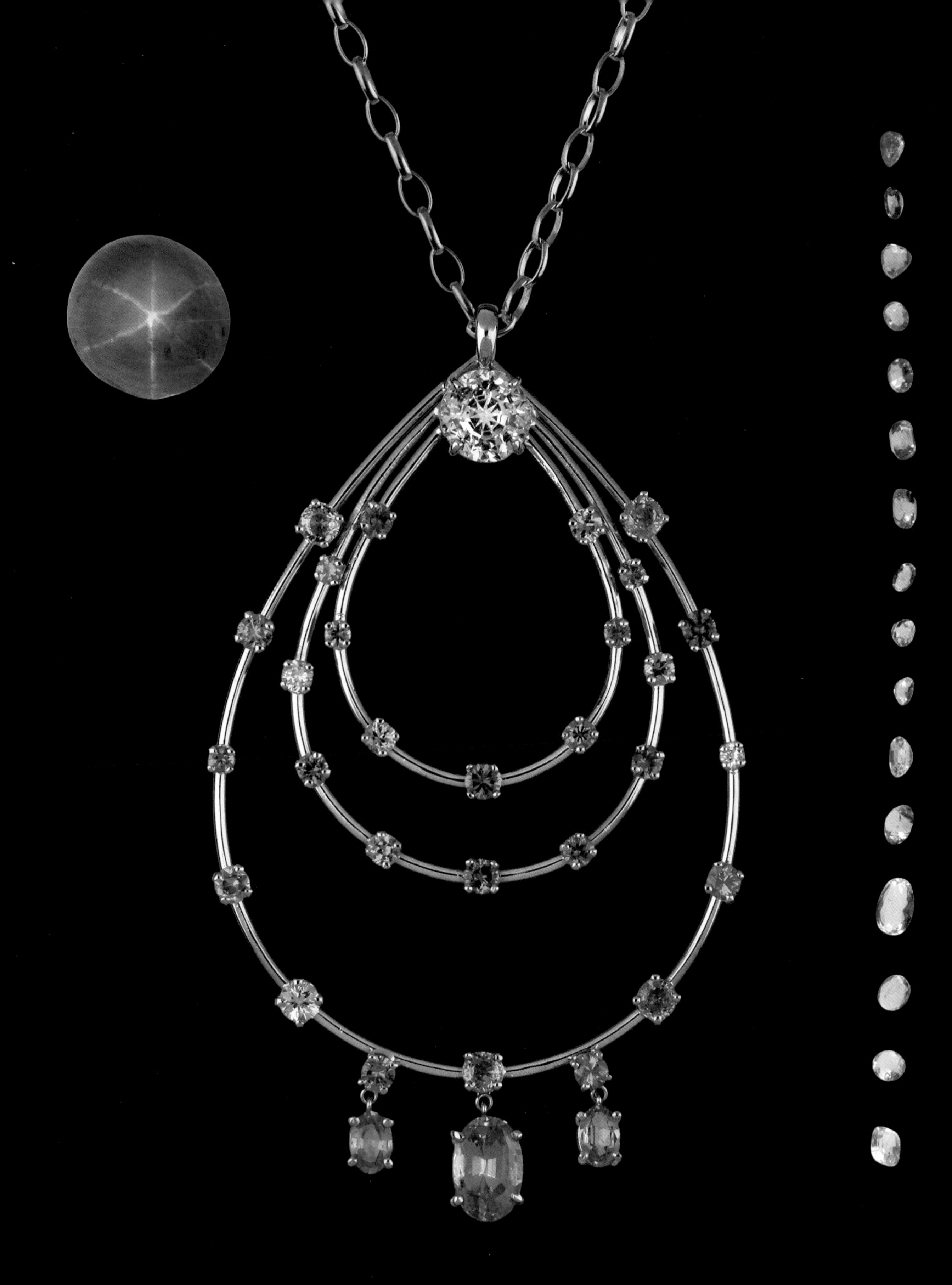

FIGURE 38.
Top left: Raw natural water-worn crystal of yellow sapphire from Sri Lanka weighing 83.9 carats and measuring 75 mm in length (FMNH H48).

FIGURE 39.
Bottom left: Collection of faceted yellow sapphires from Sri Lanka showing the variation in color hue, ranging from .87 carats to 8.25 carats in weight (FMNH H40.1–H40.7, H1298.1–H1298.3, H1740–H1742, H1800).

FIGURE 40.
Right: Modern design yellow sapphire bracelet with four sapphires from Sri Lanka mounted in solid 18-karat Gold. In oblique top view and side view to show depth of largest stone. The largest of the four sapphires is 102.2 carats. Designed by Lester Lampert in 2008 for the Grainger Hall of Gems and who named this piece "Sunrise" (FMNH H2532).

stones of over 20 carats are rare. One of the prized pieces in the Grainger Hall of Gems is an 18-karat Gold bracelet designed by Lester Lampert that contains four of The Field Museum's yellow sapphires, the largest of which is 102.2 carats (fig. 40).

Green sapphire is sometimes the result of very fine alternating bands of blue and yellow sapphire. Black, brown, olive, or gray color is often present, which reduces the value of green sapphire. The frequent olive or khaki tones lend an unattractive hue to the green, making this variety of sapphire the least popular. Some fine green sapphires come from Montana (fig. 43). Attractively colored green stones of more than 10 carats are rare and can be even more costly.

Asterism in orange, yellow, and padparadscha is extremely rare because these colors of sapphire from Sri Lanka usually lack the necessary concentration of well-defined inclusions to produce a distinct star. Nevertheless, star sapphires exist in these colors, but usually in small stones.

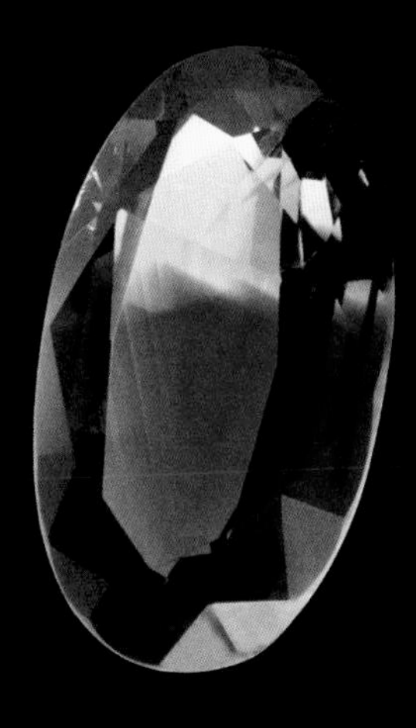

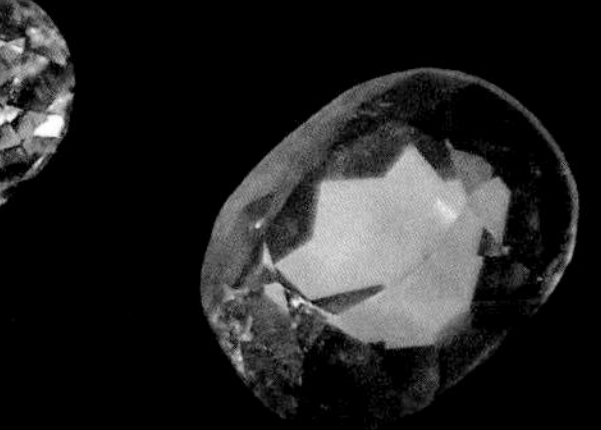

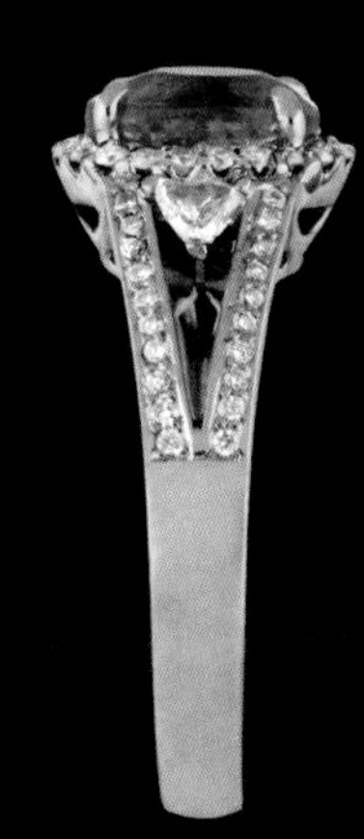

FIGURE 41.
Top: Bicolor sapphire (yellow and blue), faceted stone weighing 5.4 carats. Specimen is from Australia, measuring 14 mm in length (FMNH H53).

FIGURE 42.
Above: Padparadscha sapphire ring with 1.2-carat padparadscha from Tanzania set in 18-karat white Gold with small accent Diamonds. Side, top, and front views (FMNH H2365).

FIGURE 43.
Left: Faceted green sapphires, two from near west-central Montana, of 2.8 and 5.3 carats of weight (FMNH H1808, H1809), and one from Sapphire, Queensland, weighing 6.2 carats (FMNH H1371).

COLORLESS AND BLACK CORUNDUMS (LEUCO AND BLACK SAPPHIRE)

SYSTEM Inorganic
CLASS Oxide
GROUP Hematite
SPECIES *Corundum*
VARIETIES colorless sapphire (leuco); white star sapphire; black sapphire; black star sapphire
COMPOSITION Aluminum Oxide (Al_2O_3)
TRACE ELEMENT FOR COLOR none
HARDNESS ON MOHS SCALE 9

Colorless transparent sapphire is rare in nature and is also known as leucosapphire and also sometimes referred to as white sapphire. Because of its hardness and fine optical properties, colorless sapphire is sometimes used as alternatives to Diamonds, although it lacks the intense fire and brilliance of Diamond. In the sixteenth century, a common folklore was that colorless sapphire was a form of unripe ruby, and that if it was left to mature, it would one day become a red ruby.

Some totally colorless sapphire is made by heat-treating very pale sapphires in a way that removes all color. Natural colorless sapphire is relatively rare today, partly because it is uncommon to find it, and also because much of what has been found has been irradiated or heat-treated to turn it blue, orange, or yellow to increase its value. Sri Lanka is the major source of colorless sapphire. Asterism occurs in colorless sapphire as in other sapphire varieties and results in white star sapphire.

Synthetic colorless sapphire is used for scratch-resistant optical lenses, such as watch crystals over the dial in high-quality watches. It is also used to make scratch-resistant lenses for lasers and optical equipment as well as windows for spacecraft.

Black Corundum is technically black sapphire, but very little of this is used as gem material. Most is used as an industrial abrasive, or EMERY. When asterism is present, the result is black star sapphires, which are used by the gem trade. These are the least valuable of the star sapphire varieties in general. The largest known black star sapphire is the Black Star of Queensland, a 733-carat stone in a private collection.

FIGURE 44. Natural, double-terminated crystal of colorless sapphire, 34 mm long and weighing 32.9 carats, from Sri Lanka (FMNH H2438).

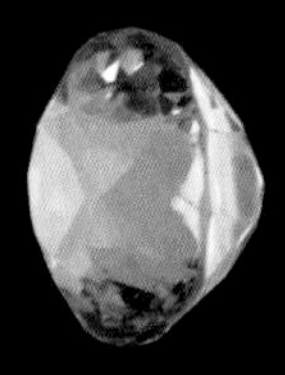
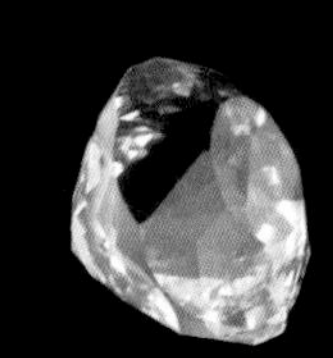

FIGURE 45.
Colorless sapphire gems from Sri Lanka, including a 2.57-carat faceted stone set in a yellow Gold ring flanked by blue sapphires, and a series of unset faceted stones ranging from 3.1 carats to 25.7 carats in weight. (Ring is FMNH H2478, unset stones are FMNH H38, H2389, H2390, H2392, H2394.)

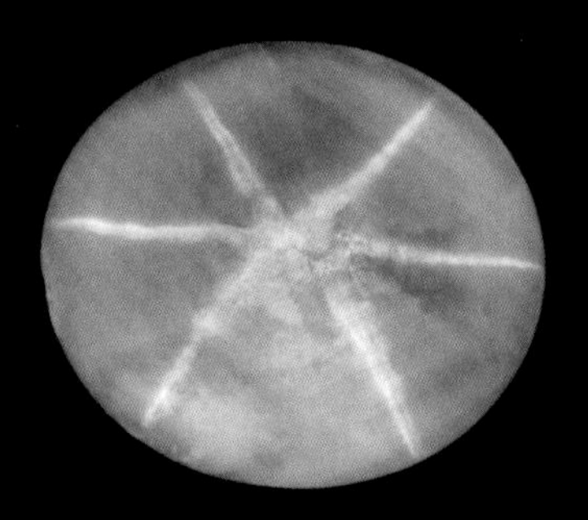
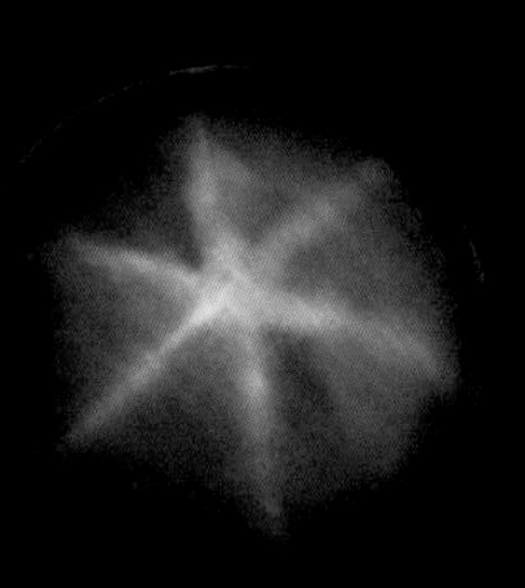

FIGURE 46.
Left: White star sapphire cabochon from Sri Lanka weighing 11.75 carats, measuring 17 × 14 × 6.5 mm (FMNH H2450).

FIGURE 47.
Right: Black star sapphire cabochon from Sri Lanka weighing 23 carats and measuring 15 × 15 × 7 mm (FMNH H2473).

Chrysoberyl

SYSTEM Inorganic

CLASS Oxide

GROUP Chrysoberyl

SPECIES *Chrysoberyl*

GEM VARIETIES color-changing Chrysoberyl (alexandrite); single-color Chrysoberyl (yellow, green, or brown); chatoyant Chrysoberyl (cat's-eye, or cymophane)

Chrysoberyl is a very hard gemstone, harder than all mineral species except Diamond and Corundum. It is a fine gemstone, with much brilliance, but lacking the fire found in Diamond. Yellow Chrysoberyl was the earliest known variety of this gem species, and in ancient times it was mistakenly thought to be a yellow variety of Beryl. The name Chrysoberyl is Greek, referring to the golden color of yellow Chrysoberyl (*chrysos*) and the Beryllium content (*beryllos*). At a Mohs hardness of 8½, Chrysoberyl is harder, has a different chemistry, and has a different crystal structure than Beryl.

Chrysoberyl gems are among the world's rarest gem varieties. Chrysoberyl varieties include the famous transparent color-changing alexandrite, the CHATOYANT cat's-eye, and fine transparent stones of yellow, green, or brown. Colorless Chrysoberyl is extremely rare. Chatoyant cat's-eye occurs primarily with the single-color varieties of yellow and green, but alexandrite also rarely occurs as chatoyant cat's-eye. Natural black Chrysoberyl is unknown to us, but there are irradiated cat's-eye stones that are black or dark gray.

Chrysoberyl is found principally in pegmatites often associated with mica, or in placer deposits of rivers and streambeds. Major mining localities for Chrysoberyl include the Ural Mountains of Russia, Sri Lanka, Myanmar, India, Zimbabwe, Tanzania, Mozambique, and Minas Gerais, Brazil. For further reading on Chrysoberyl, we recommend Bauer (1968) and O'Donoghue (2006).

COLOR-CHANGING CHRYSOBERYL (ALEXANDRITE)

SYSTEM Inorganic

CLASS Oxide

GROUP Chrysoberyl

SPECIES *Chrysoberyl*

VARIETIES alexandrite; cat's-eye alexandrite

COMPOSITION Beryllium Aluminum Oxide ($BeAl_2O_4$)

TRACE ELEMENTS FOR COLOR Chromium (Cr), Iron (Fe), and Titanium (Ti)

HARDNESS ON MOHS SCALE 8½

Alexandrite is one of the rarest gems, and it is most famous for its dramatic color change, often referred to as the ALEXANDRITE EFFECT. "Ideal" examples of the gem are typically emerald green in daylight changing to raspberry red in incandescent light or candlelight. The color-changing habit has led to the description of "emerald by day, ruby by night" for this gem. Alexandrite is named after the Russian czar Alexander II (1818–1881) and was the national stone of czarist Russia. It was first discovered in 1833 in the emerald mines of the Ural Mountains, where it was mistaken for emerald. It was eventually found to be a different mineral species because of its different chemical composition, greater hardness, and its characteristic color change. Alexandrite is much rarer than emerald. Although the best-quality alexandrites are thought to come from Russia, Sri Lanka, and Brazil, fine examples of the gem have also been found in Tanzania, India, Myanmar, Madagascar, and Zimbabwe. The color-change property of the stone tends to be less dramatic in most of the non-Russian localities.

Top-quality red-green faceted alexandrites are extremely rare, and as of 2004 "ideal" stones of more than 5 carats were selling for $100,000 per carat or more. Generally, the stronger the color change in the stone, the more valuable

FIGURE 48.
Above left: Natural alexandrite crystals on matrix from Ekaterinburg (Yekaterinburg), Russia. Specimen measures 105 mm in height and weighs 378 grams (FMNH M8644).

FIGURE 49.
Above right: Faceted alexandrite gem weighing 11.65 carats from Sri Lanka showing the green/red color-change characteristic of this gem variety. Color in daylight (*top*) and color in incandescent light (*bottom*) for the same stone. Stone is 14 mm in height. Fine alexandrites of this size and quality are extremely rare (FMNH H2537).

it is per carat. The largest faceted alexandrite is a 66-carat stone from Sri Lanka that is now in the collection of the Smithsonian Institution. Fine alexandrite gems over 10 carats are almost unheard of. One such stone in the Grainger Hall of Gems is a near-flawless 11.65-carat stone set in an 18-karat green and rose Gold ring (fig. 51). This stone exhibits a fine color change.

Some stones promoted as synthetic alexandrite are actually synthetic Corundum laced with traces of Vanadium, and not actually Chrysoberyl. Sometimes bearing the name "Czochralski alexandrite," they are more appropriately termed "simulated alexandrite" than synthetic, or better yet "synthetic alexandritic Corundum." The adjectival form of the word "alexandrite" (ALEXANDRITIC) is sometimes used to describe the color-changing property of other gem species, such as alexandritic sapphire or alexandritic Tourmaline.

There is also a rare chatoyant form of alexandrite called cat's-eye alexandrite (not to be confused with the somewhat more common non-color-changing cat's-eye discussed below). Cat's-eye alexandrite shows both color change and

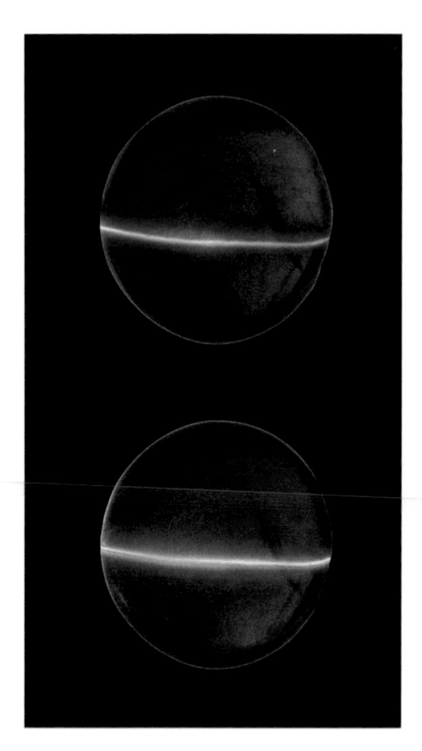

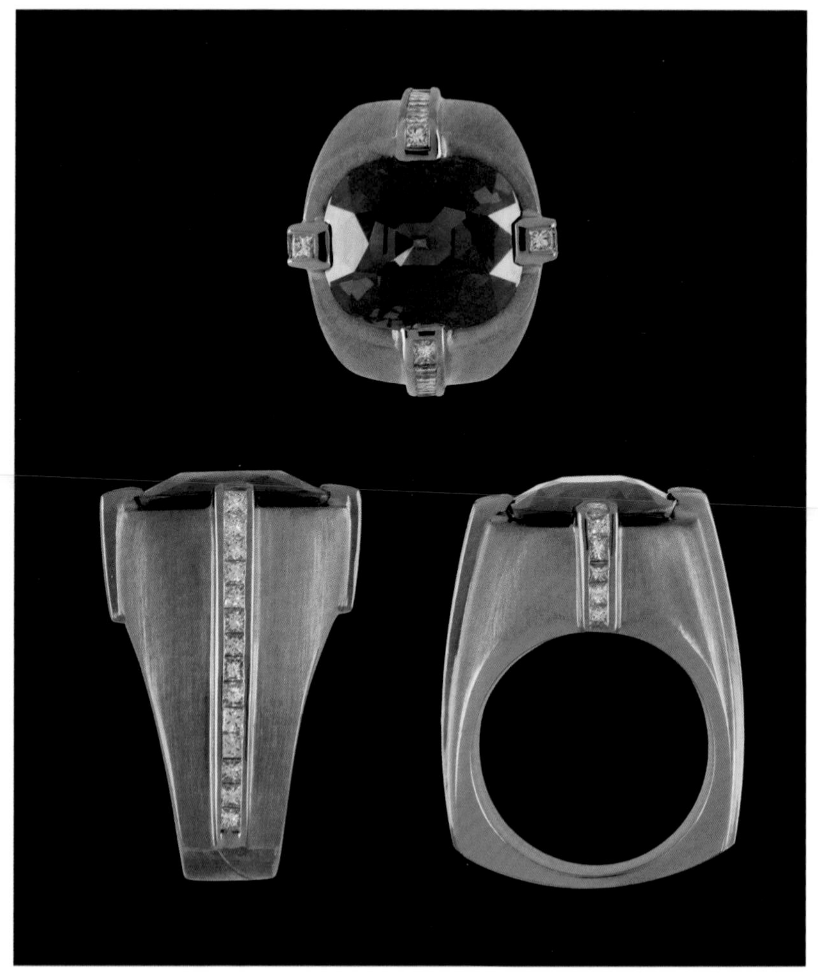

FIGURE 50.
Above: Cat's-eye alexandrite cabochon from Sri Lanka weighing 3.15 carats showing the green/red color-change characteristic of this gem variety. Color in daylight (*top*) and color in incandescent light (*bottom*) for the same stone. Stone is 8.5 × 8.5 × 3 mm (FMNH H2488).

FIGURE 51.
Right: Alexandrite ring, "Chameleon," containing the 11.65-carat gem from figure 49 set in brushed 18-karat green and rose Gold and accented with 42 princess-cut Diamonds. The rows of Diamonds divide alternating rose and green Gold sections, a theme following the color-change characteristic of the alexandrite gem. Designed by Lester Lampert in 2008 for the Grainger Hall of Gems. Top, side, and front views (FMNH H2537).

CHATOYANCY. Polished stones of cat's-eye alexandrite over 2 carats are extremely rare. Because of the chatoyancy, cat's-eye alexandrites are finished as cabochons rather than as faceted stones. A 3.15-carat cat's-eye alexandrite from Sri Lanka is in the Grainger Hall of Gems (fig. 50).

SINGLE-COLOR CHRYSOBERYL

SYSTEM Inorganic

CLASS Oxide

GROUP Chrysoberyl

SPECIES *Chrysoberyl*

VARIETIES transparent yellow Chrysoberyl; transparent green Chrysoberyl; cat's-eye Chrysoberyl (cymophane)

COMPOSITION Beryllium Aluminum Oxide ($BeAl_2O_4$)

TRACE ELEMENT FOR COLOR Iron (Fe) for yellow

HARDNESS ON MOHS SCALE 8½

FIGURE 52.
Top: Natural crystal of yellow Chrysoberyl from Espírito Santos, Brazil. Specimen weighs 213 carats and measures 30 × 30 × 35 mm (FMNH H1545).

FIGURE 53.
Bottom: A group of 10 faceted yellow Chrysoberyl gems from Sri Lanka in sizes ranging from 1.4 carats to 13.5 carats, showing color variation ranging from yellow to green (FMNH H1295, H1372, H1508, H1712, H1901, H2185, H23273, H2375–H2377).

Transparent Chrysoberyl is an excellent gemstone for faceting and jewelry. Because of its hardness, it is highly resistant to wear. Transparent Chrysoberyl is normally yellow to yellowish-green and rarely a true green (fig. 53). In the Victorian and Edwardian eras, it was referred to as "chrysolite," a name no longer used by gemologists for this species. Stones over 15 carats are rare. The largest flawless faceted Chrysoberyl known to us is the Hope Chrysoberyl in the collection of the Natural History Museum in London. This light green gem weighs 45 carats. Faceted transparent Chrysoberyl has an attractive VITREOUS LUSTER and is often PLEOCHROIC. But it is only rarely seen in jewelry due to its scarcity, and due to it being overshadowed by the more popular chatoyant

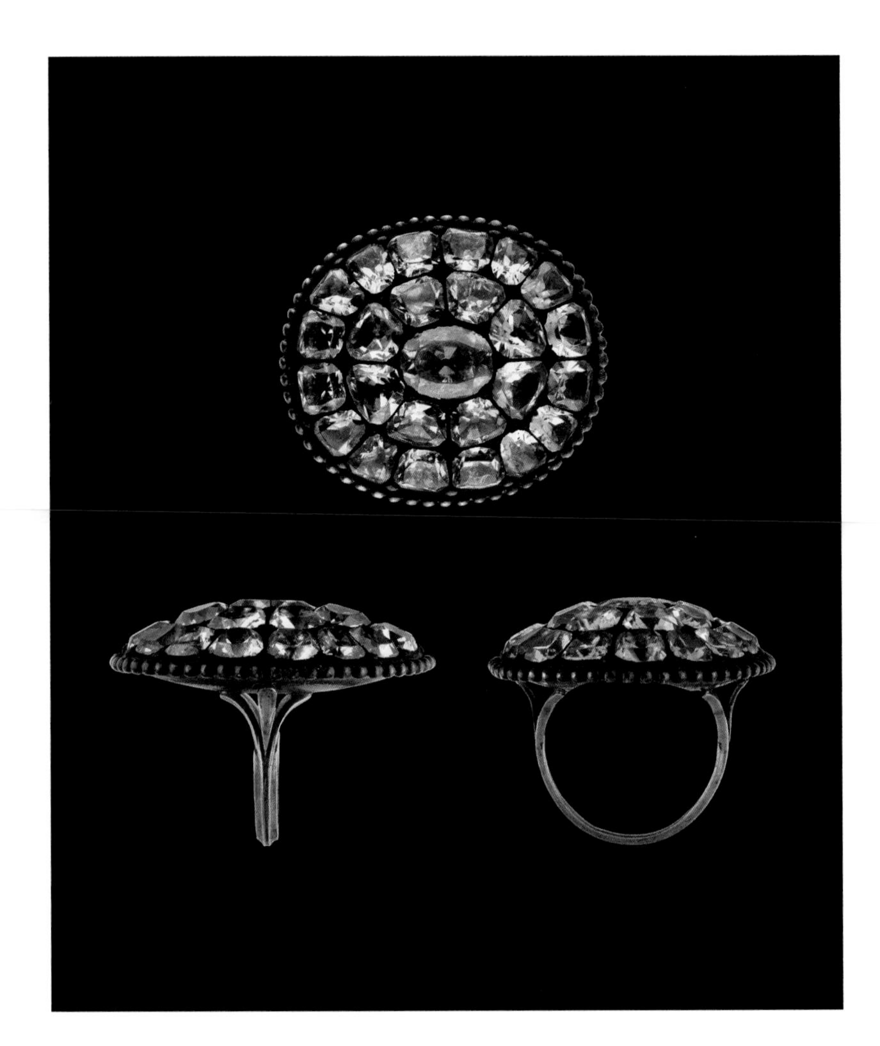

FIGURE 54.
Yellow Chrysoberyl ring from late eighteenth-century Brazil, probably of Spanish origin, with 25 faceted cushion-cut Chrysoberyls set in black-coated Gold metal. Top, side, and front views. The face of the ring measures 30 × 35 mm (FMNH H113).

form, cat's-eye. The finest transparent Chrysoberyl comes from Minas Gerais in Brazil, Russia's Ural Mountains, India, Sri Lanka, Myanmar, and Zimbabwe.

Chatoyant yellow or green Chrysoberyl is the most desirable of the non-color-changing Chrysoberyls. These are called cymophane or cat's-eye. Only a small percentage of overall Chrysoberyl production is cat's-eye. A soft honey color is considered by many gemologists to be the best color for cat's-eye, and the most valuable stones are those with a very thin, sharply defined band of light. The best cat's-eye Chrysoberyl comes from Sri Lanka, although fine material is also produced in Brazil.

Although other mineral species with single-rayed chatoyancy such as some Corundum, Elbaite, and Quartz varieties are occasionally referred to as "ruby cat's-eyes," "Tourmaline cat's-eyes," or "Quartz cat's-eyes," only chatoyant Chrysoberyl can validly bear the simple name cat's-eye. Cat's-eye is yellow, greenish-yellow, or brownish-yellow. Very rarely, cat's-eye can be reddish in

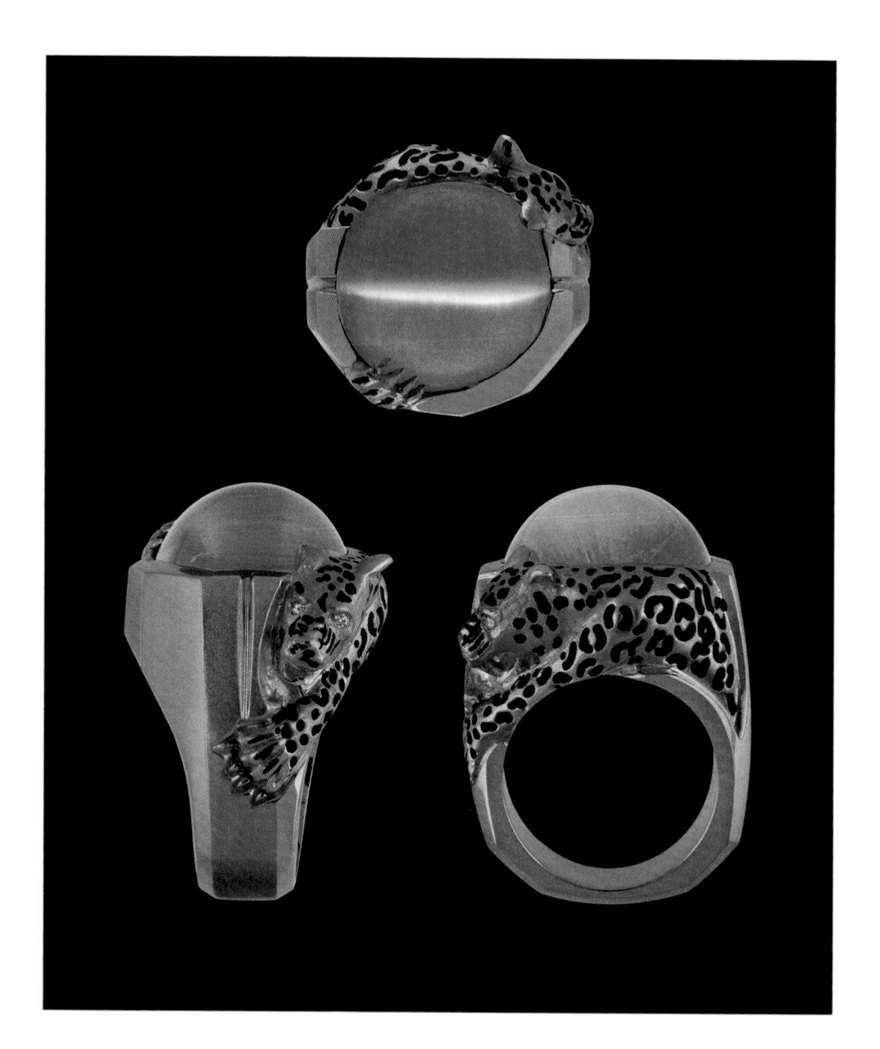

FIGURE 55.
Spectacular cat's-eye Chrysoberyl ring with a 35.5-carat cabochon from Sri Lanka. Mounted in an 18-karat Gold sculpted ring called "Eye of the Leopard." Designed by Lester Lampert in 2008 for the Grainger Hall of Gems. Top, side, and front views (FMNH H2536).

color, but these are generally stones of under a carat in weight. The chatoyancy results from fine, hair-like inclusions of Rutile, much like the chatoyancy in sapphires. As with all star and cat's-eye gems, a domed surface is needed to bring out the rays of asterism or chatoyancy. Cat's-eye is polished into cabochons rather than faceted, in order to show the chatoyant "eye" effect. Cat's-eye is to transparent Chrysoberyl as star sapphire is to transparent sapphire.

Cat's-eye gems over 10 carats are very rare. The largest known cat's-eye cabochon is the "Eye of the Lion," a 465-carat gem from Sri Lanka. This gem was cut from a rough stone of over 700 carats, discovered in the late 1800s. It is a translucent dark greenish-yellow stone with an exceptionally intense "eye" and is one of the finest gems that have come from Sri Lanka. As this book goes to press, this stone is still in private hands. There is a spectacular 35.5-carat cat's-eye Chrysoberyl in the Grainger Hall of Gems set in an 18-karat Gold sculpted ring called "Eye of the Leopard," designed by Lester Lampert (fig. 55).

Spinel

SYSTEM Inorganic

CLASS Oxide

GROUP Spinel

SPECIES *Spinel*

GEM VARIETIES red Spinel; other colored Spinel

FIGURE 56.
Natural octahedral crystal of red Spinel in white marble matrix, from Mogok, Myanmar. Entire mineral specimen is 70 mm in length (FMNH H2446).

Spinel is both a family group name for a number of mineral species, and a species name for a gemstone. Here we focus on the gemstones only, and primarily red Spinel because that is the best known. Spinel occurs in several different colors and also in black. Colorless natural Spinel was unknown until very recently, and it is still extremely rare. ASTERISM also occasionally occurs in Spinel with a four- or six-rayed star.

Synthetic Spinel has been produced since about 1910, primarily as colorless stones (to simulate Diamond or moonstone), blue stones (to simulate aquamarine, sapphire, or lapis lazuli), or red stones (to simulate ruby and pink sapphire). One variety of synthetic Spinel even shows color change from violet in daylight to more reddish in incandescent light. Because the production of synthetic Spinel is so common, there is a perception that the natural gem species is more common than it actually is. Fine natural gem-quality Spinel is very rare.

Natural Spinel crystals are sometimes sharply pointed octahedrons (see fig. 56). The name Spinel partly derives from the Latin *spina*, meaning "little thorn." Natural crystals of Spinel in MATRIX are rare. The finest natural crystals come from metamorphosed impure limestones (marble). Major gem-grade Spinel is found mainly in placer deposits of Myanmar, Sri Lanka, Madagascar, Nigeria, Afghanistan, Pakistan, Brazil, Australia, and Europe. For further reading on Spinel, we recommend Bauer (1968) and O'Donoghue (2006).

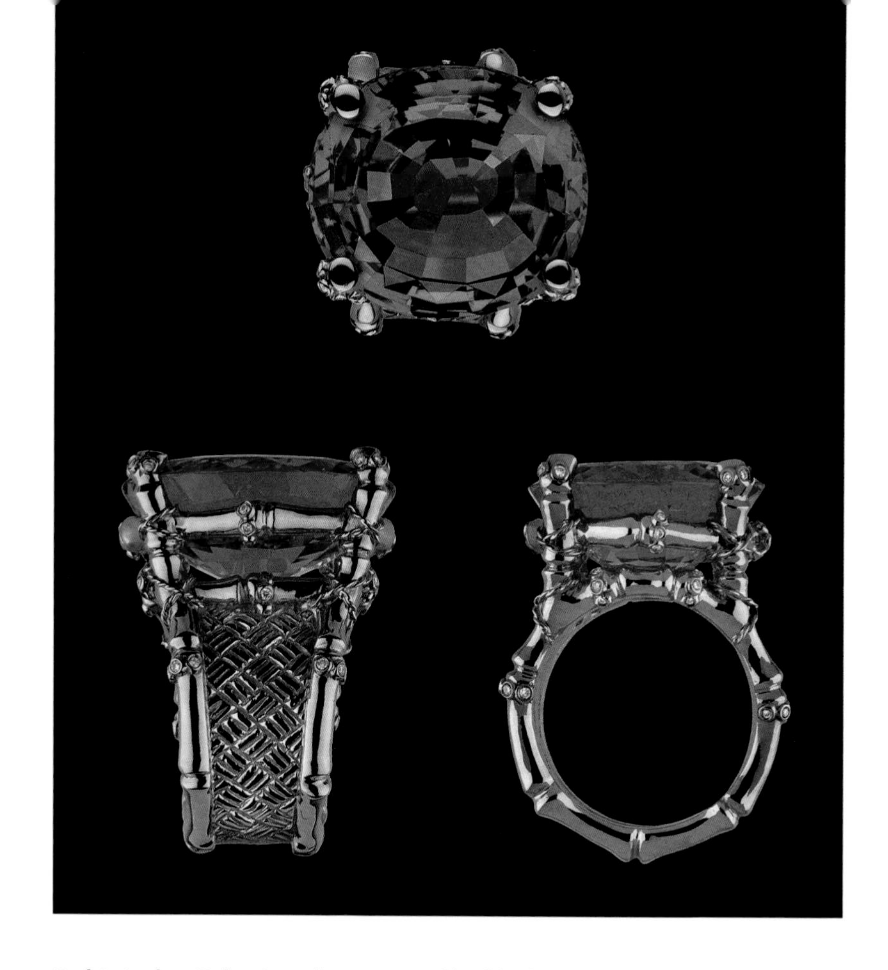

RED SPINEL

SYSTEM Inorganic

CLASS Oxide

GROUP Spinel

SPECIES *Spinel*

VARIETY red Spinel

COMPOSITION Magnesium Aluminum Oxide ($MgAl_2O_4$)

TRACE ELEMENTS FOR COLOR Chromium (Cr) and Iron (Fe)

HARDNESS ON MOHS SCALE 8

FIGURE 57. Faceted 31.9-carat red Spinel from Myanmar set in 18-karat yellow Gold "Bili-Bili" ring designed by Mish Tworkowski of Mish New York. Top, side, and front views. Spinel gem measures 22 × 20 × 10 mm. Setting © Mish, Inc. (FMNH H2521).

Red Spinel, or Balas Spinel, is a rare and highly desirable gemstone with a vitreous luster capable of high polish. The most expensive stones are characterized by a deep red color without gray or brown tones. The orange-red flame Spinels are also premium stones. Red Spinels have historically been mistaken for rubies because of a similarity of color, and because red Spinels and rubies are sometimes found together in the same water-worn gravels of placer deposits. However, Spinels are easily distinguishable from rubies by their different hardness (Mohs 9 for ruby, Mohs 8 for Spinel) and by various optical properties. Spinel was first clearly distinguished from ruby as a separate gemstone species in the eighteenth century. A famous example of misidentification includes the "Black Prince's Ruby" and the "Timur Ruby," both set in England's Crown Jewels. These two "rubies" are actually red Spinels. The fact that these stones were found to be Spinels instead of rubies did not appreciably affect their value, in part because of their historical provenance. The 361-carat red Spinel formerly known as the "Timur Ruby" is now in the collection of the queen of England and is engraved with the names of Mughal emperors who previously owned it, giving it an extremely rich pedigree. Another notable red Spinel mistaken for a ruby is the 398.7-carat stone on the Imperial Russian Crown, worn by all Russian emperors from 1762 through the abolition of the monarchy in 1917.

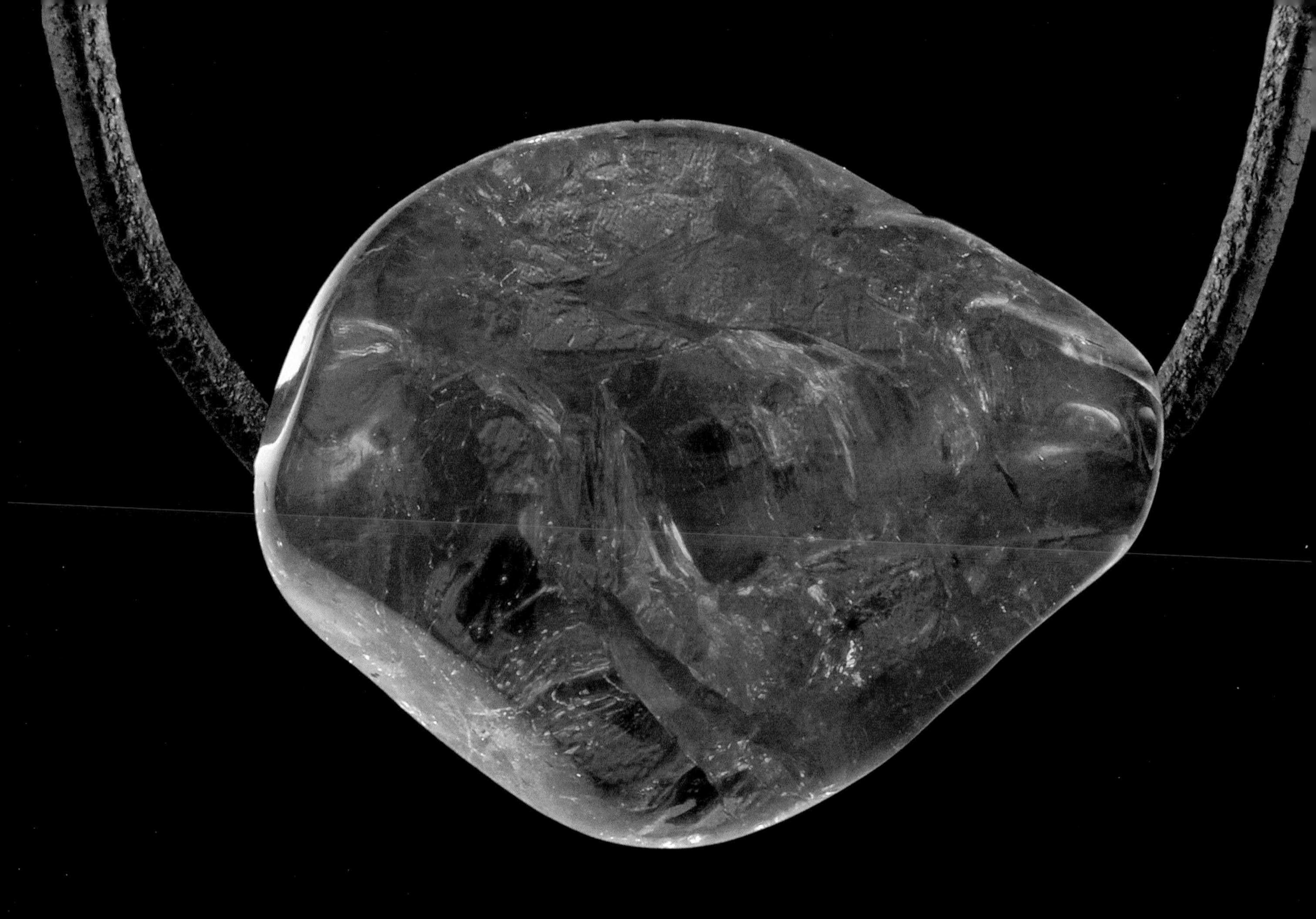

FIGURE 58.
Tumble-polished 190.6-carat red Spinel with an Urdu inscription (enlarged for inset box), from India, thought to have been used as a love token in the early to mid-twentieth century. Inscription reads "Light of the Universe" (FMNH H1659).

Even without a crown, fine red Spinels are very valuable and treasured gems. Fine red Spinels are much rarer than ruby, although they are currently less expensive due to the name-brand appeal and commercial popularity of rubies.

The largest red Spinel gem is the Samarian Spinel, an unfaceted stone of 500 carats in the Crown Jewel Collection of Iran. In 2007 a giant red Spinel crystal weighing 52 kilograms (115 pounds) was discovered in Tanzania. Although only about 3 percent of the stone was suitable for gem material, many fine stones from 5 to 30 carats were faceted from this piece.

There are some extremely fine red Spinels in the Grainger Hall of Gems, including a near-flawless 31.9-carat stone from Myanmar set in an 18-karat Gold ring by Mish Tworkowski of New York (fig. 57), and a peculiar tumble-polished 190.6-carat red Spinel with an Urdu inscription (fig. 58).

OTHER SPINEL (COLORS OTHER THAN RED)

SYSTEM Inorganic
CLASS Oxide
GROUP Spinel
SPECIES *Spinel*
VARIETIES blue Spinel (cobalt Spinel or gahnospinel); pink Spinel; purple Spinel; yellow Spinel; dark green Spinel; black Spinel (ceylonite); star Spinel
COMPOSITION gem-quality Spinels are usually Magnesium Aluminum Oxide ($MgAl_2O_4$), although gem-quality blue or purple Spinels include Zinc (Zn) and little or no Magnesium.
TRACE ELEMENTS FOR COLOR Chromium (Cr) and/or Iron (Fe) for most colors, Cobalt (Co) for blue
HARDNESS ON MOHS SCALE 8

Spinel occurs in colors other than red, but most of these gemstones are very rare. One such rare color is blue, sometimes called cobalt Spinel, gahnospinel, or sapphire Spinel. Although the luster of even the best blue stones does not quite match that of sapphire, good-quality blue Spinels are extremely rare and highly valued. Vivid blue stones without steely gray tones are the most valuable. A fine example of a top-quality blue Spinel is in the Grainger Hall of Gems: a nearly flawless 23.29-carat gem from Sri Lanka surrounded by Diamonds set in a 22-karat Platinum ring designed by Lester Lampert (fig. 59).

Spinel is found in a variety of other colors, including pale to mauve-pink or grayish light purple, orange yellow to straw-yellow Spinel (known as rubicelle or vinegar Spinel), and very dark green (not usually of gem quality). The least expensive Spinel color is opaque black, also known as ceylonite or pleonaste. Asterism also rarely occurs in black Spinel, producing four- or six-rayed stars, and black star Spinel is considered a gemstone. Colorless Spinel was unknown until recently and is extremely rare, as is pure bright green and vivid yellow Spinel. Natural color-changing (alexandritic) Spinels occur but are very rare and highly valued.

Slight variations in basic chemical composition exist for some Spinel colors, technically making them other gem species. The complexities of the Spinel family and its other species are extensive and will not be discussed in detail here.

FIGURE 59.
A rare, nearly flawless 23.29-carat blue Spinel from Sri Lanka, with deep blue color and purple overtones. Set in 22-karat Platinum ring surrounded by 14 round brilliant-cut Diamonds weighing 2.1 carats. Band is inlaid with an additional 130 tiny ideal-cut Diamonds in a tapered band. Top, side, and front views. Piece named "Duchess" was designed in 2008 for the Grainger Hall of Gems by Lester Lampert (FMNH H2541).

Quartz

SYSTEM Inorganic
CLASS Oxide
GROUP Quartz
SPECIES *Quartz*
GEM VARIETIES amethyst (purple); citrine (yellow); other (various)

As the most abundant mineral species in the earth's continental crust, Quartz seems an unlikely candidate for the "precious gem" category, yet certain varieties have been highly valued over time. Several color varieties are quite attractive (especially purple and yellow), and the fact that Quartz has no CLEAVAGE makes it resistant to chipping or cracking, and thus a sturdy stone for rings. Colorless Quartz, or rock crystal, was thought by the ancient Greeks and Romans to be ice that would not melt because it was created by the gods. This "ice" in its most transparent form has had many industrial applications in optical equipment, electronics, and fiber optics. In ancient times, transparent colorless Quartz was used to make "crystal balls" used by mystics (fig. 292). But colorless Quartz is far too common and lacking in fire to be of much use as true gem material. The only major varieties of Quartz that have been historically prized by the fine-gem trade are the transparent purple (amethyst) and the transparent yellow (citrine) varieties. Many fine-grained varieties of Quartz—such as agate, tiger's eye, and jasper—are popular lapidary source materials and beloved amateur collector stones, but they are also extremely common, making them of limited value to the fine-gem trade.

Quartz can be thought of in two major categories: MACROCRYSTALLINE varieties and MICROCRYSTALLINE varieties. The crystals of macrocrystalline Quartz

FIGURE 60.
Natural crystals of amethyst on andesite matrix from Veracruz, Mexico. Specimen is 260 mm in height (FMNH H2323).

FIGURE 61. Faceted amethyst gems from Brazil ranging in size from 34.2 to 321.5 carats each (FMNH H358, H360, H1918, H1920).

can be easily seen with the naked eye, while the crystals of microcrystalline Quartz are microscopic to submicroscopic.

Macrocrystalline Quartz includes the transparent varieties that are often faceted, such as amethyst (purple), citrine (yellow), ametrine (bicolor purple and yellow), smoky Quartz (gray-black), rose Quartz (faint pink), prasiolite (pale light green), and rock crystal (colorless). Chatoyant varieties of macrocrystalline Quartz include a pale gray-bluish cat's-eye Quartz, star Quartz, and tiger's eye (or tiger-eye).

Microcrystalline Quartz includes the translucent to opaque varieties that are made into cabochons, polished slices, or carvings. These varieties include agate (banded chalcedony of many varieties), carnelian (red to yellow chalcedony), onyx (black or black with white banding), sardonyx (onyx with red or orange layers mixed in), chrysoprase (apple-green chalcedony), jasper (various colors), bloodstone (green chalcedony with red speckles), chert, and flint. Some of these varieties are extremely common.

Various forms of simulated Quartz products are also used in costume jewelry, such as goldstone, which is a brown glass containing flakes of Copper to simulate reddish-brown aventurine. Quartz has been synthesized since the mid-nineteenth century, and since the mid-twentieth century flawless crystals of 1 pound or more have been manufactured. Most synthesized Quartz is colorless and used for technical purposes, although some smoky Quartz, green Quartz, and poor-quality amethyst are also synthesized.

The distribution of Quartz is worldwide. The only varieties that we discuss in detail here are the two varieties most commonly used in fine-gem pieces: amethyst and citrine. For further reading on Quartz, we recommend Sinkankas (1959), Bauer (1968), and O'Donoghue (1987, 2006).

PURPLE QUARTZ (AMETHYST)

SYSTEM Inorganic

CLASS Oxide

GROUP Quartz

SPECIES *Quartz*

VARIETY amethyst

COMPOSITION

Silicon Dioxide (SiO_2)

TRACE ELEMENT FOR COLOR

Iron (Fe)

HARDNESS ON MOHS SCALE 7

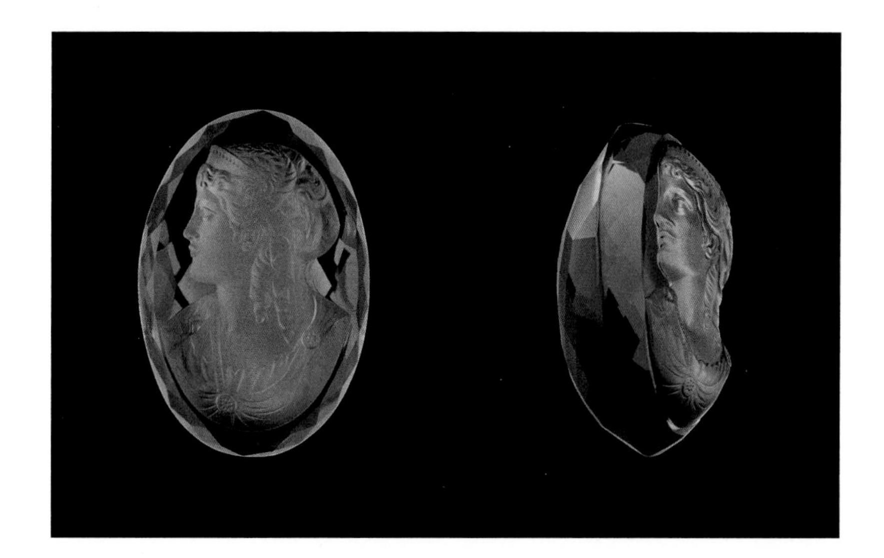

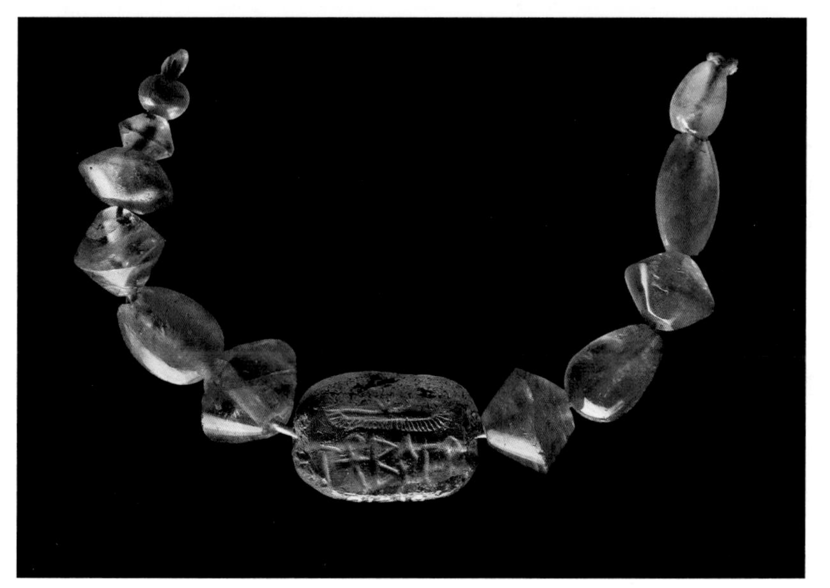

FIGURE 62.
Top: Faceted 106.7-carat amethyst from Bahia, Brazil, with the image of a woman carved into the table facet sometime prior to 1923. Top view (*left*) and oblique side view showing depth of carved image (*right*). Stone is 20 × 25 × 37 mm (FMNH H363).

FIGURE 63.
Bottom: Ancient amethyst necklace. Center stone is a scarab from the Middle Kingdom of Egypt, ca. 2000 BC, which was recarved in South Arabian script during the eighth or ninth century BC (FMNH A31218).

The gemological history and folklore of amethyst is extensive and interesting. Pieces of amethyst jewelry in the Grainger Hall of Gems date back as far as 2000 BC (fig. 63). The name amethyst comes from the Greek *a-methustos* ("not intoxicated"), which refers to the ancient belief that amethyst prevented drunkenness. This gem was a favored adornment of the drinking cups of the ancient Romans and Greeks. Leonardo da Vinci wrote that amethyst increased intelligence and dispelled evil thoughts.

Amethyst occurs in colors ranging from lilac or mauve to deep purple. The current favored color is "Deep Siberian" or "Deep Russian," an intense violet hue with traces of blue and red flashes. Another, less desirable color is a pale pinkish-

FIGURE 64.
Amethyst pendant with 56-carat faceted amethyst from Brazil and small cultured Pearls set in 14-karat yellow Gold (FMNH H2267).

lavender or lilac shade sometimes referred to as "Rose de France." Some amethyst can be heat-treated to change it to citrine (yellow Quartz), and from certain localities it can also be heat-treated to change to prasiolite (pale green Quartz).

Amethyst was much more highly valued in the past, in a class with sapphire, ruby, Diamond, and emerald. But in the mid-nineteenth century, vast deposits of fine amethyst were discovered in Brazil, collapsing the market. Amethyst remains relatively inexpensive to this day, except for the finest Russian material. It can make exceptionally beautiful jewelry, and several fine pieces can be found in the Grainger Hall of Gems (fig. 64).

Amethyst has been worn in rings by ranking members of the Roman Catholic Church since the Middle Ages. In Tibet, amethyst is also considered sacred in Buddhism. In ancient times, rings of amethyst set in bronze were worn as charms against evil. Throughout history, amethyst has also been worn as a symbol of royalty and is included in the British Crown Jewels. Amethyst is sometimes found lining the cavities of huge **GEODES** or volcanic **VUGS** (see fig. 10); one such amethyst-lined geode discovered in Brazil in 1900 weighed 8 tons and measured 33 × 10 × 16 feet.

Fine-quality amethyst comes from the Ural Mountains of Russia, Mexico (fig. 60), Sri Lanka, Zambia, and Madagascar. Some localities in the United States also produce a few fine gem-quality specimens (e.g., Maine, Colorado, Arizona, North Carolina, and New Hampshire), but not in abundance. Highly included or fractured amethyst is very common in some regions of the United States.

YELLOW QUARTZ (CITRINE)

SYSTEM Inorganic
CLASS Oxide
GROUP Quartz
SPECIES *Quartz*
VARIETIES citrine; ametrine
COMPOSITION
Silicon Dioxide (SiO_2)
TRACE ELEMENT FOR COLOR
Iron (Fe)
HARDNESS ON MOHS SCALE 7

The name citrine is derived from the word "citrus" and is reflective of its yellow color. Although it has sometimes been referred to as "Oriental Topaz," it is not Topaz, and that name should be avoided. Citrine is easily distinguished from Topaz because it is softer (Mohs hardness 7 compared to 8 for Topaz). Citrine ranges in color from lemon-yellow to reddish-brown, but the most desirable color is a dark golden or orangish-yellow, without a significant brown hue. Light lemon-yellow Quartz, called "lemon Quartz," is frequently Quartz that has been artificially irradiated to give it its lemon-yellow hue. Natural, untreated, gem-quality citrine is currently very rare, much rarer than amethyst; and most of the citrine in today's gem market is treated and enhanced in one way or another. Some amethyst, mainly Brazilian amethyst, will turn light yellow at 880°F and dark yellow to red-brown at around 1,025°F. Some smoky Quartz will turn yellow at about 390°F. A trained specialist can usually distinguish between natural and heat-treated citrine because heat-treated stones have subtle strips of yellow, and the yellow of natural stones is usually cloudier and more uniform (except for some of the Bolivian citrine).

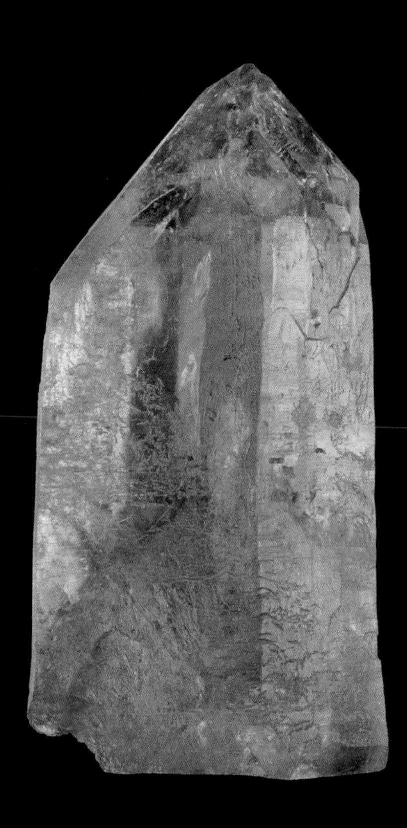

FIGURE 65.
Natural citrine crystal of pale yellow color from Minas Gerais, Brazil. Specimen weighs 473 grams (2,365 carats) and is 125 mm in height (FMNH H2596).

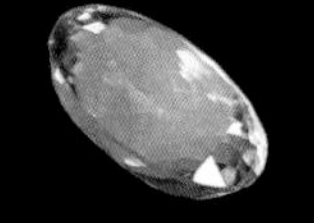

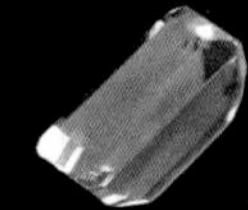

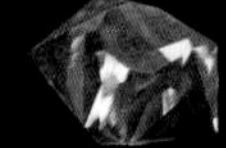

FIGURE 66.
Faceted citrine gems ranging from 10.8 to 122.0 carats, from Brazil (FMNH H322, H1455, H1655, H2171, H2470).

FIGURE 67.
Above: Carved citrines set in a pair of 18-karat Gold earrings together with amethyst, Tourmalines, and Diamonds (FMNH H2388).

FIGURE 68.
Left: Faceted ametrines from the Anahi mine, Sandoval Province, Bolivia; two faceted gems of 14.8 and 17.8 carats in weight (FMNH H2468, H2469).

Citrine is an old gemstone, used since at least the fourth century BC in Greece. It was once thought to symbolize joy and happiness. Although Brazil is the main source of gem-quality citrine today, it is also found in Madagascar, Spain, Sri Lanka, and Russia.

Ametrine is a variety of Quartz that has become popular again recently, but its use as a gemstone goes back to the seventeenth century. This gemstone is part citrine and part amethyst with a distinct border in between (fig. 68) and is found primarily in the Anahi mine of Bolivia. Stones generally must be 6 or more carats in weight in order to see the color transition clearly. This variety is generally faceted in a rectangular shape so that half the stone is yellow and the other half is purple, but many other patterns of faceting are popular. Ametrine is also sometimes cut in a way to blend the two colors, resulting in a mix of purple, yellow, and peach hues throughout the stone.

OTHER QUARTZ

SYSTEM Inorganic

CLASS Oxide

GROUP Quartz

SPECIES *Quartz*

VARIETIES pink Quartz (rose Quartz); pale green Quartz (prasiolite); brown to black Quartz (smoky Quartz); star Quartz; cat's-eye Quartz; tiger's eye; colorless Quartz (rock crystal); rutilated Quartz (sagenitic Quartz); tourmalinated Quartz; aventurine Quartz; jasper; onyx; agate (banded chalcedony); carnelian; chrysoprase

COMPOSITION Silicon Dioxide (SiO_2)

TRACE ELEMENTS FOR COLOR Iron (Fe), Chromium (Cr), Nickel (Ni), Aluminum (Al) (depending on color)

HARDNESS ON MOHS SCALE 7

In addition to amethyst and citrine, there are many other varieties of Quartz that are attractive and popular. But because most are very common, they are rarely used in fine jewelry and not considered precious gemstones. Therefore, we talk only briefly about those varieties here.

Rose Quartz is a light pink to peach-colored Quartz used primarily for beads or decorative carvings. Rose Quartz is usually light and cloudy in appearance without the rich pink hue of morganite or kunzite. Massive rose Quartz is extremely common, but crystals of rose Quartz are rare. Faceting-grade transparent rose Quartz is also very uncommon and is usually somewhat milky translucent rather than completely transparent (fig. 69). The best material comes from Madagascar, Brazil, and Russia.

Smoky Quartz ranges from brown to grayish-black in color and is very common (fig. 70). This is the least valuable of the colored Quartz faceting varieties. Stones with good clarity are also common and used for faceting. Smoky Quartz can be made artificially by irradiating colorless Quartz. The darkest smoky Quartz, that which is almost perfectly black in thick pieces, sometimes goes by the name "morion."

Prasiolite is the name most often used for transparent green Quartz (fig. 71). This gem variety is very rare in nature; and where it is known to occur naturally (rarely Brazil and Poland), it is usually a pale light hue of green lacking the rich chrome green hue of emerald, tsavorite, demantoid, or true hiddenite. Most green Quartz in the gem market today is artificially created by heat-treating and/or irradiating natural Quartz, particularly amethyst. Sometimes, in fact, treated green Quartz is referred to as "greened amethyst." There is also synthetic green Quartz that is occasionally used for inexpensive jewelry. Prasiolite

FIGURE 69.
Rose Quartz. Large faceted transparent stone of 129 carats from Oxford County, Maine, surrounded by a necklace of small faceted beads. Faceted gem is 15 × 29 × 40 mm (FMNH H372, H2460).

FIGURE 70.
Natural crystal of smoky Quartz from Silesia, Poland, weighing 1,030.5 carats and measuring 80 mm in height (FMNH M1124). Faceted gems from Spain weighing 27.6 and 35.2 carats (FMNH H353.2 AND H353.5).

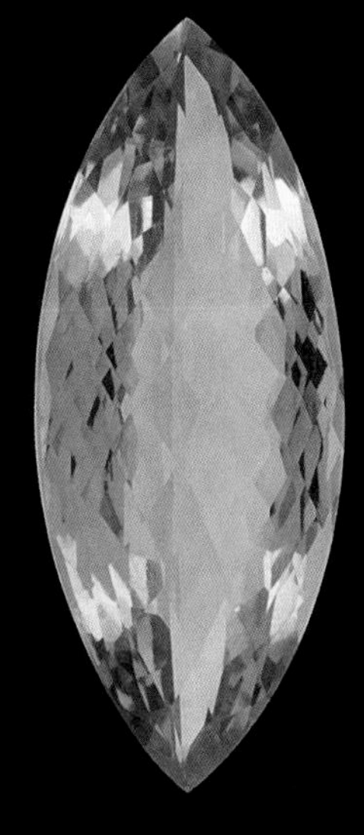

FIGURE 71.
Prasiolite (transparent green Quartz) from Bahia, Brazil, faceted with a long marquise cut. Gem measures 33.5 × 14 × 10 mm and weighs 24.65 carats (FMNH H2467).

FIGURE 72.
Left: Star Quartz from Ontario, Canada, cut and polished into a 13 × 13 × 6.5 mm cabochon weighing 7.6 carats (FMNH H2082). *Right:* Cat's-eye Quartz from Hof, Bavaria, Germany, cut and polished into a cabochon measuring 24 × 24 × 12 mm (FMNH H2077).

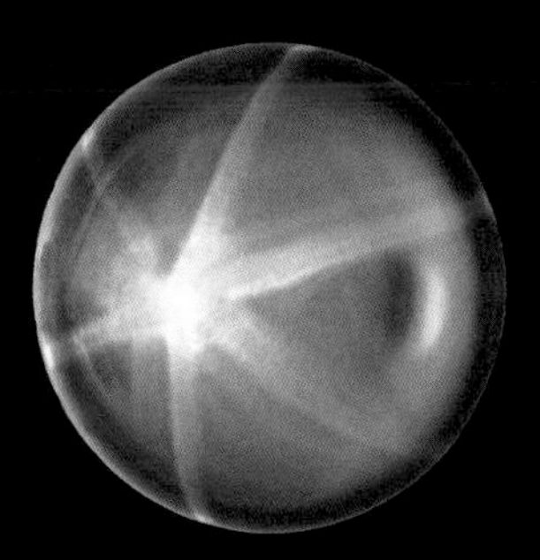

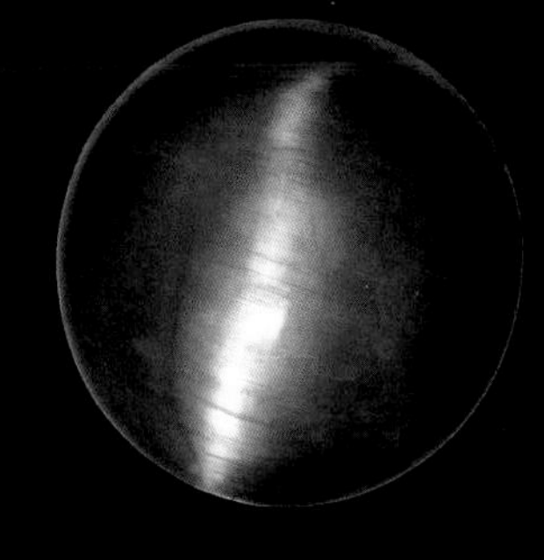

FIGURE 73.
Tiger's-eye Quartz from South Africa. Cabochons of yellow and blue tiger's eye, measuring 38 mm and 42 mm in height (FMNH H1966, H1969).

can be faceted, but because of its pale light, unexceptional color, it is not generally used for fine jewelry.

Star Quartz and cat's-eye Quartz are the scarcest and most valuable forms of the species. These varieties are cut into cabochons to show their chatoyancy and asterism. The cat's-eye and star varieties of Quartz come primarily from India, Sri Lanka, Brazil, Germany, and Ontario, Canada (fig. 72).

In addition to cat's-eye Quartz, there is another chatoyant variety called tiger's eye, or tiger-eye. Tiger-eye is fibrous-textured Quartz that usually occurs in yellow but also occasionally in blue or blue-and-yellow combinations (fig. 73). Red tiger-eye has most always been heat-treated to achieve that color. Tiger-eye was much more valuable during the nineteenth century until huge deposits were found in South Africa, sending the value from $11,200 per pound to 25 cents per pound. Today it remains of little value because of its great abundance and is a prime example of how scarcity (or the lack of it) can affect the gem market. Tiger-eye still comes primarily from Africa.

FIGURE 74.
Natural crystal of rutilated (sagenitic) Quartz with surface polished to display the large Rutile needle inclusions within that define this variety of Quartz. This piece, from Bahia, Brazil, weighs 560 grams (2,800 carats) and measures 156 × 86 × 34 mm (FMNH H2228). To the right is a polished cabochon of rutilated Quartz of 57 carats, from Brazil, measuring 30 × 23 × 11.5 mm (FMNH H2084).

Colorless Quartz, or rock crystal, is extremely common and has important uses for its optical properties, but not as a gemstone. Colorless Quartz that has fine, needle-like inclusions of Rutile is called rutilated Quartz, sagenite, or hairstone, and a variety with needle-like inclusions of black Tourmaline is called tourmalinated Quartz. The rutilated and tourmalinated varieties are sometimes made into cabochons for jewelry, and if the inclusions are well proportioned and oriented within the finished stone, the stone makes an attractive pendant (fig. 74).

Aventurine is Quartz that includes small inclusions of Muscovite mica that give Muscovite a silvery green or bluish sheen. Orange and brown varieties of aventurine colored by traces of Hematite or Goethite are also known, but the green variety is the most popular. Aventurine is sometimes made into cabochons for inexpensive jewelry, but it is common enough that it is also used for carvings or even building stone. The major source of aventurine is India (fig. 75).

Jasper is a red, yellow, or brown form of microcrystalline Quartz that was widely used as a gem by ancient societies of Greece, Rome, and Persia (fig. 75). It is actually a variety of chert, and although it can occasionally be similarly colored to carnelian, it is less translucent. On Minoan Crete, jasper was used to produce wax seals more than 3,800 years ago. Jasper is fairly common in the New World and rarely used in fine jewelry.

A special type of green jasper with red spots of Iron Oxide is called bloodstone (fig. 75). Bloodstone, also known as heliotrope, is sometimes called "the martyr's stone" because in the Middle Ages it was often used to carve scenes of the Crucifixion and Christian martyrs. There is a legend that says that bloodstone was first formed when drops of Christ's blood fell and stained green jasper at the foot of the cross. Bloodstone is mined in India, Australia, and the United States.

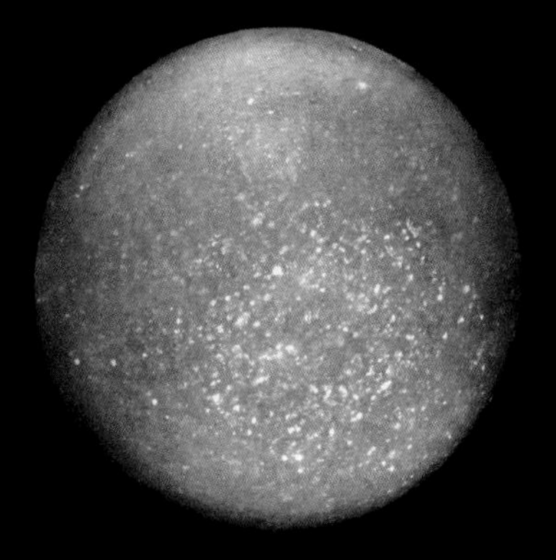

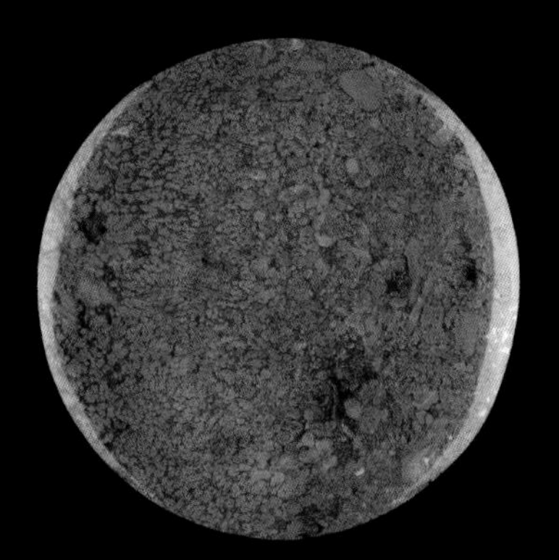

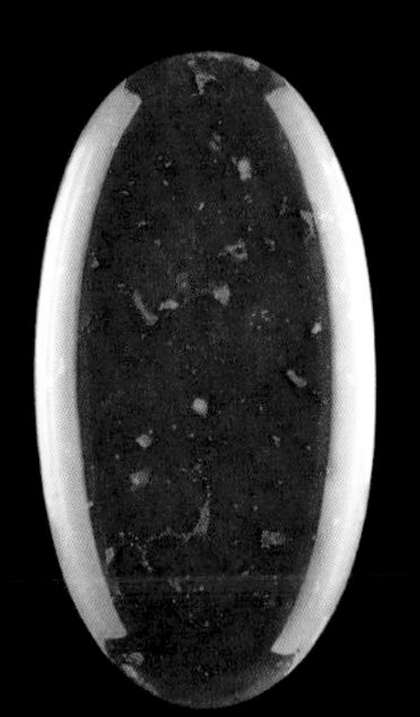

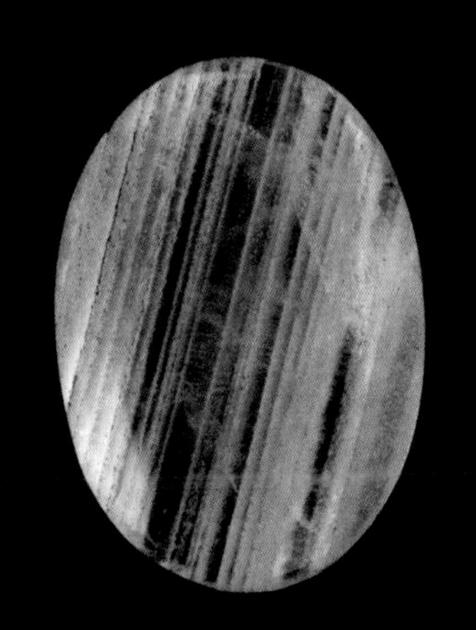

FIGURE 75.
Left: Four varieties of microcrystalline and cryptocrystalline Quartz cut and polished into cabochons. Aventurine Quartz from Madras, India (*top left*), measuring 29 × 29 × 11 mm (FMNH H366). Jasper from Oregon (*top right*), measuring 40 × 40 × 4 mm (FMNH H1953). Bloodstone from India (*bottom left*), measuring 32 × 17 × 5 mm (FMNH H1979). Onyx from Brazil (*bottom right*), measuring 18 × 13 × 4 mm (FMNH H1327).

FIGURE 76.
Below: Agate, a common but extremely popular type of lapidary stone. Cabochons of Brazilian agate (*left*) and Mexican lace agate (*right*). Pieces are 43 mm and 63 mm in height (FMNH H1936, H1980).

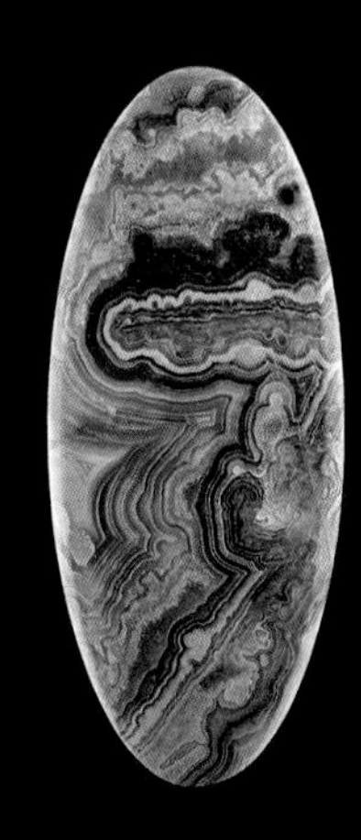

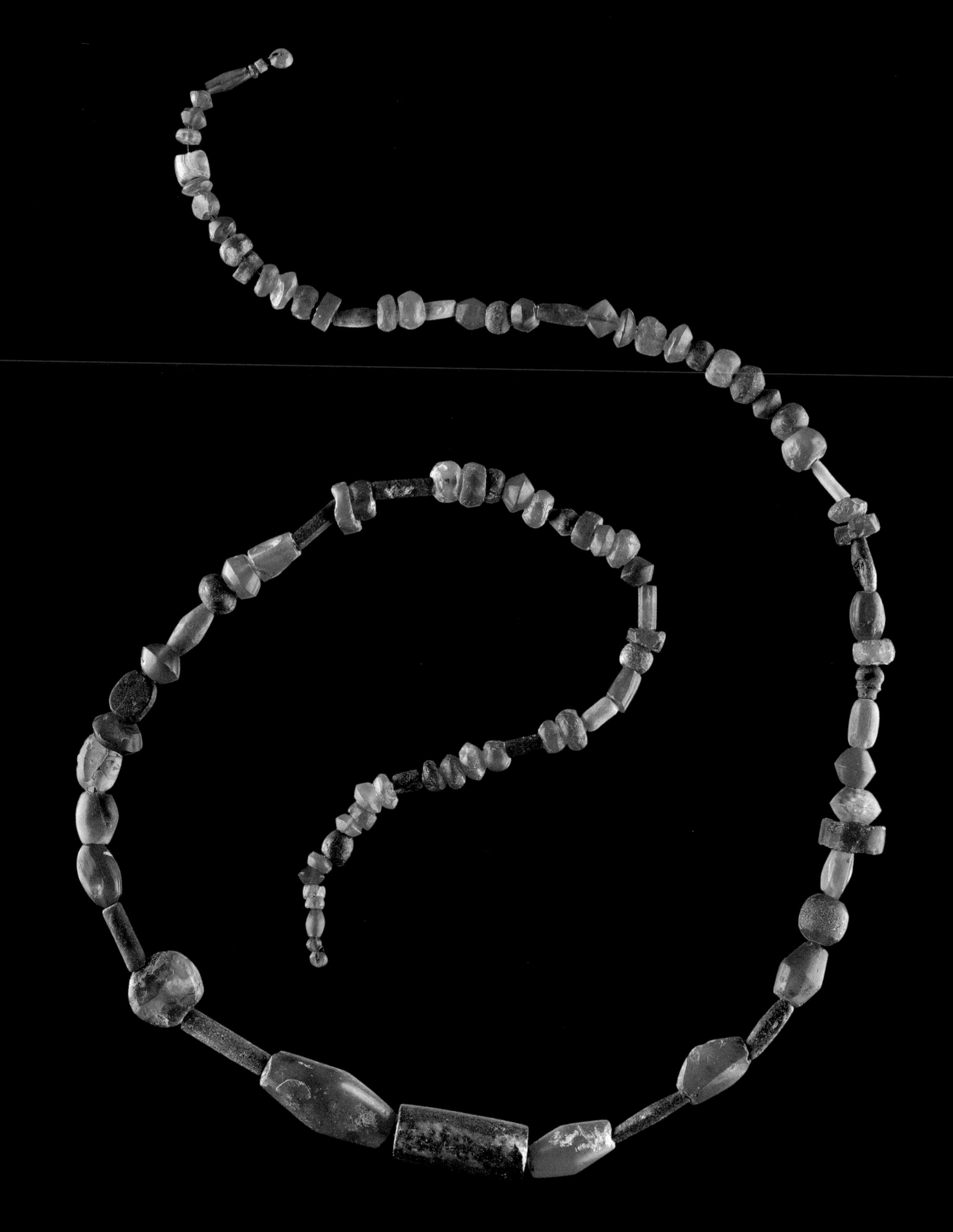

FIGURE 77.
Facing page: Ancient necklace from the Kish site (dated about 2500–2300 BC), in what is now Iraq. The necklace contains carnelian beads, probably mined from Indus Valley in what is now Pakistan and part of India, strung together with chalcedony, amethyst, agate, and lapis from various other regions (FMNH A230030). The contrasting colored stones were typical of ancient Mesopotamian jewelry. Carnelian and lapis had deep metaphysical significance to the ancient Mesopotamians.

Onyx can be pure black (a form still used in some fine jewelry) or banded CRYPTOCRYSTALLINE Quartz. Banded onyx is very similar to agate, but the banding is either less curved than that of most agate, or straight (fig. 75). The term "Mexican onyx" generally refers to a form of banded Calcite and is not onyx at all. The majority of inexpensive carvings sold as "onyx" today, particularly from Mexico, consist of Calcium Carbonate and not true Quartz onyx.

Agate is a microcrystalline variety of Quartz and is a banded form of chalcedony that occurs in an amazing variety of colors and patterns. It is found in many localities all over the world. Most types of agate have a distinctive banding pattern, except for moss agate (or landscape agate), which is formed by dendritic Manganese Oxide or Iron Oxide inclusions in translucent chalcedony, and fire agate, an iridescent form from Arizona and Mexico. There are dozens of different agate types, often named after regional geographic areas or features (e.g., Lake Superior agate, Montana moss agate, Mexican lace agate, Brazilian agate, Botswana agate, Laguna agate, Queensland agate, etc.). One of the largest producers of banded agate today is Brazil, which has exported thousands of tons of agates to the Unites States. Banded agate is also frequently dyed or stained to enhance or change its color. Agate is made into cabochons for use in inexpensive jewelry. Many a geologist and gemologist first became interested in rocks and minerals by searching for agates in their early years (present first author included).

Carnelian (also spelled cornelian) is a translucent, reddish-orange variety of chalcedony. The red color comes from traces of Iron Oxide. Carnelian was widely used by ancient Greeks and Romans for seal rings because it does not stick to sealing wax. It was also commonly used in ancient jewelry (fig. 77). It is closely related to another reddish-brown chalcedony called sard, which differs only slightly in color. The finest carnelian comes from India. Much of the carnelian on the market today is artificially stained chalcedony from Brazil.

Chrysoprase (also called chrysophrase) is a form of cryptocrystalline Quartz that is apple-green to deep green in color. It was used in jewelry by the ancient Greeks and Romans. The color is the result of traces of Nickel and can sometimes fade in sunlight. Because of its glassy luster and its opaque apple-green color, it is sometimes confused with Jadeite. The finest chrysoprase comes from Poland, Czechoslovakia, and Queensland, Australia. Chrysoprase also is found in Brazil, California, and Russia.

Opal

SYSTEM Inorganic

CLASS Oxide

GROUP Opal

SPECIES *Opal*

GEM VARIETIES precious black Opal; precious white Opal; precious fire Opal

The PLAY OF COLOR from a high-quality precious Opal, sometimes referred to as "nature's fireworks," can be breathtaking. Precious Opal has long been revered as one of the most beautiful of gems. Precious Opal permeates the ancient aboriginal legends of Australia, where the "creator" of everything came down to earth on a rainbow to bring a message of peace, and wherever his foot touched the ground, Opals formed. Opal's great beauty was described 2,000 years ago by the great Roman philosopher Pliny the Elder, who said of Opals, "There is in them a softer fire than in the carbuncle, there is the brilliant purple of amethyst; there is the sea-green of the emerald—all shining together in incredible union."

In Roman times, Opal was thought to bring good luck and fortune, but in the early nineteenth century, the popularity of Opals waned for several decades because they were believed to bring bad luck to the owner. This brought a corresponding decrease in the popularity and value of the gem. Reasons for the start of this myth have been due to numerous factors, ranging from a rumored connection to cholera in nineteenth-century Spain, to jealous Diamond traders trying to eliminate competing gemstones from the marketplace. Eventually Opal's bad luck reputation subsided, and today they are again among the world's most desirable gems.

Precious Opal is hard, hydrated silica, containing 3 to 25 percent water molecules. The process of precious Opal formation is not entirely clear. It is thought that Opal starts out as a concentrated silica gel that accumulates in cavities and spaces within a host rock, eventually hardening into gemstone. The silica gel is believed to be derived from silica-rich waters that percolate through the bedrock over long periods of time, perhaps millions of years, concentrating silica in the spaces and cavities. In IGNEOUS host rocks, the spaces within the rock are the result of gas bubbles frozen in volcanic MAGMA or LAVA, or cracks or fissures in the rock. In SEDIMENTARY rocks, the spaces are sometimes formed by fossils inside the rocks that are dissolved away by acidic groundwater flowing through the rock, leaving behind a mold of the original fossil. The resulting space or mold is then filled with silica gel that eventually becomes precious Opal. The opalized fossil is referred to as a REPLACEMENT FOSSIL; that is, a fossil that is replaced with a different mineral, in this case Opal. Some of these opalized fossils are remarkable in their anatomical detail and fiery beauty. Several of these are illustrated and discussed below in the section on precious white Opal.

Most Opal is common Opal, which lacks the play of color and is not used as gemstone material. Precious Opal is that Opal which exhibits play of color. There are three types of precious Opal discussed in this book: black Opal, white Opal, and fire Opal. Each of these is defined by the overall color of the body tone or base forming the background for the play of color in the stone. The base for the black Opal is black or dark gray, although the best black Opals show mostly play of color and very little black or gray. For white Opal, the base is milky white to translucent white, and for fire Opal the base is transparent or translucent orange to reddish-orange. Another type of Opal that is sometimes used for gems that can resemble black or white Opal is called boulder Opal. This type of Opal consists of a very thin layer of precious Opal coating the surface of dark, opaque ironstone, and it can be quite beautiful. The coating of precious Opal in boulder Opal is generally far too thin to be completely removed from the ironstone. The major producer of boulder Opal is Quilpie, Queensland.

Because of the water content, precious Opal can sometimes dry out and crack with time. For this reason, uncut precious Opal is sometimes stored in water. Care must be taken when grinding and polishing Opals for use in jewelry, because excessive heat may evaporate the water locked inside the stone and diminish or eliminate the play of color. Opal can also be destroyed by acids or strong alkalis. Ultrasonic cleaners and strong solvents should also be avoided with precious Opal. Nevertheless, if properly cared for, precious Opal can last centuries. One precious Opal in the Grainger Hall of Gems, the "Sun-god Opal," is thought to have been mined over 400 years ago (fig. 82).

Opal's beauty is best shown in CABOCHONS, so Opal is not normally faceted for jewelry, with the exception of transparent fire Opal, which typically lacks play of color but has an attractive orange hue. Faceting this material can bring out its fiery orange brilliance and produce flame-like reflections from the inner surfaces of the facets.

Opal is sometimes impregnated with plastic to enhance its appearance and stability. This decreases the value of the gem. Other methods of enhancement include backing a translucent precious white Opal with a thin piece of black stone such as onyx, jade, jet, or black glass. The two pieces are tightly cemented together, creating the appearance of the more expensive black Opal, and this composite is called a DOUBLET. Some doublets are of very high quality and used in fine jewelery (see fig. 83). Boulder Opal cabochons resemble doublets because they consist of a very thin layer of precious Opal covering a dark, opaque ironstone, but boulder Opal is formed in nature rather than being a man-made composite. In some cases, a man-made doublet receives a thin covering of a colorless stone such as clear Quartz or glass for surface strength and to give the top of the stone a curved cabochon surface. This type of composite is referred to as a TRIPLET. Doublets and triplets are, of course, normally worth much less than completely natural black Opals.

Most precious Opal mined since the early twentieth century has come from Australia, which produces over 95 percent of the world's supply. Opal is the national stone of Australia. Other sources of precious Opal include Mexico, the United States (Idaho), Brazil, Slovakia, the Czech Republic, and South Africa.

Here we discuss three major forms of precious Opal: precious black Opal, precious white Opal, and fire Opal. For further reading on Opal, we recommend Leechman (1978), Bauer (1968), and O'Donoghue (2006).

PRECIOUS BLACK OPAL

SYSTEM Inorganic

CLASS Oxide

GROUP Opal

SPECIES *Opal*

VARIETY precious black Opal

COMPOSITION hydrated silica gel ($SiO_2 \cdot nH_2O$)

TRACE ELEMENT FOR COLOR none

HARDNESS ON MOHS SCALE 5½–6½

Black Opal is the variety of precious Opal with the most brilliant colors and is the most valuable variety. It is not really a black stone, because a truly black stone would show no play of color. Instead it is a stone that has a black to dark gray transparent body tone, or background color, that highlights the multicolored play of color in the stone. If the body of the Opal is completely transparent with play of color in it, it is sometimes referred to as crystal Opal. The most valuable black Opal is usually that which contains bright red, blue, green, yellow, and orange flash. Most black Opals show only blue or blue and green, and are of significantly less value. Opals that show only a single-hued flash of color are called flash Opal. The highest-grade black Opals can sell for more than $40,000 per carat and show nothing but colors when set in jewelry. Because of its extreme value and sensitivity to heat, many of the highest-quality black Opals are set into jewelry as free-form stones, so as not to lose carat weight or risk damage during the finishing process. Black Opals are normally cut with a very low cabochon surface, because black Opal tends to be much thinner than other types of Opal.

Nearly all black Opal comes from one of two localities in South Australia: the famous Lightning Ridge locality, first mined in 1903, or the Mintabie locality. Fine black Opal ranks with the world's most expensive gems. There are many extremely fine pieces of black Opal in the Grainger Hall of Gems (figs. 78–80).

FIGURE 78. Precious black Opal from Lightning Ridge, New South Wales, Australia. Natural rough piece of 147.9 carats with polished surface measuring 55 × 37 × 8 mm (FMNH H456).

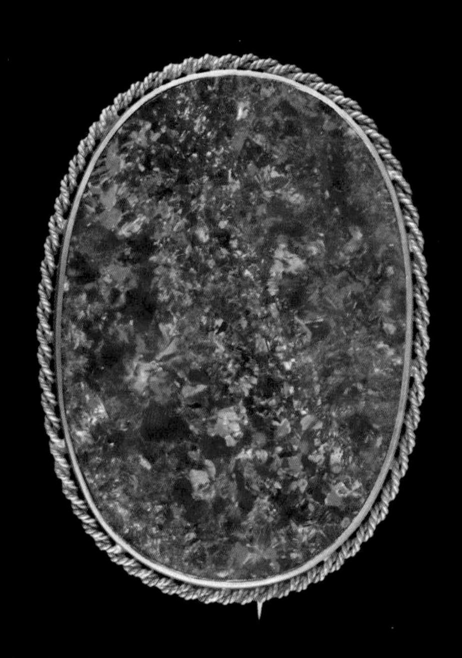

FIGURE 79.
Precious black Opal cabochon measuring 48 × 34 × 4 mm oval, weighing 87 carats, mounted in an 18-karat yellow Gold brooch (FMNH H1633).

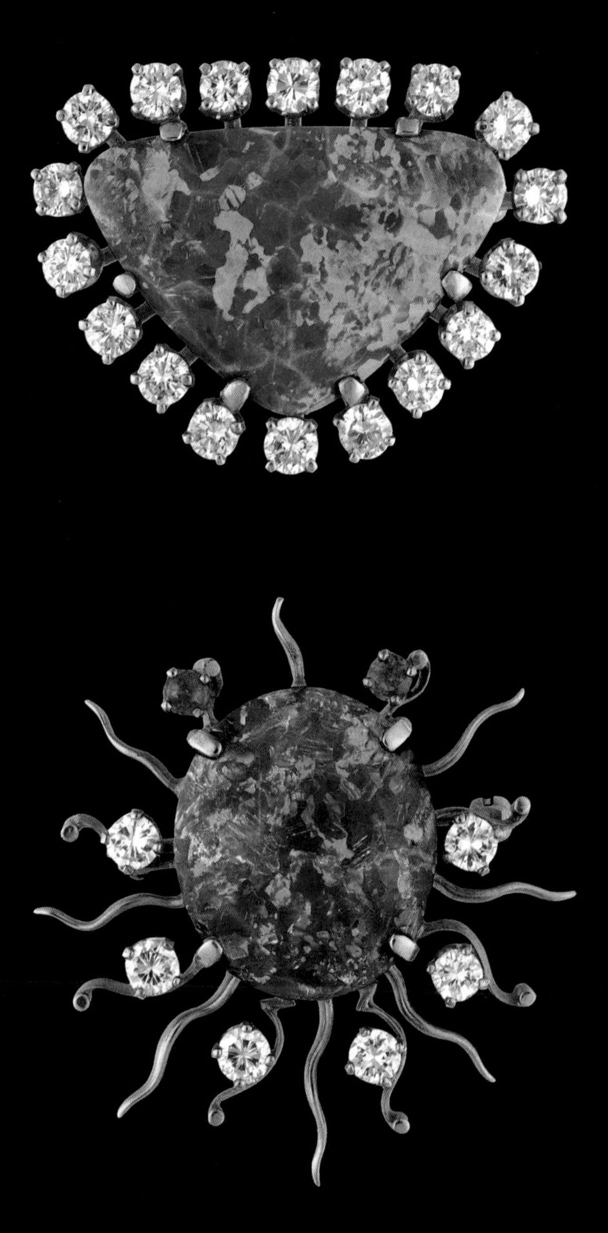

FIGURE 80.
Two precious black Opal pins made in the early twentieth century from Australia. *Top right:* Triangular-shaped cabochon mounted in 14-karat yellow Gold surrounded by 18 brilliant-cut Diamonds. Opal cabochon measures 24 mm across the top (FMNH H1517). *Bottom right:* Oval cabochon measuring 16 × 19 mm, mounted in 14-karat Gold with Diamonds and two rubies (FMNH H1516).

PRECIOUS WHITE OPAL

SYSTEM Inorganic

CLASS Oxide

GROUP Opal

SPECIES *Opal*

VARIETY precious white Opal

COMPOSITION hydrated silica gel ($SiO_2 \cdot nH_2O$)

TRACE ELEMENT FOR COLOR none

HARDNESS ON MOHS SCALE 5½–6½

White Opal (sometimes called milk Opal) is the most common of the precious Opals, but the highest-quality white Opal is both rare and valuable. It has a white background or base color containing the play of color. As with black Opal, the most desirable (expensive) white Opals contain bright red color in addition to the blue and green. The best stones have a nearly transparent base with solid multicolor flash. When the milky body of the Opal is transparent or translucent, it is sometimes referred to as crystal Opal or water Opal.

Some white Opal is treated to resemble black Opal, using a series of heated solutions including heated glucose (sugar) and sulfuric acid. This turns the white base to dark gray. A gemologist can always detect an Opal treated in this way.

Opalized fossils are formed mainly in white Opal. It is sometimes a struggle between paleontologists and gem dealers to obtain some of the opalized Cretaceous marine reptiles and dinosaurs found in Australia. The paleontologists want the skeletons to go into a museum intact, while gem dealers will often want to break the skeletons up for gems. The gem quality of some skeletons can sometimes be very high. A vertebra from one such fossil reptile, a plesiosaur, is on display in the Grainger Hall of Gems, weighing 822.5 carats (fig. 86). Fossil snails and clams that have been replaced with white precious Opal also are known (fig. 85).

White Opal is found in many more regions than black Opal. It is found in Australia, Slovakia, the Czech Republic, Hungary, Brazil, Mexico, South Africa, Idaho, Washington, Oregon, and Nevada. White Opal is sometimes found in very large pieces (fig. 81). One of the most famous gems in the Grainger Hall of Gems is the Aztec "Sun-god Opal." It is a 35-carat precious white Opal that has been carved into the shape of a human face. This stone was thought to have been mined in Mexico by the Aztecs in the sixteenth century. Later it showed up in the Middle East, where it was mounted in a Gold setting and remained in a Persian temple for several centuries. It eventually came into the possession of the famous gem collector Philip Hope in the mid-nineteenth century and eventually found its way to The Field Museum in 1893, where it has remained ever since (fig. 82).

FIGURE 81.
Precious white Opal rough from White Cliffs, New South Wales, with face polished to show fiery play of color. The stone is 613.5 carats and measures 62 × 77 × 22 mm (FMNH H442).

FIGURE 82.
The Aztec "Sun-god Opal," a 35-carat precious white Opal carved into the shape of a human face. This Opal was mined in Mexico by the Aztecs in the sixteenth century and eventually found its way to The Field Museum in 1893, where it has remained ever since. *Top:* Front view showing the play of color. *Bottom:* Oblique view with angled lighting to highlight the delicately carved surface (FMNH H447).

FIGURE 83.
Top: Large white Opal cabochon doublet of 32 × 23 mm set in 14-karat Gold, surrounded by 29 round brilliant-cut demantoid Garnets and 33 round brilliant-cut Diamonds. Late nineteenth century by an unknown designer (FMNH H2224).

FIGURE 84.
Bottom: Precious white Opal cabochon measuring 22 × 14 mm, set in 14-karat yellow Gold with rubies and round brilliant-cut Diamonds. Late nineteenth century by an unknown designer (FMNH H1518).

FIGURE 85.
Precious white Opal as natural replacement fossils for clam and snail shells from White Cliffs, New South Wales, Australia. Opal clams are 32.8 and 4.6 carats, and snails are 9.8 and 33.5 carats (FMNH H464, H1930–H1932).

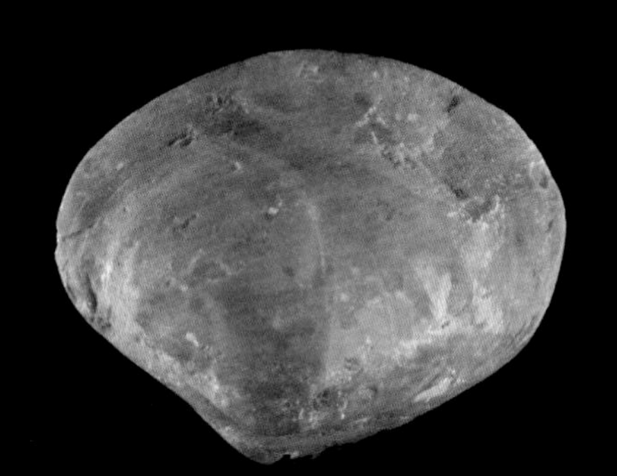

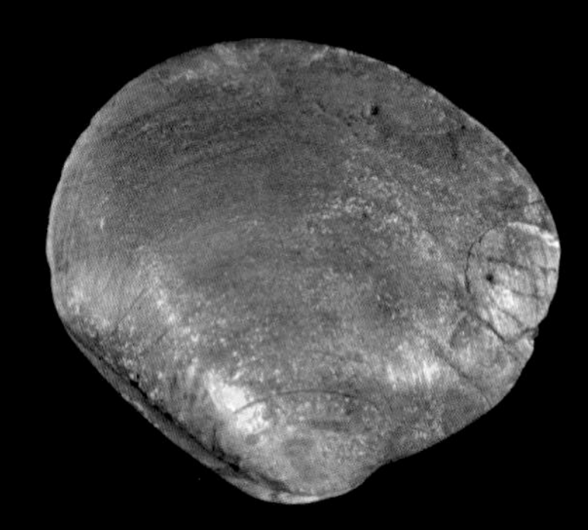

FIGURE 86.
Below left: Precious white Opal replacement of a fossil vertebra from a 115-million-year-old plesiosaur. This specimen, from White Cliffs, New South Wales, Australia, weighs 822.5 carats and measures 60 × 50 × 45 mm. It is very rare to find gem-quality Opal fossils like this intact, because gem cutters usually cut them up for production of gems for jewelry (FMNH H443). *Below right:* Artist's mural reconstruction of swimming plesiosaurs from Late Jurassic time (approximately 150 million years before present) by Charles R. Knight. Mural is in The Field Museum's Evolving Planet exhibit (FMNH 6564).

FIRE OPAL

SYSTEM Inorganic

CLASS Oxide

GROUP Opal

SPECIES *Opal*

VARIETIES fire Opal, precious fire Opal

COMPOSITION hydrated silica gel ($SiO_2 \cdot nH_2O$)

TRACE ELEMENT FOR COLOR none

HARDNESS ON MOHS SCALE 5½–6½

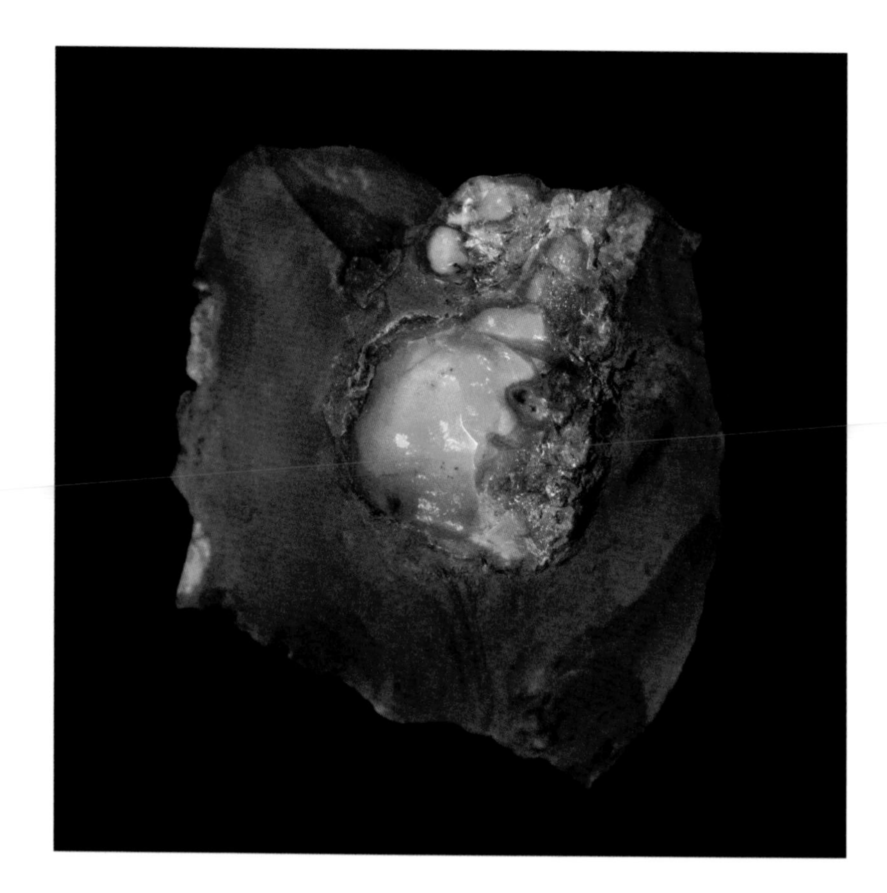

FIGURE 87. Natural raw nodule of precious fire Opal in matrix from Querétaro, Mexico, measuring 90 × 90 mm. Opal nodule measures 32 × 30 mm (FMNH M8568).

Fire Opal gets its name from the radiant orange to orange-red color of its body tone. Fire Opal that has play of color is treated like other precious Opals and made into cabochons to show off the multicolored flash within. Transparent fire Opal that lacks play of color does not qualify in the "precious" category of Opal; but if the body tone is of good orange color and is transparent, it is sometimes faceted for rings or other jewelry (fig. 89) or made into beads to be strung for necklaces. Transparent fire Opal with good play of color is by far the most valuable form (e.g., fig. 90). The most common color flashes in the precious fire Opal are yellow, red, and green, while blue is much rarer. Fire Opal is often found filling small vugs in volcanic rock. It is one of the least durable forms of precious Opal, and it is easily fractured. It is thought that the drier the environment, the more durable the fire Opal found there.

Fire Opals were treasured by the Mayan and Aztec cultures, who called it *quetzalitzlipyollitli*, or "stone of the bird of paradise," and used it in rituals. Although known from a number of regions around the world, the most significant deposits of fire Opal are in the Mexican highlands in a region containing many extinct volcanoes. Precious fire Opal also is found in Honduras, Guatemala, the United States, Canada, Australia, Ethiopia, Turkey, and Brazil,

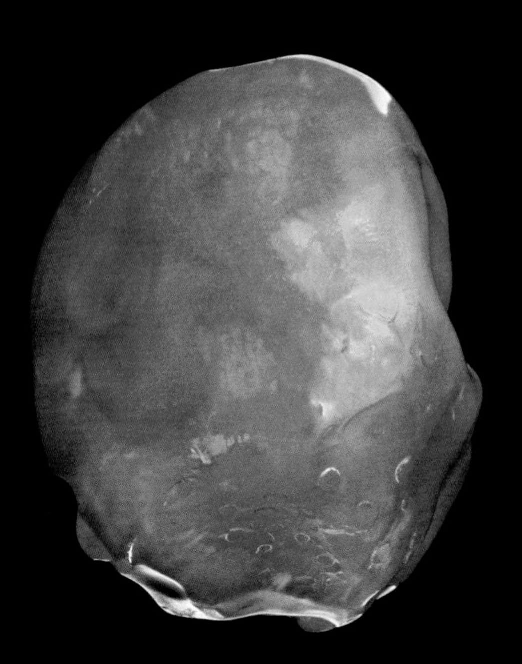

FIGURE 88.
Left: A nodular fire Opal of the type shown in figure 87 from Querétaro, Mexico, that has been removed from matrix weighing 163.4 carats and measuring 30 × 31 × 25 mm (FMNH H2208).

FIGURE 89.
Below: Fire Opal ring and large faceted stone from Mexico showing little or no play of color, but with sufficient transparency to be facetable for jewelry. Stone in ring is 2.63 carats and set in 14-karat white Gold with 24 small round brilliant-cut Diamonds. Top, side, and front views (FMNH H2472). Unset stone is 9.04 carats, measuring 15 × 15 × 11 mm (FMNH H1683).

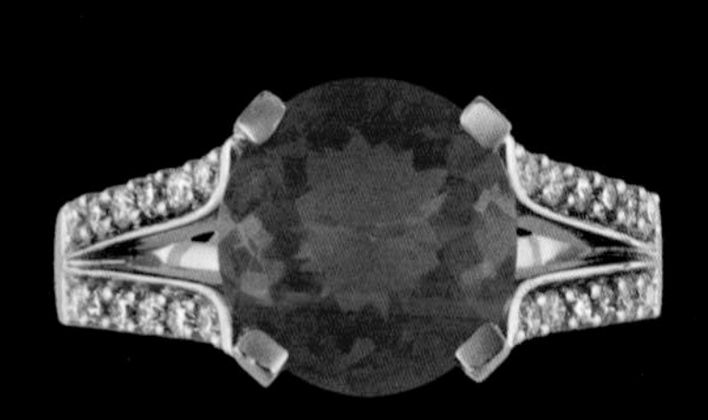

FIGURE 90.
A 21.5-carat precious fire Opal of rare quality, from Querétaro, Mexico, that shows not only the characteristic orange base color and good transparency, but also excellent play of color with deep iridescent flashes of green, yellow, red, and blue color throughout the stone. Opal is a teardrop-shaped cabochon measuring 23 × 17 × 12 mm set in 18-karat Gold with 39 ideal-cut Diamonds. Designed by Maureen Lampert of Lester Lampert, Inc., 2009, for the Grainger Hall of Gems. This piece was named "Caliente" by the designer (FMNH H2588).

although fire Opals from Brazil do not normally have play of color. One of the outstanding gems in the Grainger Hall of Gems is a 21.5-carat precious fire Opal with good transparency, fine orange base color, and excellent play of color; a rare combination for such a large stone. This stone, cut and polished as a teardrop cabochon, was set in 18-karat yellow Gold. It was designed by Maureen Lampert of Lester Lampert, Inc., in 2009 for the Grainger Hall of Gems and is called "Caliente" (fig. 90).

Topaz

SYSTEM Inorganic

CLASS Silicate

GROUP Topaz

SPECIES *Topaz*

GEM VARIETIES blue Topaz; imperial Topaz; other colors and colorless Topaz

FIGURE 91.
Huge, faceting-grade colorless Topaz crystal, from Minas Gerais, Brazil. Specimen is approximately 210,000 carats (over 90 pounds), measuring 440 mm in length. Such stones are often irradiated or heat-treated to produce blue Topaz for jewelry (FMNH H998).

Topaz is literally a giant among gemstones; some faceting-grade crystals are weighed in pounds rather than carats. It is also the hardest of all Silicate minerals (Mohs 8). Its icy clarity and durable resistance to scratching make it an ideal gem that would be much more valuable if it were not for the abundance of colorless, brown, and light blue varieties. Topaz was a rare and highly sought-after stone in the Middle Ages and for centuries after, until vast deposits of the gemstone were discovered in Brazil in the nineteenth century. Today blue Topaz is a popular, beautiful, and relatively inexpensive gem often used in fine jewelry. The name Topaz is thought by some to be derived from the Sanskrit word meaning "fire," referring to the golden-yellow variety. Others believe the origin of the name is after the Island of Topazos (St. John's Island), where the yellow Olivine gem chrysolite is found. In ancient times, the name "Topaz" was simply a name used for just about any yellow gemstone; nearly all yellow stones, including citrine and chrysolite, were once called Topaz. Then in the early eighteenth century, the name was more rigorously applied exclusively to minerals with a composition of Aluminum Fluoro-Hydroxy-Silicate (true Topaz by today's definition), a species that occurs in several different colors.

Of all colors of Topaz, colorless and blue are the most common, and dark pink and orange-red are the most valuable. Red stones are extremely rare, and one of the world's largest is in the Grainger Hall of Gems (see fig. 96). Very light blue Topaz can be irradiated with gamma rays and heat-treated to produce a deeper blue Topaz that is visually difficult to distinguish from aquamarine without magnification. But Topaz is still easily distinguishable upon closer examination because it is generally harder than aquamarine (Mohs 8 compared to 7½–8 for aquamarine) and has characteristic teardrop-shaped inclusions that do not occur in aquamarine. A recent trend in the jewelry trade is an iridescent-colored stone called "Mystic Topaz" that displays many colors. The unusual color effect is not natural and is the result of applying a thin coating of Titanium Oxide to the surface of the stone. Although Topaz is a hard gemstone, it is also somewhat brittle and can crack or shatter with a hard blow. Therefore, Topaz gems should be protected from hard knocks.

Topaz is found in placer deposits, pegmatites, and volcanic rocks. Topaz is also found in pegmatites or cavities of hardened lava as found in western Utah. Clear, faceting-grade colorless blue or pale yellow Topaz from Brazil (Minas Gerais) is famous for its sometimes gigantic proportions. One crystal of gem-grade colorless Topaz in the collection of The Field Museum is 210,000 carats (over 90 pounds) measuring 44 × 23 × 20 centimeters (fig. 91). The largest Topaz gemstones can occasionally weigh over 100 pounds. Topaz is found in Mexico, the United States, Sri Lanka, Japan, Siberia, Russia, Nigeria, Zaire, and numerous other places. For further reading about Topaz, we recommend Hoover (1992), O'Donoghue (2006), and Bauer (1968).

BLUE TOPAZ

SYSTEM Inorganic

CLASS Silicate

GROUP Topaz

SPECIES *Topaz*

VARIETY blue Topaz

COMPOSITION

Aluminum Fluoro-Hydroxy-Silicate [$Al_2SiO_4(F,OH)_2$]

TRACE ELEMENT FOR COLOR

Iron (Fe)

HARDNESS ON MOHS SCALE 8

FIGURE 92.
Right: Natural blue faceting-grade Topaz crystals on matrix from near Alabashka, in the Ural Mountain region of Russia. Specimen is 100 mm in height (FMNH H218).

FIGURE 93.
Below: The "Chalmers Topaz," a faceted blue Topaz from Minas Gerais, Brazil, weighing 5,899.5 carats, measuring 131 × 85 × 71 mm. This is one of the world's largest faceted Topaz gems (FMNH H1597).

Blue Topaz is one of the more common varieties of Topaz, primarily because colorless Topaz can be turned blue by HEAT TREATMENT and IRRADIATION using gamma rays. Very light blue Topaz can also be enhanced to a deeper blue by the same treatments. The blue color of treated stones is generally stable (i.e., does not fade with time), and therefore treated stones are widely accepted in the gem trade. In fact, most blue Topaz in the gem trade today is treated material. Its exceptional hardness of Mohs 8, affordability, and beauty make it a highly desirable stone for the gem industry.

Blue Topaz resembles aquamarine, but it is generally harder than aquamarine (Topaz Mohs 8, aquamarine Mohs 7½–8) and has certain optical properties that are easily discernible by any gemologist. Blue Topaz is worth considerably less today than aquamarine because it is more available. Blue Topaz, like other varieties of the species, is frequently found with few or no inclusions, and as a gem it is a characteristically EYE-CLEAN stone (see table 3). The Greeks believed that blue Topaz could increase strength and make its wearer invisible, while the Romans believed it could improve eyesight. It is the state gemstone of Texas. The largest faceted gem in the Grainger Hall of Gems is the "Chalmers Topaz," a pale blue Topaz with a teardrop cut that is 5,899.5 carats, or over 2½ pounds (fig. 93). Most of the largest gem-quality crystals come from Minas Gerais, Brazil.

FIGURE 94.
An emerald-cut blue Topaz from Bahia, Brazil, weighing 15.18 carats, mounted in a 14-karat Gold pendant. Designed by Teri Wallace of Mineral Search, Inc., in 1995 and named "Ensenada" by the designer. Topaz gem is 16 × 12 × 8 mm (FMNH H2337).

PINK TO RED TOPAZ (IMPERIAL TOPAZ)

SYSTEM Inorganic

CLASS Silicate

GROUP Topaz

SPECIES *Topaz*

VARIETIES pink imperial Topaz; orange imperial Topaz; red imperial Topaz

COMPOSITION Aluminum Fluoro-Hydroxy-Silicate [$Al_2SiO_4(F,OH)_2$]

TRACE ELEMENTS FOR COLOR Iron (Fe) and Chromium (Cr)

HARDNESS ON MOHS SCALE 8

FIGURE 95. Natural crystal of deep pink or red imperial Topaz in matrix from the Mardan district, Pakistan. Specimen measures 39 mm in height (FMNH H2447).

Imperial Topaz is the rarest and most expensive variety of Topaz. It occurs in orange, pink, and red, the rarest of which is red. The name imperial Topaz is thought to have originated in nineteenth-century Russia, in the Ural Mountain mines where it was first discovered. At that time, pink and red Topaz were said to have been reserved exclusively for the family of the czar, which is why it is called "imperial" Topaz. Today these rare gems come mainly from Minas Gerais, Brazil. Good red Topaz is extremely rare. The largest known gem-grade red Topaz in a museum is a 97.45-carat untreated gem in the Grainger Hall of Gems. It is a beautifully clear stone that has been finished into a rectangular step-cut. This is a unique piece not only for its size, but for the fact that it is almost completely inclusion-free. In 2008 this stone was set in 14-karat rose Gold in a piece designed by Lester Lampert (fig. 96).

Some yellow Topaz can be heat-treated to turn pink, but natural stones can generally be distinguished from heat-treated stones by professional gemologists. As with other varieties of Topaz, great care must be exercised when cutting and polishing red Topaz, because it is relatively brittle. Imperial Topaz usually looks best under incandescent light, as opposed to blue Topaz, which looks best under daylight or fluorescent light. Before buying an expensive Topaz, it is best to look at it under both daylight and incandescent light.

FIGURE 96.
One of the world's largest ruby Topaz gems, weighing 97.45 carats and set in 14-karat rose Gold. Surrounded with 140 small ideal-cut Diamonds set in three-dimensional Gold "flames." Height of Topaz is 30 mm. The stone, originally from Minas Gerais, Brazil, was set into this original design named "Blaze" by Lester Lampert in 2008. Designed for the Grainger Hall of Gems (FMNH H2531).

OTHER TOPAZ

SYSTEM Inorganic

CLASS Silicate

GROUP Topaz

SPECIES *Topaz*

VARIETIES champagne Topaz (light brown Topaz); yellow Topaz; green Topaz; colorless Topaz

COMPOSITION Aluminum Fluoro-Hydroxy-Silicate [$Al_2SiO_4(F,OH)_2$]

TRACE ELEMENT FOR COLOR unknown, but some color due to external radiation over geologic time

HARDNESS ON MOHS SCALE 8

In addition to blue and imperial Topaz, there are a number of other varieties including light brown or tan-colored champagne Topaz, a relatively common variety. Yellow Topaz is common in its pale form, but much scarcer as a deeper golden shade sometimes referred to as sherry Topaz. One of the prize pieces of the Grainger Hall of Gems is a magnificent faceted golden Topaz of 54.33 carats from Brazil that is set in a 22-karat Gold ring designed by Eva Regan (fig. 100). Yellow, golden, and champagne varieties of Topaz have been mined in Germany since the Middle Ages.

Some yellow Topaz is heat-treated to form pink Topaz. Pale green Topaz is rare but generally has a lifeless appearance. "Hyacinth" is a name once applied to the scarce dark orange variety of Topaz, but the term is no longer used for Topaz. We restrict the use of the name hyacinth for a variety of Zircon discussed in a later chapter.

Colorless Topaz is the most common variety of transparent Topaz by far. When cut and polished, these stones have good luster but lack the fire of colorless Diamond. Colorless Topaz pebbles from ALLUVIAL DEPOSITS in northeastern Brazil are called *Pingos d'agoa* ("drops of water") by the local population.

FIGURE 97. Natural crystals of champagne Topaz on white Feldspar matrix from Pakistan. Specimen is 150 mm wide (FMNH H2351).

FIGURE 98.
Natural double-terminated raw crystals of champagne Topaz from Thomas Mountain, Utah. Specimen is 58.8 carats and measures 51 mm in length (FMNH M9112).

FIGURE 99.
Rectangular-faceted golden Topaz of 16.62 carats from Minas Gerais, Brazil, measuring 20 × 11 × 8 mm (FMNH H190).

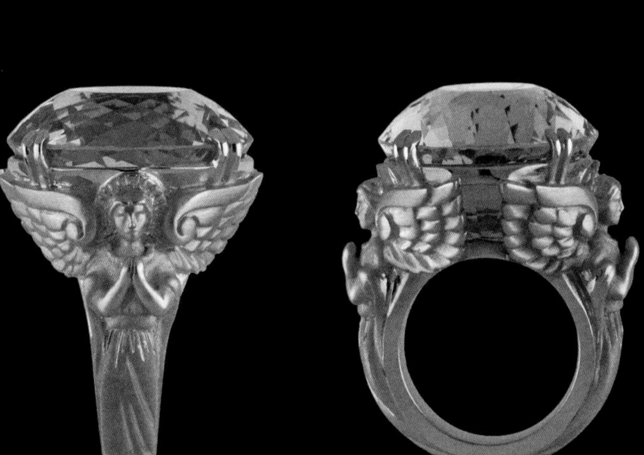

FIGURE 100.
A magnificent faceted golden Topaz of 54.33 carats from Minas Gerais, Brazil, set in 22-karat yellow Gold with intricately carved angels flanking each side of the Topaz. Each angel seems to stand "guard" with wings outstretched to embrace the Topaz. Christened "Guardian" by the designer, the Gold has a sand-blasted bright finish. Designed by Eva Regan of Lester Lampert, Inc., in 2008 for the Grainger Hall of Gems. Top, side, and front views (FMNH H2530).

FIGURE 101.
Right: Natural crystal of colorless Topaz (*left*) next to colorless Quartz crystal (*right*) on white Albite matrix from Pakistan. Topaz crystal is 965 mm height (FMNH H2463).

FIGURE 102.
Above: Faceted 93.7-carat colorless Topaz from the Ural Mountain region of Russia, in top, side, and oblique views. Stone measures 27 × 25 × 19 mm (FMNH H209).

Clear crystals of colorless Topaz can be very large, occasionally exceeding 100 pounds in weight. One clear faceting-grade colorless Topaz in The Field Museum collection weighs 210,000 carats, or about 90 pounds (fig. 91), and larger stones are known. As a gemstone, colorless and near-colorless Topaz is used mainly in the production of blue Topaz by means of irradiation and heat treatment. A notable exception is a piece of colorless Topaz jewelry designed by Ellie Thompson in the Grainger Hall of Gems (fig. 103).

One of the most famous colorless Topaz gems was a stone mined in Brazil during the eighteenth century. It was originally thought to be a Diamond and was faceted into a 1,680-carat gem. The gem was christened the "Breganza Diamond" and was set into the Portuguese Crown Jewels. It was much later determined to be a colorless Topaz. Colorless Topaz is known from most all localities that produce colored Topaz, including Brazil, Germany, Sri Lanka, the United States, Pakistan, Russia, Japan, Nigeria, Zaire, and Namibia.

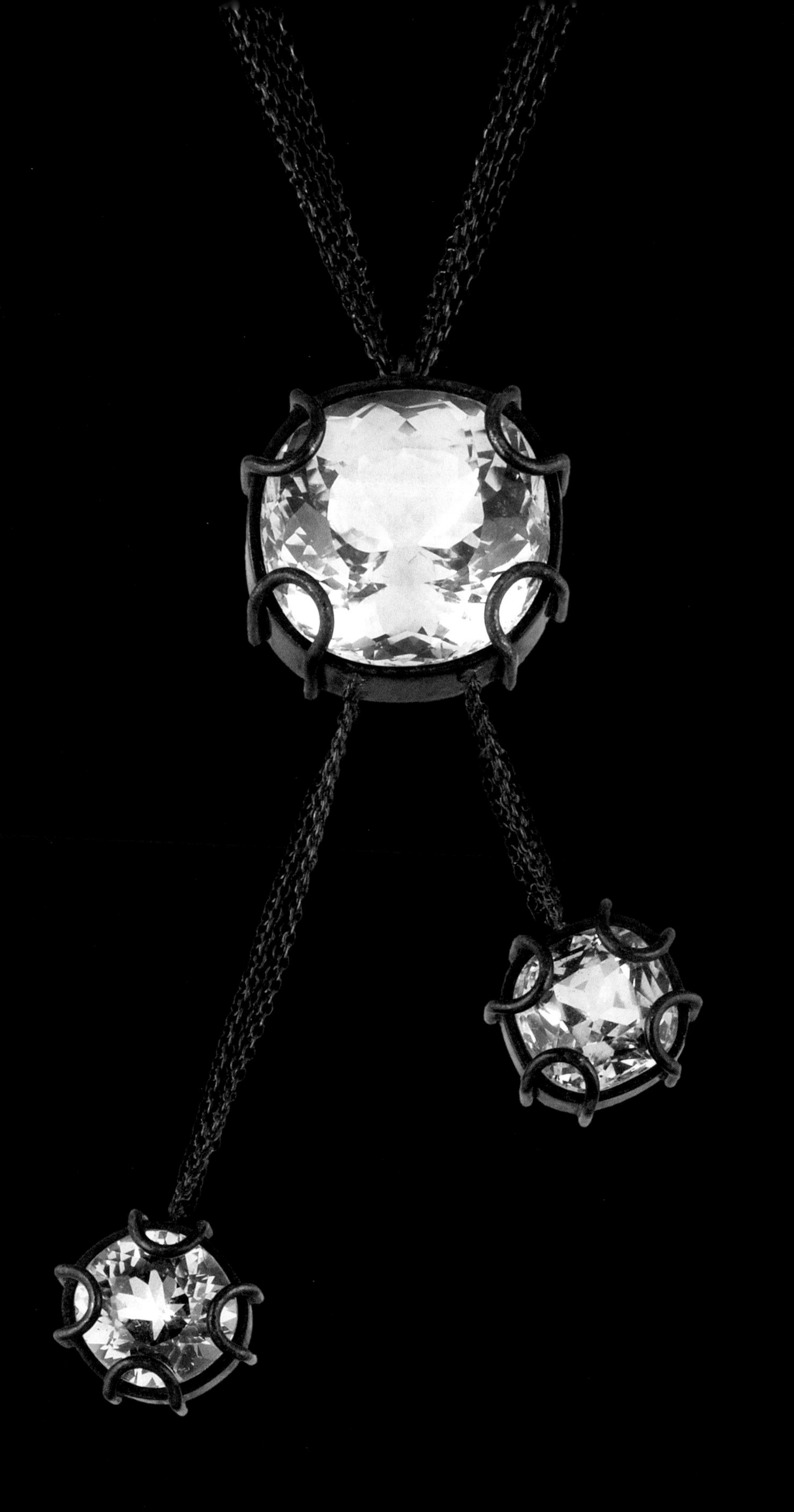

FIGURE 103.
Colorless Topaz necklace with stones set in blackened Silver. The three faceted Topaz gems weigh 13.6, 16.5, and 61.1 carats. Designed by Ellie Thompson of Ellie Thompson & Co. in 2008 for the Grainger Hall of Gems and named "Midnight" by the designer (FMNH H2544).

Beryl

SYSTEM Inorganic

CLASS Silicate

GROUP Beryl

SPECIES *Beryl*

GEM VARIETIES emerald (deep green); bixbite (red); aquamarine (blue); heliodor (yellow-green); morganite (pink); pale green Beryl; goshenite (colorless)

From the bright blue charm of aquamarine to the intense green fire of emerald, from the soft pink glow of morganite to the rich yellow hue of heliodor, and from the blood-red color of bixbite to the colorless clarity of goshenite—all of these gem varieties are forms of a single gem species: Beryl. Colorless goshenite is often the purest form of Beryl, while the other Beryl varieties are colored by the presence of minute trace elements or impurities.

"Massive" Beryl (non-gem-quality Beryl) is a primary ore of the rare metal Beryllium, which is used in ALLOYS for making metal stronger and more flexible. Some crystals of massive Beryl are enormous. A single Beryl crystal discovered in Malakialina, Madagascar, was 18 meters long, 3½ meters wide, and weighed over 400 tons.

Transparent gemstone-quality Beryls, particularly aquamarine and pale green Beryl, are also occasionally found as large crystals in pegmatite vugs. Near-flawless gem-quality aquamarine crystals can occasionally reach a meter in length. Flawless emerald crystals, on the other hand, are never large because emerald often grows in mica schists or as inclusions in Calcite deposits, tending to be more flawed than other varieties. Bixbite is also nonexistent as large flawless stones, and fine-quality faceted bixbite gems over 1 carat are extremely rare. Some Beryl showing CHATOYANCY or ASTERISM is known, although it is rare. Cat's-eye emeralds and aquamarines have been reported from Colombia and Brazil, and black star Beryl is known from Mozambique.

The occurrence of Beryl is widespread globally, but some varieties, such as bixbite, are mined from only a single small locality. For further reading on Beryl, we recommend Sinkankas (1981), O'Donoghue (2006), and Bauer (1968).

RICH GREEN BERYL (EMERALD)

SYSTEM Inorganic

CLASS Silicate

GROUP Beryl

SPECIES *Beryl*

VARIETY emerald

COMPOSITION Beryllium Aluminum Silicate $[Be_3Al_2(SiO_3)_6]$

TRACE ELEMENTS FOR COLOR Chromium (Cr) and possibly Vanadium (V) in some cases

HARDNESS ON MOHS SCALE 7½–8

As ruby is to red, so emerald is to green. Usually only the rich intense green Beryls with traces of Chromium are considered to be fine emeralds. Some dark green emeralds are also colored with traces of Vanadium. The pale green Beryls, often colored with the trace element Vanadium, do not qualify to bear the name emerald and are considered to be either simply green Beryls, greenish aquamarines, or greenish heliodors. Top-quality emeralds are rare and generally more valuable than colorless Diamond. The finest-quality stones of 15 carats or larger can sell for as much as $100,000 per carat today.

Nearly all emeralds have INCLUSIONS (cracks and fissures), and they are thus graded mostly on the basis of color. Unlike most other varieties of Beryl, large flawless emeralds are basically nonexistent. The inclusions in emeralds are sometimes referred to as the *jardin* (garden), and they can be used by specialists to help identify the geographic origin of individual stones. Emeralds are also brittle, so care must be taken with emerald rings not to knock them against a hard surface.

Rich Green Beryl (Emerald)

FIGURE 104.
Natural emerald crystals on black mica schist matrix from Lagham Province, Afghanistan. Entire specimen is 120 × 90 × 80 mm (FMNH H23218).

OILING is a common enhancement technique to hide cracks and fissures in natural emeralds. This involves soaking an emerald in oil such as cedarwood oil. The oil is absorbed into cracks and exposed fractures and hides imperfections for a time by filling them. After several years, the oil decomposes and must be reapplied. The treatment when fresh is effective enough that it has today become a standard procedure used by many of the world's main emerald mining and processing centers. Most newly mined emeralds sold today have been oiled. This is one reason it is best not to immerse emerald jewelry in strong solvents that dissolve oil, such as nail polish remover (acetone). Another method of filling fractures in emerald is to use epoxies or resins, such as synthetic polyester epoxy resin. But even this is often an impermanent emerald filler. Usually an experienced gemologist can tell whether or not an emerald has been filled with oil or resin.

Emerald mining has a long history going back at least as early as 2000 BC. The ancient Egyptians mined emeralds near the Red Sea thousands of years ago. Legend has it that Cleopatra was an avid collector of emeralds. The ancient Romans called all green gemstones "emerald" and believed that richer hues had ripened from lighter greens, so the paler Beryls were considered to

FIGURE 105.
Natural emerald crystals. *Top:* an 877.5-carat crystal from Bahia, Brazil, measuring 89 × 37 mm. *Bottom:* two double-terminated crystals of 151.8 and 152.6 carats, 70 and 60 mm in length, from Alexander County, North Carolina (FMNH H95, H85, H86).

be immature. Egypt is no longer a productive source of emerald, and the major mines of today are in Colombia, Brazil, India, Zimbabwe, Zambia, South Africa, Tanzania, Russia, Australia, and Pakistan.

Gem-quality emeralds have also been mined in the United States in North Carolina, near and at the localities that produce hiddenite. The Aztecs and Incas of South America regarded emeralds as holy. The maharajas and maharanis of India also highly valued the stones. One Indian emerald, the "Mogul Emerald," weighs 217.8 carats, is 10 centimeters tall, and dates back to 1695. It is inscribed with prayer texts and was auctioned by Christie's of London in 2001 for $2.2 million. Fine emeralds are an important part of the crown or national jewel collections of many nations. A very rare cat's-eye emerald is also known from Colombia and Brazil. These stones can wholesale for thousands of dollars per carat.

The Grainger Hall of Gems features some exceptional natural emerald crystals, including an 877.5-carat crystal from Bahia, Brazil, and two fine crystals of 151.8 and 152.6 carats from Alexander County, North Carolina (fig. 105). The Grainger Hall of Gems also has some fine pieces of emerald jewelry, including a necklace containing 25 carats of African emeralds mounted with 238 round brilliant-cut Diamonds (fig. 106).

FIGURE 106.
Emerald and Diamond necklace set in 18-karat Gold, with 18 African emeralds totaling 25 carats and 238 round brilliant-cut Diamonds totaling 14.75 carats (FMNH H2259).

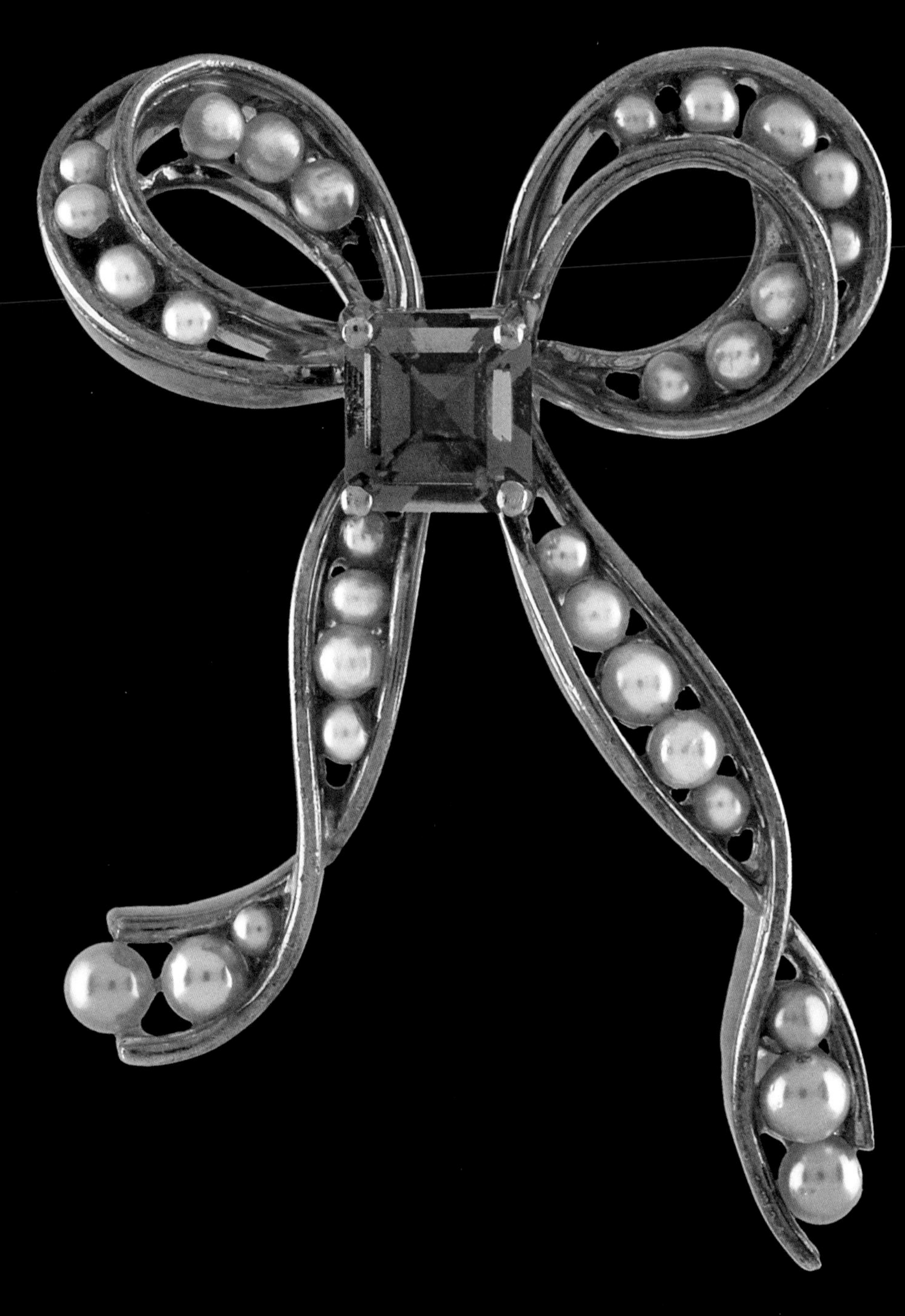

FIGURE 107.
Antique brooch, with an unusually clean, step-cut African emerald of 1.06 carats set into 18-karat yellow Gold with 30 seed Pearls (FMNH H2482).

RED BERYL (BIXBITE)

SYSTEM Inorganic

CLASS Silicate

GROUP Beryl

SPECIES *Beryl*

VARIETY bixbite

COMPOSITION

Beryllium Aluminum Silicate

$[Be_3Al_2(SiO_3)_6]$

TRACE ELEMENT FOR COLOR

Manganese (Mn)

HARDNESS ON MOHS SCALE

7½–8

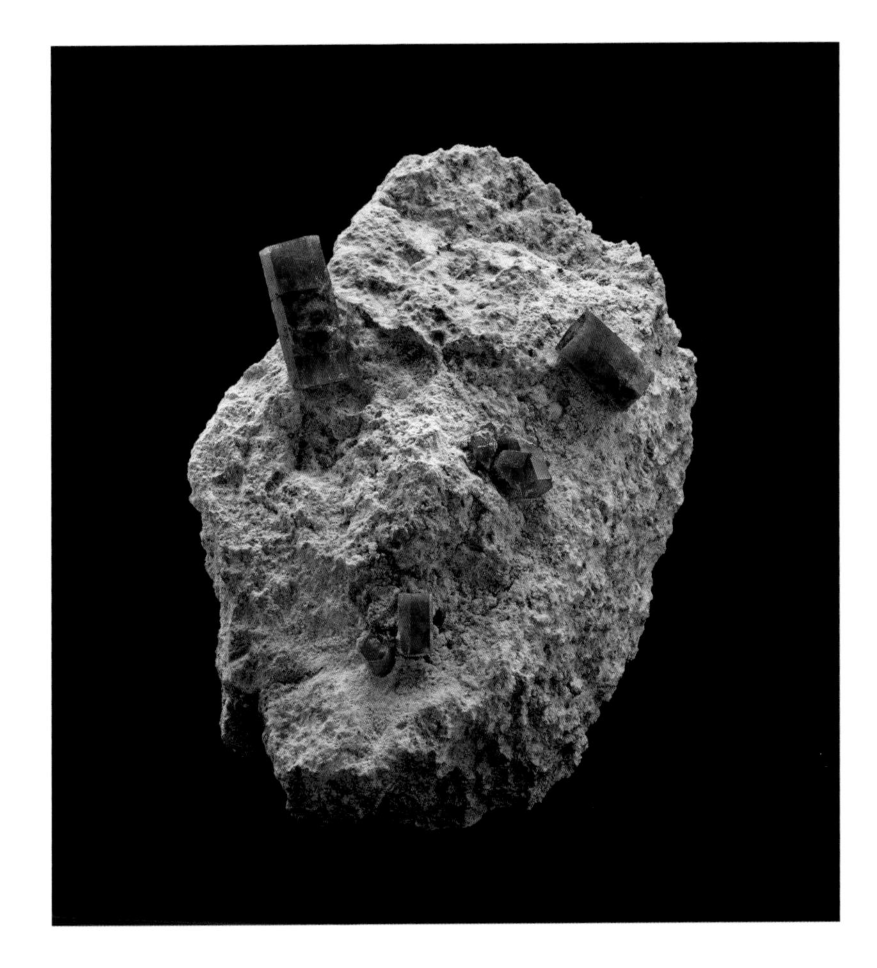

FIGURE 108. Natural double-terminated crystals of bixbite (red Beryl) on rhyolite matrix, from the Wah Wah Mountains of Utah. Specimen measures 140 mm in height, and the largest crystal measures 27 mm in length (FMNH M2330).

Bixbite is the rarest of the Beryls and a little-known variety in the precious-gem market. Hundreds of times rarer than even emerald, large gem-quality stones are elusive and usually unavailable. To date the largest faceted red Beryl is only about 8 carats, the second largest is 4 carats, and any top-quality faceted stone of a carat or more is a true rarity. It is so rare, in fact, that its scarcity has kept the popularity (and hence the value) of this gem from being much greater than it currently is. Even so, the largest stones of over 1 carat can sell for up to $20,000 per carat. Like emerald, bixbite is almost always heavily included and fractured.

Bixbite usually occurs as small crystals in a soft volcanic rock called rhyolite (fig. 108). It is a relative newcomer to the gem trade, first discovered in 1904 by a mineral collector named Maynard Bixby in the Thomas Mountains of central Utah. The variety was named after him. Gem-quality stones were later discovered in the Wah Wah Mountains of Utah in the 1950s. The Wah Wah localities are where all gem-quality stones are mined today. Regular mining of gem-quality bixbite began in the mid-1970s. Bixbite has also been reported

FIGURE 109.
Right: Rare fine bixbite ring called "Berylicious," with three faceted bixbites including a 1.73-carat stone flanked by two .51-carat stones. Set in 18-karat yellow Gold with 24 Diamond baguettes lining the band. Top, side, and front views. Designed by Lester Lampert in 2008 for the Grainger Hall of Gems. Height of ring is 25 mm. The bixbites are from the Wah Wah Mountains in Utah. *Above:* close-up of the larger central stone before it was set in the ring (FMNH H2539).

from New Mexico, but these deposits are not of gem quality so far. The only mineable deposits of bixbite anywhere in the world today remain the deposits of the Wah Wah Mountains.

The best bixbite is a raspberry color and has few obvious inclusions (although no large stones are flawless). The average faceted stone is only about 2/10th of a carat in weight. It is estimated that currently only about 10 to 20 crystals of faceting-grade bixbite are produced each year that will yield faceted stones of 1 carat or more. Even the richest red Beryl deposits yield less than 2 carats of red Beryl per ton of ore. Recently, synthetic red Beryls have been produced, but an experienced gemologist can easily distinguish between natural and synthetic bixbite. Bixbite is one of the rarest and most unique gemstones of North America.

There is a fine bixbite ring designed by Lester Lampert in the Grainger Hall of Gems with 2.75 carats of faceted bixbite set in 18-karat yellow Gold with 24 Diamonds (fig. 109). There is also an extremely fine specimen of natural bixbite crystals in MATRIX in the exhibition (fig. 108).

BLUE BERYL (AQUAMARINE)

SYSTEM Inorganic

CLASS Silicate

GROUP Beryl

SPECIES *Beryl*

VARIETIES aquamarine; cat's-eye aquamarine

COMPOSITION

Beryllium Aluminum Silicate $[Be_3Al_2(SiO_3)_6]$

TRACE ELEMENT FOR COLOR

Iron (Fe)

HARDNESS ON MOHS SCALE

7½–8

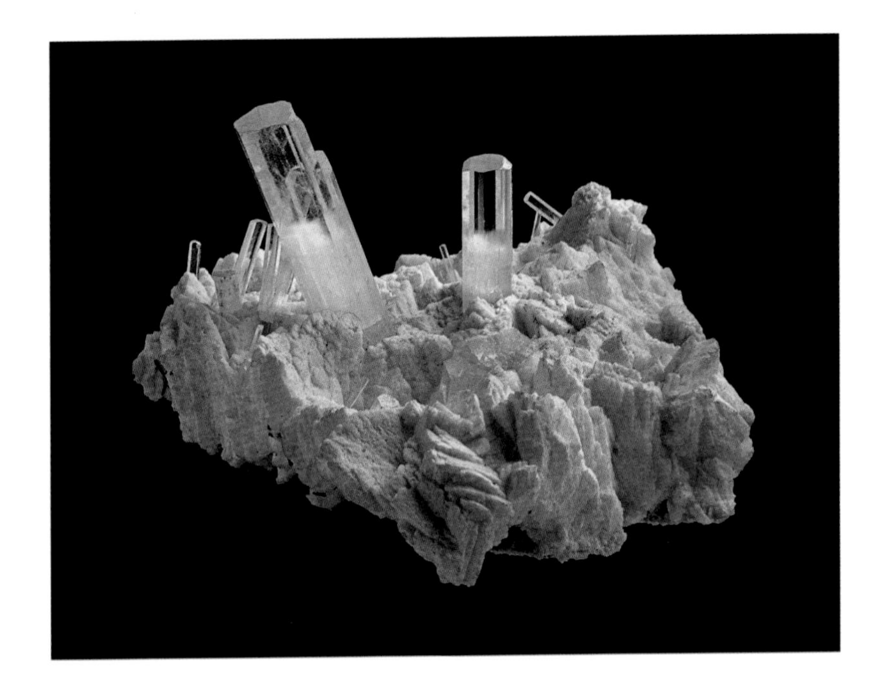

FIGURE 110.
Natural aquamarine crystals in white Albite Feldspar matrix, from the Gilgit district, Pakistan. Length of largest crystal is 70 mm (FMNH H2322).

There is no gem that rivals the cool sea-blue color and clarity of a high-quality aquamarine. Even the natural crystals are truly outstanding in their beauty (fig. 110). The name aquamarine refers to the water (Latin *aqua*) of the sea (Latin *mare*). A rare variety of deep blue aquamarine called maxixe is known from the Maxixe mine in Minas Gerais, Brazil. The intense color of maxixe is thought to be derived from natural radiation, but the color eventually fades from long-term exposure to daylight. The color can be revived through artificial irradiation but will eventually fade once again. The best aquamarines by today's standards are natural stones ranging from sky blue to dark blue. Most ideally colored aquamarines in the gem trade today are heat-treated. By heating poor-colored aquamarines to 725°–850°F, the yellowish and greenish tints can be eliminated, and the blue deepened. The change is permanent.

Aquamarine, unlike emerald and bixbite, commonly occurs with few or no noticeable inclusions. It is also PLEOCHROIC, sometimes appearing blue or colorless as the stone is viewed from different angles, or blue and greenish-blue at different angles. The PLEOCHROISM is one of the optical differences between aquamarine and blue Topaz.

Gem-quality near-flawless aquamarine crystals can be very large, sometimes reaching lengths of 1 meter or more. The largest known gem-quality aquamarine was found in 1910 in Minas Gerais, Brazil. It weighed over 240 pounds and was cut into more than 100,000 carats of gems.

Aquamarine is more costly than blue Topaz, but less expensive than either emerald or bixbite. Extra-fine-quality aquamarines wholesale for a few hundred

FIGURE 111.
Faceted 137-carat aquamarine from Oxford County, Maine. This is one of the largest known gem aquamarines cut from a North American gemstone. Gem is 36 × 36 mm (FMNH H148).

FIGURE 112.
The "Crane Aquamarine," a flawless 341-carat faceted gem from Brazil measuring 60 × 37 × 22 mm that was part of the founding gem collection in The Field Museum in 1893 (FMNH H149).

dollars per carat. Lower-grade and paler color stones sell for substantially less. Large, EYE-CLEAN stones are readily available. Aquamarine's great beauty and an abundance of eye-clean stones make this a popular colored gem. It is the birthstone for March. A rare form of CHATOYANT aquamarine also exists, both in star and cat's-eye varieties.

The best aquamarine deposits today are in Brazil, but other important deposits occur in Pakistan, Australia, Myanmar, China, India, Madagascar, Mozambique, Nigeria, Namibia, Zimbabwe, and the United States (Maine and California). The largest faceted high-quality aquamarine from a North American locality known to us is a near-flawless untreated 137-carat stone with good color from Stoneham, Maine, in the Grainger Hall of Gems (fig. 111). The Grainger Hall of Gems also contains a number of other large faceted aquamarines, including the flawless 341-carat "Crane Aquamarine" (fig. 112), and some fine aquamarine gems set in jewelry (figs. 113, 114).

FIGURE 113.
Victorian-era 22-karat Gold brooch with a faceted oval aquamarine. Piece is 36 mm in height (FMNH H2451).

FIGURE 114.
An exquisite 148.5-carat aquamarine from Minas Gerais set in a Platinum pin with Gold accents and white Diamonds, called "The Schlumberger Bow." Aqua gem measures 30 × 28 × 20 mm. Designed by Jean Schlumberger for Tiffany & Co. in the late twentieth century. The Field Museum aquamarine was set by Tiffany's in the Schlumberger design for the Grainger Hall of Gems in 2009 (FMNH H2590).

YELLOW BERYL (HELIODOR)

SYSTEM Inorganic

CLASS Silicate

GROUP Beryl

SPECIES *Beryl*

VARIETY heliodor

COMPOSITION

Beryllium Aluminum Silicate $[Be_3Al_2(SiO_3)_6]$

TRACE ELEMENT FOR COLOR

Iron (Fe)

HARDNESS ON MOHS SCALE

7½–8

FIGURE 115.
Natural crystals of heliodor on Feldspar matrix from the Zelatoya Vata Mine, Tajikistan. Largest crystal is 68 mm in length (FMNH H2332).

The lemon-yellow, greenish-yellow, or golden-yellow heliodor is somewhat scarcer than its cousin aquamarine but is currently less valuable. There is some confusion in the gem literature about which shades of yellow are included in the variety heliodor. Here we include golden-yellow Beryl in the heliodor category, but some authors list this as a separate variety. The name heliodor is from the Greek *Helios* (sun) and *doron* (gift), and means "gift from the sun." This stone sometimes looks similar to yellow Chrysoberyl, but it is not as hard (Mohs 7½–8 compared to 8½ for Chrysoberyl) and is of a different chemical composition. Heliodor was discovered in Namibia in 1910, although it is also found in Brazil, Russia, Tajikistan, and Madagascar. The finest stones are said to come from the Ural Mountains of Russia. Heliodor is often found associated with aquamarine in pegmatitic vugs. Stones with rich yellow color are fine yet relatively inexpensive gemstones, often found with very few inclusions. Heliodor gems of 20 carats or more are very scarce. The largest known faceted heliodor is a 2,054-carat stone in the collection of the Smithsonian Institution.

The Grainger Hall of Gems has an extremely fine natural crystal specimen of heliodor from Tajikistan (fig. 115) and a beautiful pendant designed for the Hall of Gems in 2008 by Ellie Thompson using one of the museum's fine faceted heliodors (fig. 116).

PINK BERYL (MORGANITE)

SYSTEM Inorganic

CLASS Silicate

GROUP Beryl

SPECIES *Beryl*

VARIETY morganite

COMPOSITION

Beryllium Aluminum Silicate $[Be_3Al_2(SiO_3)_6]$

TRACE ELEMENT FOR COLOR

Manganese (Mn)

HARDNESS ON MOHS SCALE

7½–8

FIGURE 116.
Facing page: Heliodor pendant designed by Ellie Thompson of Ellie Thompson & Co. in 2009 for the Grainger Hall of Gems. The largest heliodor is from Tajikistan and weighs 8.3 carats. It is set in 18-karat yellow Gold with 100 small Diamonds and 14 small heliodors. Named "Sunlit Diamonds and Fringe" by the designer (FMNH H2513).

FIGURE 117.
Above right: Natural crystal of morganite on bluish-white Albite Feldspar matrix from the Pala mining district of Southern California. Specimen is 125 mm at widest diameter, and morganite crystal measures 68 × 68 × 30 mm (FMNH H1803).

Morganite is well known today as the pink variety of Beryl, but it is a relatively recent addition to the list of precious gems. It has been known by the name of morganite only since 1911, when the famous gemologist George F. Kunz named this gem variety after the famous banker and mineral collector J. P. Morgan.

Morganite is found in various shades of pink, ranging from light violet pink to deep pink. The deep pink variety of morganite is the most valuable, and most morganites tend to be pale. The pink color of pale morganites can be intensified and the yellow or orange tinges removed with heat treatment of 725° to 730°F. Morganite crystals tend not to be as elongate as most other varieties of Beryl but are instead tabular (flat, tablet-shaped) (fig. 117). The first morganites were described from California, and the gem has since been discovered in Madagascar, Brazil, Afghanistan, Italy, Mozambique, Namibia, Zimbabwe, and Pakistan. Large stones over 20 carats are readily available, and morganite is currently slightly less valuable than aquamarine. The largest known faceted morganite is a 598.7-carat cushion-shaped stone from Madagascar in the collection of the British Museum of Natural History.

There are some large faceted morganite gems in the Grainger Hall of Gems, including a pendant designed by Marc Scherer (fig. 118).

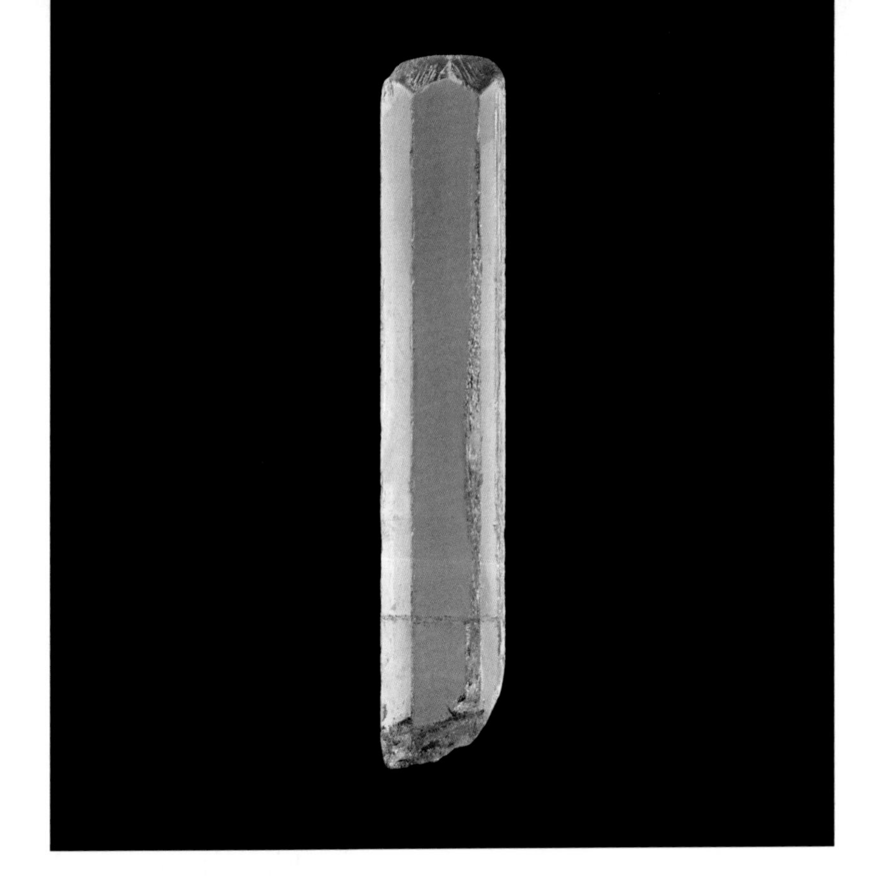

PALE GREEN BERYL

SYSTEM Inorganic

CLASS Silicate

GROUP Beryl

SPECIES *Beryl*

VARIETY pale green Beryl

COMPOSITION

Beryllium Aluminum Silicate

[$Be_3Al_2(SiO_3)_6$]

TRACE ELEMENT FOR COLOR

Iron (Fe) or Vanadium (V)

HARDNESS ON MOHS SCALE

7½–8

Going under the names pale green Beryl, light green Beryl, or sometimes simply green Beryl, this is one of the most abundant and least expensive varieties of Beryl, even as EYE-CLEAN, gem-grade material. Sometimes marketed as "light emerald" or "green aquamarine," the color is far too pale to legitimately be called emerald and too green to be aquamarine. For a Beryl to be considered true emerald, the color must be deep, intense green, blue green, or grass green. To be an aquamarine, the color should be blue.

Like aquamarine, very large, eye-clean crystals of pale green Beryl are not rare, and some large, eye-clean crystals are even used for carving. In the Grainger Hall of Gems, there is a large pale green Beryl carving of Marshall Field (the museum's main founder). This piece weighs 1,424 carats and is carved from nearly flawless faceting material (fig. 123). Pale green Beryl is also sometimes included with aquamarine, because heat treatment of some stones (those with traces of Iron) will turn them blue. Care must be taken when heat-treating pale green Beryl and greenish aquamarine, because overheating may result in the disappearance of all color. There is a necklace of fine pale green Beryls and amethysts set in 18-karat white Gold in the Grainger Hall of Gems (fig. 122).

Pale green Beryl occurs in most all of the localities that produce aquamarine, including Brazil, the United States, Pakistan, Australia, Myanmar, China, India, Madagascar, and various localities in Africa.

FIGURE 118.
Facing page: Morganite and Diamond pendant with a 42.8-carat faceted morganite from Minas Gerais, Brazil, set in 18-karat white Gold with 200 small round brilliant-cut Diamonds. Designed by Marc Scherer of Marc & Co. in 2008 for the Grainger Hall of Gems. Named "Morning Rose" by the designer (FMNH H2522).

FIGURE 119.
Above right: Natural crystal of eye-clean, pale green Beryl from Minas Gerais, Brazil. Crystal is 188.5 carats and measures 90 mm in length (FMNH H183).

FIGURE 120.
Large flawless, faceted pale green Beryl from the Siberian region of Russia, weighing 339.8 carats and measuring 55 × 40 × 21 mm (FMNH H162).

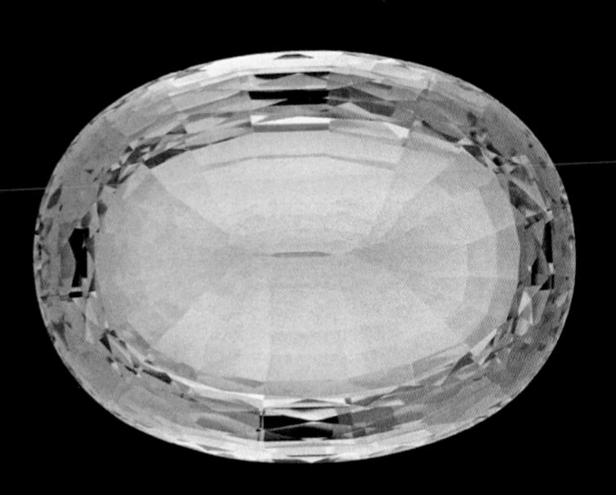

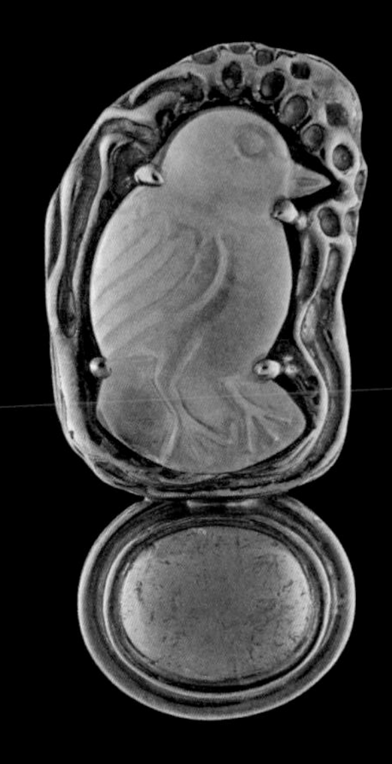

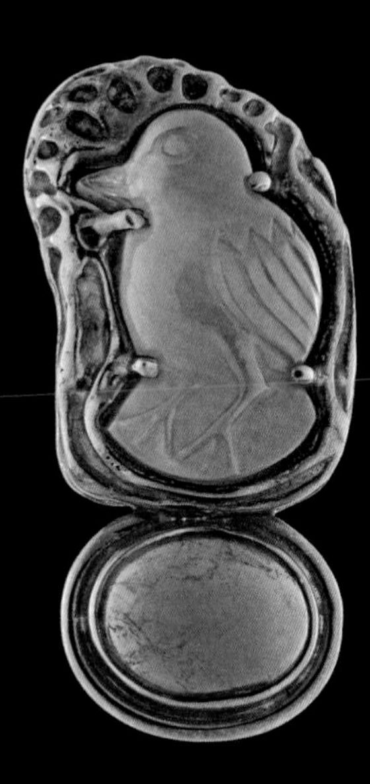

FIGURE 121.
Above: Earrings of pale green Beryl and pink Tourmaline set in 18-karat yellow Gold. Beryl is carved into the image of a bird. Designed by Elizabeth Gage, late twentieth century (FMNH H2456).

FIGURE 123.
Left: Pale green Beryl carving of Marshall Field sculpted by artist Ute Bernhardt, weighing 1,424 carats and measuring 85 × 90 × 20 mm. Cut from a larger unusually eye-clean crystal recovered from Sri Lanka (FMNH H1752).

FIGURE 122.
Necklace with 106 carats of faceted pale green Beryls and 118 carats of faceted amethysts set in 18-karat white Gold (FMNH H2349).

COLORLESS BERYL (GOSHENITE)

SYSTEM Inorganic

CLASS Silicate

GROUP Beryl

SPECIES *Beryl*

VARIETY goshenite

COMPOSITION

Beryllium Aluminum Silicate $[Be_3Al_2(SiO_3)_6]$

TRACE ELEMENT FOR COLOR

none

HARDNESS ON MOHS SCALE

7½–8

FIGURE 124.
Natural crystal of goshenite (colorless Beryl) on matrix of white Feldspar and mica from the Gilgit district of Pakistan. Specimen is 120 mm in length (FMNH H2437).

FIGURE 125.
Facing page: Goshenite and Diamond necklace named "Empress," with three large faceted goshenite Beryls from Death Valley, California. Goshenite gems weigh 3, 3.8, and 13.1 carats and are set in 14-karat white Gold. They are surrounded in Etruscan style by 280 single-cut Diamonds and 12 old mine-cut Diamonds together weighing 3.55 carats. Pendant is suspended with a black silk cord with two Diamond rondel sections. Designed by Lester Lampert in 2008 for the Grainger Hall of Gems (FMNH H2538).

Goshenite is, in theory, Beryl in its purest form. Because there are generally no trace elements in the stone, goshenite is completely colorless. It is sometimes referred to as the "mother of gemstones" because the addition of various trace elements in nature transforms it into emerald, aquamarine, morganite, heliodor, or bixbite.

Goshenite is named after the town of Goshen, Massachusetts, one of the localities where it is found. It is also found in Canada, Brazil, Russia, China, Mexico, the United States, and Pakistan.

It is said in some Internet and printed sources that goshenite and very pale-colored Beryl were once used for magnifying lenses and spectacles during the Middle Ages or even earlier. We were unable to verify this with any documented evidence. Goshenite is occasionally used as a Diamond SIMULANT (imitation) by placing silver foil behind a cut stone and setting it into a piece of jewelry. Similarly, it has been used to imitate emerald with a backing of green foil. As a faceted gemstone, goshenite has exceptional brilliance.

Goshenite, like other forms of Beryl, is brittle and should be handled carefully. It should not be exposed to wide temperature fluctuations. It is frequently found as clear, eye-clean material, so pieces with visible inclusions are not used in jewelry. There is a fine goshenite and Diamond necklace named "Empress" in the Grainger Hall of Gems with faceted goshenite from Death Valley, California. This piece was designed by Lester Lampert in 2008 (fig. 125).

Cordierite

SYSTEM Inorganic

CLASS Silicate

GROUP Cordierite

SPECIES *Cordierite*

GEM VARIETIES blue Cordierite (iolite); colorless Cordierite

COMPOSITION

Magnesium Aluminum Silicate $[(Mg,Fe)_2Al_4Si_5O_{18}]$

TRACE ELEMENT FOR COLOR

Iron (Fe)

HARDNESS ON MOHS SCALE

7–7½

FIGURE 126.
Top: Natural piece of rough faceting-grade blue Cordierite (iolite) from Tsihombe, Madagascar. Specimen measures 45 mm in length (FMNH M18008). *Bottom:* Three faceted colorless Cordierite gems from Sri Lanka weighing 3.9–5.7 carats each (FMNH H84, H2034, H2035). It is the blue variety of Cordierite (iolite) that is most valued by the gem and jewelry trade (see fig. 127).

FIGURE 127.
Facing page: Faceted iolite gems of two carats each from Tanzania set in 18-karat white Gold earrings with 58 ideal-cut Diamonds weighing 2.19 carats total. Designed by Lester Lampert in 2009 for the Grainger Hall of Gems and named "Soirée" by the designer (FMNH H2593).

Iolite—also known as "water sapphire," "lynx sapphire," or "dichroite"—is the sapphire-blue variety of the mineral Cordierite. It has been used in jewelry since the eighteenth century, and the name is derived from the Greek *io*, meaning "violet flower." It resembles violet-blue tanzanite and sapphire, although it is much softer than sapphire, and much harder than tanzanite. There is also a variety called bloodshot iolite from Sri Lanka that shows a reddish color from Hematite inclusions. Iolite is also much more common than either sapphire or tanzanite and therefore much less expensive, currently wholesaling for well under $100 per carat. Gems over 10 carats are scarcer and more valuable. Iolite is strongly pleochroic, making it necessary to properly orient the stone for cutting in order to get the best show of color. It is a very brittle stone and can be difficult for gem cutters to work with. The mineral Cordierite also occurs as colorless faceting-grade material, but it is the more common blue iolite variety that is in demand by the gem industry.

Gem-quality iolite is generally found in veins in METAMORPHIC ROCKS, and iolite gemstones are generally irregular-shaped masses rather than well-formed crystals (fig. 126). It is also found in alluvial gravels as water-worn pebbles. Iolite is generally cut as a rectangular step-cut, with the table oriented to view the optimum color, although darkly colored and scarce cat's-eye iolite is usually cut as a cabochon. The best iolite gem material is found in Sri Lanka, Myanmar, India, and Madagascar. There is a pair of fine iolite earrings in the Grainger Hall of Gems designed by Lester Lampert in 2009 (fig. 127). For further reading on iolite, we recommend Bauer (1968) and O'Donoghue (2006).

Phenakite

SYSTEM Inorganic

CLASS Silicate

GROUP Willemite

SPECIES *Phenakite*

GEM VARIETIES colorless Phenakite; cat's-eye Phenakite; star Phenakite

COMPOSITION

Beryllium Silicate (Be_2SiO_4)

TRACE ELEMENT FOR COLOR

none

HARDNESS ON MOHS SCALE

7½–8

FIGURE 128.
Natural crystals of Phenakite on matrix from Minas Gerais, Brazil. Specimen is 70 mm in width (FMNH M11123).

Phenakite (also called phenacite) is "the gemstone that isn't." Usually included in any book on gemstones, it is harder than Quartz, and is very rare in faceting grade. Well-cut Phenakite is brilliant, with a silvery appearance. This mineral would seem to be an excellent candidate for the gem trade given its hardness, scarcity, and brilliance, but Phenakite has never been popular as a gemstone. The colorless, faceting-grade material resembles colorless sapphire when cut and polished, but is softer (Mohs 7½–8 for Phenakite vs. Mohs 9 for sapphire). It can also resemble Quartz, but is harder (Mohs 7 for Quartz). In fact, the name Phenakite is derived from the Greek *phenakos*, meaning "deceiver," because it was once often confused with Quartz. Phenakite is very rarely found with a yellow, pale rose, or brown hue, but the color often fades after years of exposure to daylight. Although more brilliant and lustrous than Quartz, Phenakite lacks the fire of Diamond.

Phenakite crystals form in pegmatitic vugs with other gems such as Beryl, Topaz, and Chrysoberyl. Some of the largest faceted Phenakite gems we know of are two examples in the Natural History Museum of London, weighing 34 and 43 carats, and a recently discovered 1,470-carat stone from Sri Lanka that was cut into a 569-carat faceted oval, currently in a private collection. Gem-quality Phenakite comes primarily from the Ural Mountain region of southern Russia, Brazil, Sri Lanka, Zimbabwe, Tanzania, Norway, and the United States (Colorado and New Hampshire). Translucent Phenakite can sometimes be cut as a cabochon to show cat's-eye chatoyancy. Four-rayed star Phenakite has also been reported from Sri Lanka.

FIGURE 129.
Phenakite from the Ural Mountain region of Russia faceted into gems weighing from 14.2 to 19.1 carats each. Largest stone measures 21 × 17 × 10 mm (FMNH H223–H225, H227).

TOURMALINE GROUP

SYSTEM Inorganic

CLASS Silicate

GROUP Tourmaline

ALKALI SPECIES *Elbaite* Tourmaline; *Dravite* Tourmaline; *Schorl* Tourmaline

CALCIC SPECIES *Liddicoatite* Tourmaline; *Uvite* Tourmaline

The name Tourmaline is derived from the Sinhalese (Sri Lankan) word *tourmali*, meaning "gemstone." In ancient times, Tourmalines were generally mistaken for other gems known at the time, such as emerald, sapphire, and ruby. In fact, the Russian empress Catherine the Great had a large crown jewel she thought to be a ruby that was later found to be a rubellite Tourmaline. Tourmaline was first recognized as a distinct mineral in the eighteenth century. Since that time, the complexity of the Tourmaline group has increased greatly to include fourteen species, five of which include popular gem varieties that are discussed below. Of the five gem species of Tourmaline, most gem varieties are contained within the species Elbaite.

Tourmaline, like Garnet, is more than a single-gem species. It is what is referred to as a group, consisting of at least fourteen species, each with a different chemical composition, making it chemically one of the most complicated groups of Silicate minerals. The group is further complicated in that some species can combine and form as **BLENDS** (Elbaite and Dravite, for example), much like Garnet and some Feldspar species. Chemical compositions given here for Tourmaline gem varieties are idealized and in actuality are somewhat variable. But all Tourmaline species share a common crystal structure and contain the principal components of silica, Aluminum, and Boron. Natural crystals are usually columnar and roughly triangular in cross section. In the gem trade, Tourmalines are often classified by color rather than by composition to simplify classifications for this extremely complex mineral group.

Five Tourmaline species have been used as gemstones. These include *Elbaite* Tourmaline (or Lithium Tourmaline), which is the most diverse and includes the most valuable Tourmaline gemstones; *Liddicoatite* Tourmaline (or Calcium-Lithium Tourmaline), which is a relatively rare species; *Dravite* Tourmaline (or Sodium Magnesium Tourmaline), which includes one brown variety occasionally used as a gem and a very valuable rich green variety colored by traces of Chromium called "chrome Tourmaline" or "chromdravite"; and *Schorl* Tourmaline (or Iron-rich Tourmaline), an opaque black variety formerly used in mourning jewelry. More recently, *Uvite* Tourmaline (Calcium Magnesium Tourmaline) gems are becoming more widely available on the market. One variety of Uvite sometimes called "chrome Uvite" is a rich green variety usually colored by traces of Vanadium. This variety, like chromdravite, can be very valuable in faceting-grade quality. Among the five Tourmaline species, Elbaite varieties are the most widely used as gems. Gem-quality Tourmalines are usually found in pegmatite vugs, and the crystals show characteristic "phonograph record"–type striations parallel to the length of long crystals. Exceptions are Dravite, which usually occurs in marble and often exhibits smoother crystal faces, and Uvite, which tends to form stubby crystals.

Tourmaline crystals are strongly **PIEZOELECTRIC**, which is the ability to pro-

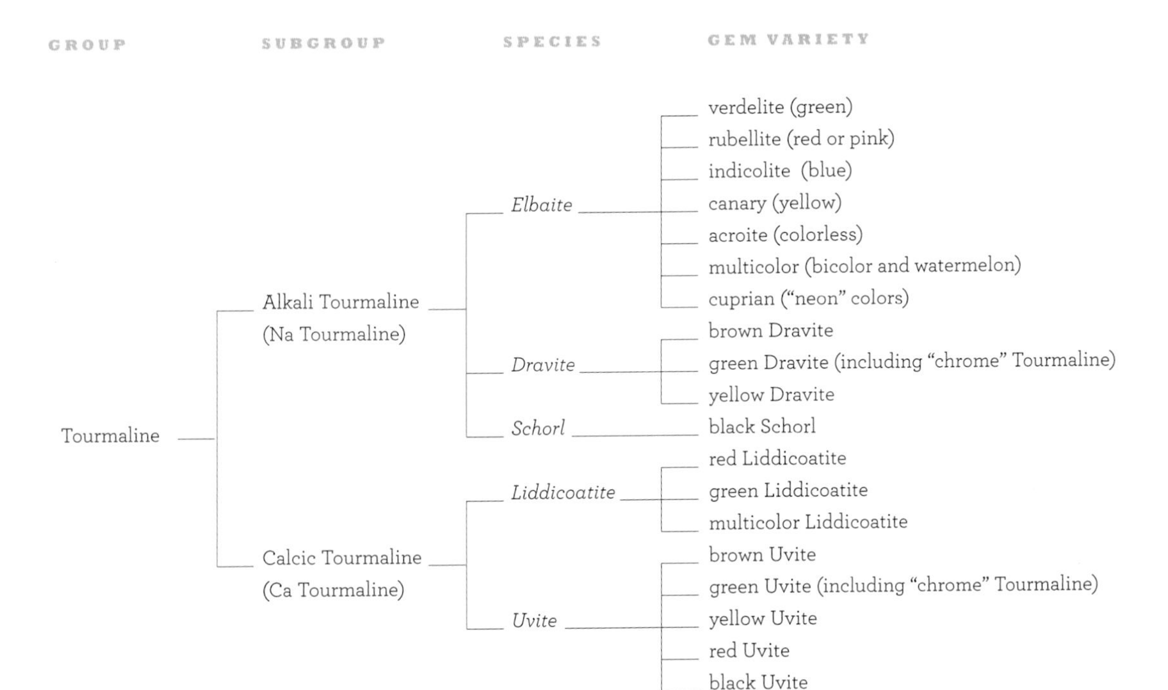

FIGURE 130. Classification of Tourmaline gemstone varieties discussed in this book.

duce an electric charge when compressed. This property gives Tourmaline value for industrial use, such as the construction of pressure gauges. Tourmalines are also PYROELECTRIC, which is the ability to become electrically charged due to changing temperature in the crystal (although Schorl apparently lacks this trait). This property led to their use in the Netherlands as *aschentrekkers*—a tool used to pull ashes and dust out of a pipe—long before science could explain this phenomenon. When one end of a Tourmaline crystal is heated, it becomes charged at the end, attracting ashes and dust to stick to it so they can be removed. There is a disadvantage to the piezoelectric and pyroelectric properties of Tourmaline: the gems tend to collect more dust than other gems because of the electrical charges that develop over time. Consequently, Tourmalines require more frequent cleaning in museum exhibition cases than other gems. Beyond its value to industry, Tourmaline is a major group that contains many beautiful gemstone varieties. One of the Tourmaline species, Elbaite, exhibits the widest ranges of colors for any gem species. For additional reading on Tourmaline, we recommend Dietrich (1985), Sinkankas (1959), and Bauer (1968).

Elbaite Tourmaline

SYSTEM Inorganic

CLASS Silicate

GROUP Tourmaline

SUBGROUP Alkali Tourmaline

SPECIES *Elbaite*

GEM VARIETIES green Elbaite (verdelite); pink to red Elbaite (rubellite); blue Elbaite (indicolite); yellow Elbaite (canary); multicolored Elbaite (bicolor, tricolor, watermelon); colorless Elbaite (achroite); "neon" colors of blue, green, purple (cuprian Elbaite)

Elbaite Tourmaline, sometimes also referred to as Lithium Tourmaline or lithia Tourmaline, is among the most colorful of all gem species. Every color of the rainbow is represented by one of the numerous varieties, and it is a favorite of colored-gem connoisseurs and collectors. Some stones even exhibit two or three different colors (watermelon, bicolor, and tricolor Tourmalines). Almost any color is possible in Elbaite, but we discuss only the major ones here. Transparent-colored stones of Elbaite also show strong pleochroism, that is, their color changes when viewed at different angles. Also, as in Corundum, Quartz, Chrysoberyl, and other species, some colors of Elbaite can show asterism or chatoyancy, usually due to inclusions of fine parallel needles of Rutile.

Elbaite is named after the island of Elba, Italy, where it was first discovered. Most Elbaite today comes from Minas Gerais, Brazil. Excellent crystals also have come from California, Maine, South Africa, Nigeria, Mozambique, Namibia, Myanmar, East Africa, Madagascar, Pakistan, Afghanistan, Sri Lanka, and elsewhere. Major varieties of Elbaite include green verdelite (the most common variety), pink or red rubellite (scarce), blue indicolite (rare), yellow canary (rare), colorless achroite (rare), and bicolor Elbaite, usually pink and green but occasionally other colors as well. High-quality, near-flawless bicolor and tricolor stones of over 20 carats are quite scarce and command high prices. High-quality blue-and-yellow bicolor stones of any size are extremely rare. In the early 1900s, California and Maine were the largest producers of gem-quality Elbaite. Now the largest producers are Brazil and Africa.

GREEN ELBAITE TOURMALINE (VERDELITE)

SYSTEM Inorganic

CLASS Silicate

GROUP Tourmaline

SUBGROUP Alkali Tourmaline

SPECIES *Elbaite*

VARIETIES verdelite; cat's-eye verdelite

COMPOSITION complex Borosilicate $[Na(Li,Al)_3Al_6(BO_3)_3[Si_6O_{18}](OH)_4]$

TRACE ELEMENTS FOR COLOR Iron (Fe) and Titanium (Ti)

HARDNESS ON MOHS SCALE 7–7½

Green Elbaite, or verdelite, is the most common color variety of Elbaite, and the one most often associated with Tourmaline. It is both popular and abundant, and used extensively for jewelry. Here we include all green Elbaite as verdelite. The name verdelite is derived from the Latin *verde*, meaning "green," and the Greek *lithos*, meaning "stone." Color hues of verdelite can range from light yellowish-green to deep dark green. "Chrome Toumaline," the intense emerald-green variety, is a form of Uvite or Dravite and not Elbaite. Gem-quality green Elbaite also occurs as another variety, cuprian Elbaite (discussed in another section below), which is a bright "neon" hue that is also very expensive. Prior to the eighteenth century, certain shades of green Elbaite were often mistaken for emerald. Occasionally, verdelite from Brazil is mistakenly referred to as "Brazilian emerald." Some verdelite stones tend to be very dark in color to near opaque, which decreases their value significantly. The best stones are those that retain their transparency in incandescent light. Chatoyant green cat's-eye Elbaite is also known.

Green Elbaite is often found as clear, eye-clean material with few inclusions

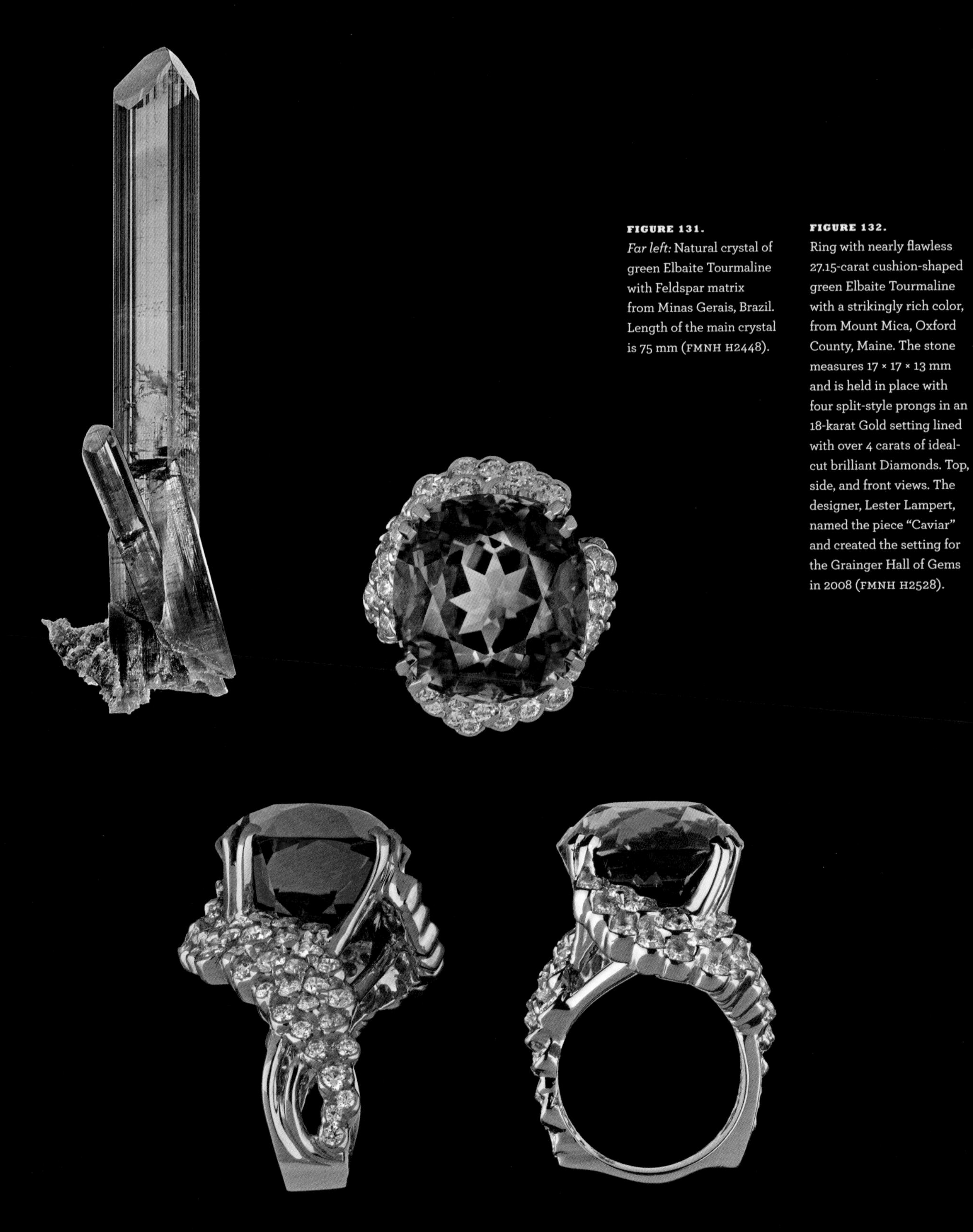

FIGURE 131.
Far left: Natural crystal of green Elbaite Tourmaline with Feldspar matrix from Minas Gerais, Brazil. Length of the main crystal is 75 mm (FMNH H2448).

FIGURE 132.
Ring with nearly flawless 27.15-carat cushion-shaped green Elbaite Tourmaline with a strikingly rich color, from Mount Mica, Oxford County, Maine. The stone measures 17 × 17 × 13 mm and is held in place with four split-style prongs in an 18-karat Gold setting lined with over 4 carats of ideal-cut brilliant Diamonds. Top, side, and front views. The designer, Lester Lampert, named the piece "Caviar" and created the setting for the Grainger Hall of Gems in 2008 (FMNH H2528).

FIGURE 133.
Three faceted 1-carat green Elbaite Tourmalines set in a 14-karat rose and yellow Gold brooch by Tiffany & Co., mid-twentieth century (FMNH H2552).

(fig. 131). It comes primarily from Brazil, but also comes from South Africa, East Africa, California, Maine, Pakistan, Sri Lanka, and many other localities around the world. The Grainger Hall of Gems has an exceptionally fine 27.15-carat cushion-shaped green Elbaite gem named "Caviar," which comes from Mount Mica, Maine, set in 18-karat Gold by Lester Lampert (fig. 132).

PINK TO RED ELBAITE TOURMALINE (RUBELLITE)

SYSTEM Inorganic

CLASS Silicate

GROUP Tourmaline

SUBGROUP Alkali Tourmaline

SPECIES *Elbaite*

VARIETIES rubellite; pink Tourmaline; color-change rubellite

COMPOSITION complex Borosilicate $[Na(Li,Al)_3Al_6(BO_3)_3[Si_6O_{18}](OH)_4]$

TRACE ELEMENT FOR COLOR Manganese (Mn)

HARDNESS ON MOHS SCALE 7–7½

Rubellite is one of the most beautiful hues of Elbaite, occasionally even mistaken for fine ruby. Some hues of rubellite are occasionally referred to as "Siberian ruby," a name that is misleading and best not used. Color hues range from ruby red to shocking pink. The name rubellite comes from the Latin *rubellus*, or "reddish." The most valuable forms of rubellite by far are those that are red in color without brown MODIFIERS, and such stones that are also free of noticeable inclusions are extremely rare. One of the finest rubellite Tourmalines we know of is a clear 55.4-carat gem from Minas Gerais, Brazil, set in 22-karat yellow Gold and 14-karat rose Gold in the Grainger Hall of Gems (fig. 135). Rubellite gems tend to have more inclusions than verdelite. Rubellite also occurs in a chatoyant form (cat's-eye rubellite).

FIGURE 134. Natural red Elbaite Tourmaline (rubellite) crystals in lepidote matrix from Minas Gerais, Brazil. Width of specimen is 120 mm (FMNH H2320).

FIGURE 135.
Facing page: Rubellite Tourmaline of 55.4 carats from Minas Gerais, Brazil, set in Etruscan-style necklace, named "Incredible" by the designer. Stone is set in 22-karat yellow Gold and 14-karat rose Gold with a hammered finish. Pendant is 44 mm in width. Designed by Lester Lampert in 2008 for the Grainger Hall of Gems (FMNH H2527).

FIGURE 136.
Natural crystal of pink Elbaite Tourmaline from Mesa Grande, San Diego County, California. Double-terminated crystal measures 85 x 20 mm (FMNH M11319).

Some authors, following the trend of the gem trade, distinguish between rubellite and pink Tourmaline and consider only the deep ruby-red shades to qualify as rubellite. This is not unlike the controversy that has existed for ruby and pink sapphire. We prefer the simpler system here and consider all shades of red Elbaite, including the lighter shades, to be rubellite. Pink varieties of large sizes, between 20 and 50 carats, are not very rare, but deep red stones over 20 carats are. A rare form of rubellite also exists that shows an ALEXANDRITIC change from reddish to orangish when going from daylight to incandescent light.

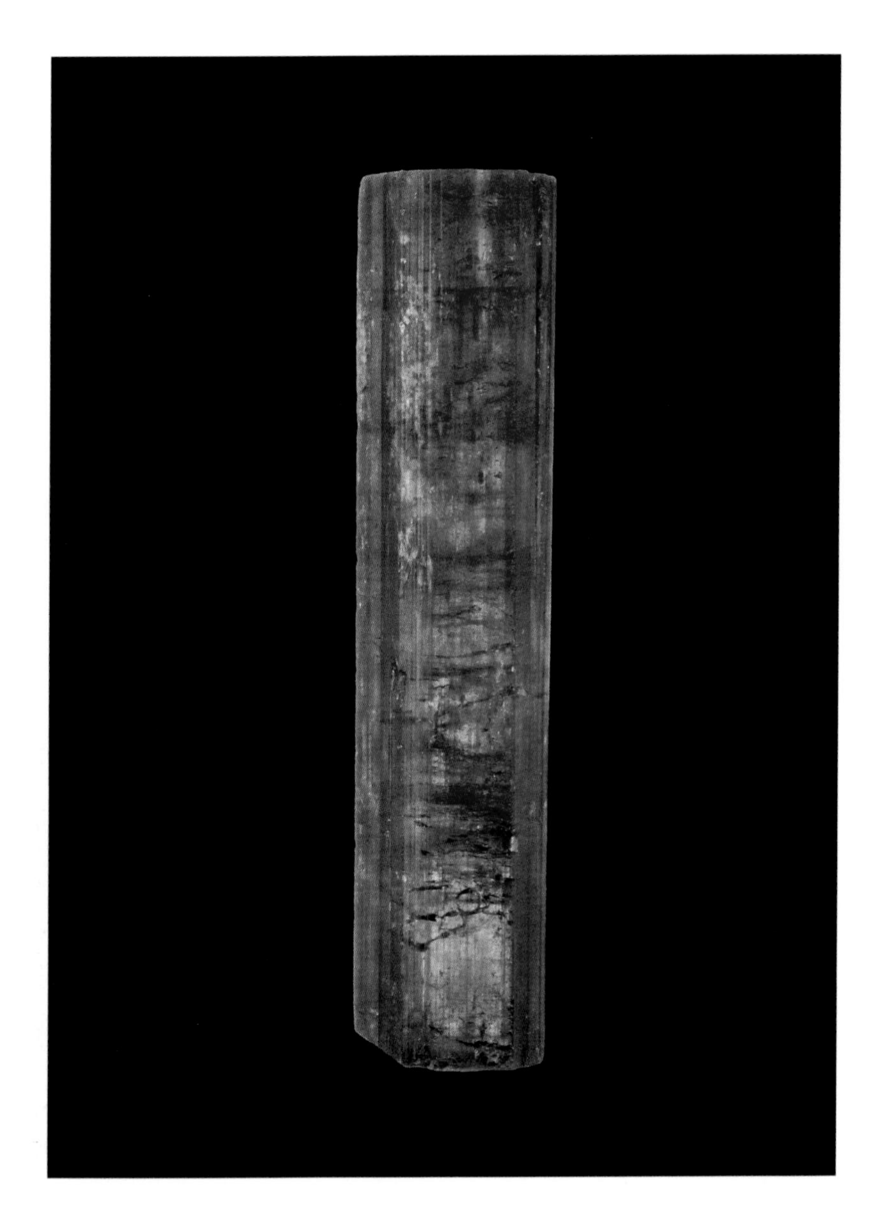

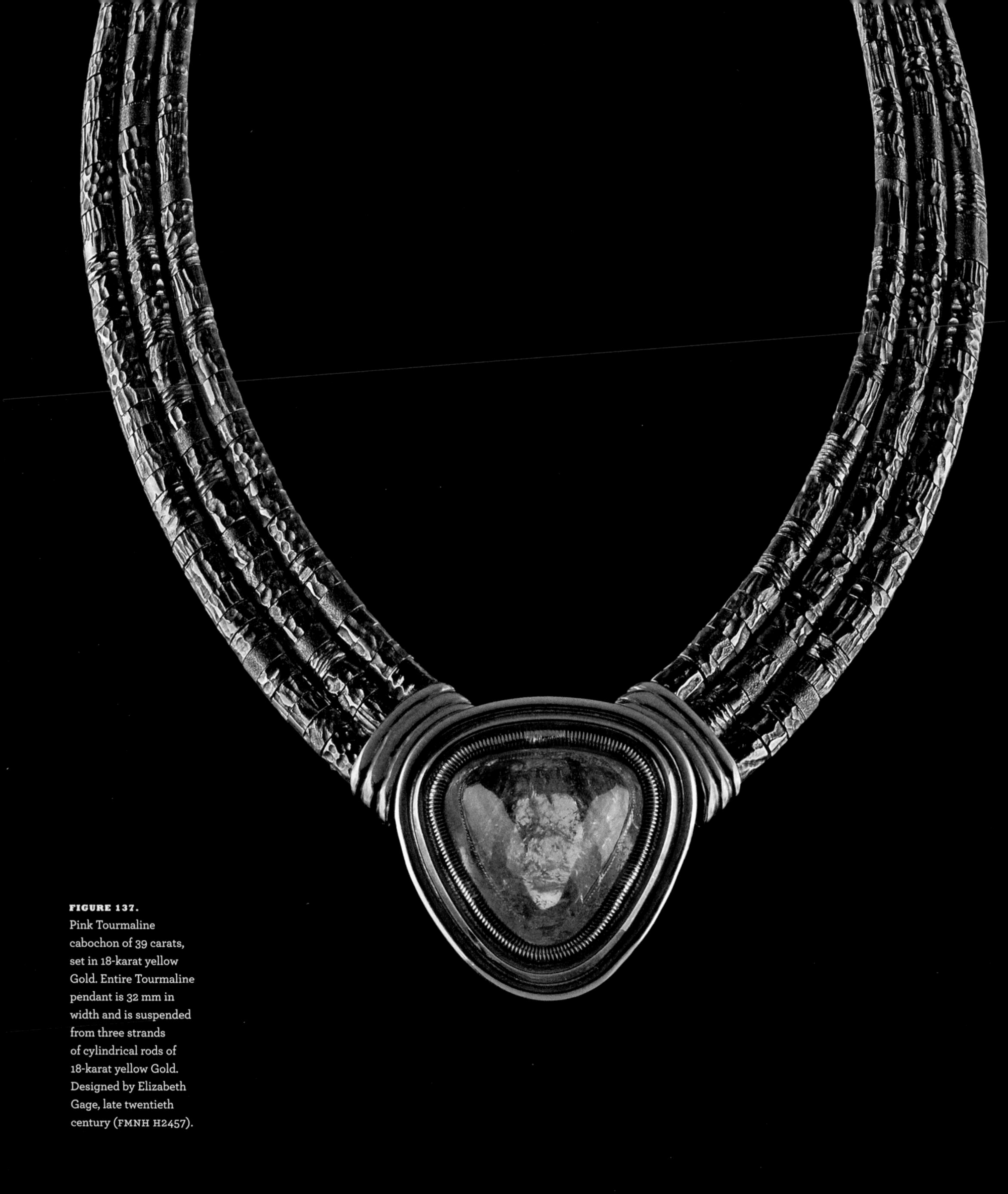

FIGURE 137.
Pink Tourmaline cabochon of 39 carats, set in 18-karat yellow Gold. Entire Tourmaline pendant is 32 mm in width and is suspended from three strands of cylindrical rods of 18-karat yellow Gold. Designed by Elizabeth Gage, late twentieth century (FMNH H2457).

FIGURE 138.
Flower brooch containing three faceted pink Tourmalines from Mozambique of 20.6, 7.5, and 7.4 carats, set in 18-karat yellow Gold and Platinum with green tsavorite Garnet and colorless Diamond accent stones. Designed by Marc Scherer of Marc and Co. in 2008 for the Grainger Hall of Gems and named "Spring Repose" by the designer (FMNH H2523).

Rubellite has been a very popular gemstone for well over a century. Empress dowager Tzu Hsi, the last empress of China, evidently loved pink rubellite, as she purchased nearly a ton of it from the Himalaya mine in California. After she died in 1908, she was laid to rest in her tomb with her head on a round pillow carved from pink Tourmaline. Some of the rubellite she had carved can be seen in the Fallbrook Mineral Museum in Southern California.

Like most Tourmaline varieties, rubellite usually forms in pegmatites. Rubellite crystals can reach up to a meter in length, but large crystals are always internally flawed and heavily included. Large eye-clean rubellite gems are very rare. Brazil is now the world's largest producer of rubellite, but other major localities include Madagascar, Myanmar, Afghanistan, Kunar, Nuristan, Madagascar, Russia, East Africa, and the United States (Maine and California). Other fine rubellite gems in the Grainger Hall of Gems include a pink cabochon of 39 carats set in 18-karat yellow Gold designed by Elizabeth Gage (fig. 137). Another pink Tourmaline piece of note is a flower brooch designed by Marc Scherer containing three fine faceted pink Tourmalines (fig. 138).

BLUE ELBAITE TOURMALINE (INDICOLITE)

SYSTEM Inorganic
CLASS Silicate
GROUP Tourmaline
SUBGROUP Alkali Tourmaline
SPECIES *Elbaite*
VARIETY indicolite
COMPOSITION complex Borosilicate $[Na(Li,Al)_3Al_6(BO_3)_3[Si_6O_{18}](OH)_4]$
TRACE ELEMENT FOR COLOR Iron (Fe)
HARDNESS ON MOHS SCALE 7–7½

Blue Elbaite Tourmaline, or indicolite, is the rarest and most expensive variety of Tourmaline other than blue cuprian Elbaite (discussed in a later section). Indicolite, or indigolite, is usually a deep blue color resembling sapphire and, like verdelite, gem specimens are often free of any major inclusions. Indicolite is highly pleocroic, and it appears darker if viewed down the long axis of the crystal than if viewed from the side of the crystal. This property is taken into consideration by gem cutters when orienting a stone for faceting in order to optimize the color hue. Very dark indicolite is often heat-treated to lighten its color. Pure blue hues of indicolite are more valuable and scarce than blue-green hues. A rare type of indicolite is known to have a lavender hue, and such stones are sometimes called siberite, after Siberia, where it was first discovered. Indicolite is known, although rarely, in bicolor stones with rubellite or verdelite, and even more rarely in chatoyant form as cat's-eye indicolite. Some authors also include a bright "neon blue" Tourmaline, called cuprian Tourmaline, in indicolite; but cuprian Tourmaline has Copper in its chemical structure and is classified as a different variety known as cuprian Elbaite Tourmaline.

Important localities for indicolite include Russia, Brazil, Madagascar, and the United States. Indicolite is sometimes inappropriately offered in the gem trade as "Brazilian sapphire," a name also occasionally given to blue Topaz. Indicolite is softer than either sapphire or Topaz, has a different chemistry and a different crystal structure. There is an extremely fine indicolite ring in the Grainger Hall of Gems with a 8.68-carat stone set in 18-karat Gold (fig. 140).

FIGURE 139.
Far left: Natural crystals of blue Elbaite Tourmaline (indicolite) on lepidolite and Albite matrix from Minas Gerais, Brazil. Specimen is 160 mm in height (FMNH H2319).

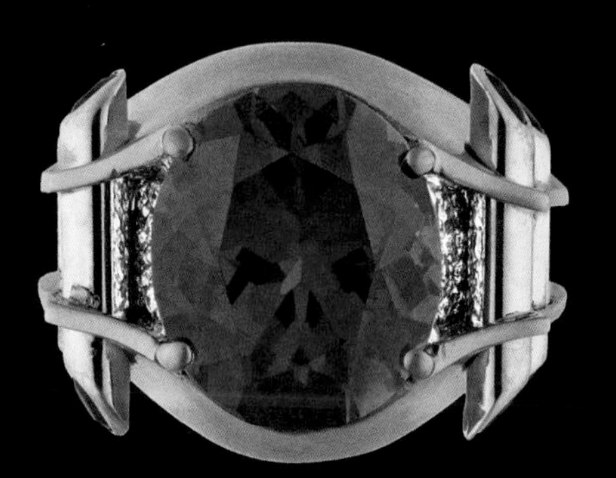

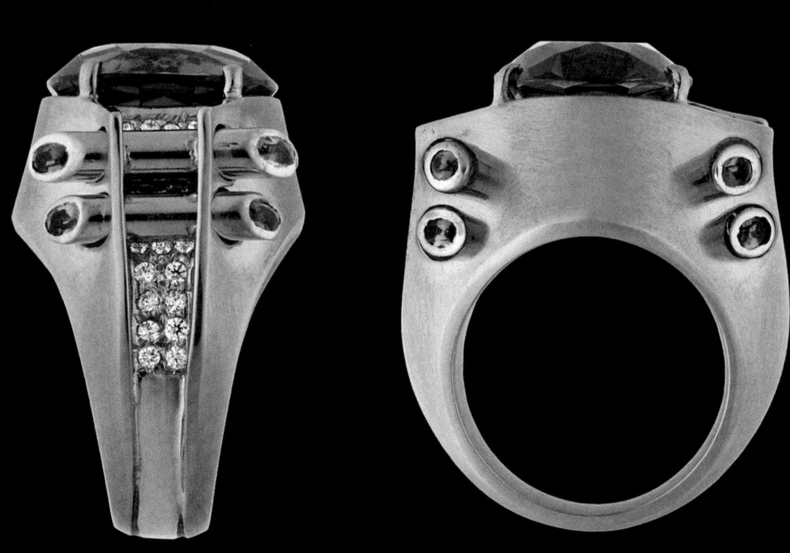

FIGURE 140.
Ring with a near-flawless 8.68-carat deep blue Elbaite Tourmaline (indicolite) from Minas Gerais, Brazil. The stone measures 15 × 13 × 8 mm and is set in a ring of white and 18-karat yellow Gold with 6 smaller indicolite gems and 56 ideal-cut Diamonds. The designer, David Lampert of Lester Lampert, Inc., named the piece "Azzurro" and created the setting in 2009 for the Grainger Hall of Gems (FMNH H2586).

YELLOW ELBAITE TOURMALINE (CANARY)

SYSTEM Inorganic

CLASS Silicate

GROUP Tourmaline

SUBGROUP Alkali Tourmaline

SPECIES *Elbaite*

VARIETY yellow

COMPOSITION complex Borosilicate $[Na(Li,Al)_3Al_6(BO_3)_3[Si_6O_{18}](OH)_4]$

TRACE ELEMENTS FOR COLOR

Manganese (Mn) and Titanium (Ti)

HARDNESS ON MOHS SCALE

7–7½

Until fairly recently, yellow Tourmaline was quite scarce, and most of it had tinges of green or brown. Then in the twenty-first century, deposits of "canary" radiant yellow Tourmaline were discovered in Malawi, southern East Africa. Since the mid-1990s, commercial quantities of bright yellow stones have been excavated, although the most intense, pure yellow stones are extremely rare at this time. Some of the Malawi material also has a slight brownish tinge; but this can be removed by heat-treating the stone with temperatures of about 700°C (1,300°F). This process turns the secondary color of the stone (the brown component) yellow to match the primary yellow color. The heat-treated stones are completely stable, and the pure yellow color appears to be irreversible. Coincidentally, imported rough Malawi stones occasionally have a noticeable lemon smell to go with their lemon-yellow color. This is because the miners often boil the crystals in water and lemon juice to dissolve a black coating that covers the stones when newly removed from the ground. The process does not alter the color of the stone. The acidic lemon juice simply helps dissolve the alkaline coating. The Tumbuku mine in Malawi was said to have employees whose sole job is to procure and squeeze lemons for the process. Stonecutters in the United States and elsewhere who have worked with Malawi canary material say it gives off a faint lemon odor when it is cut and polished.

Large yellow stones are extremely rare in Malawi, and over 90 percent of all stones weigh less than 1 carat. Although the bright yellow "canary" stones come primarily from the Kabelubelu and Tumbuku areas of Malawi, yellow Tourmaline also comes from Sri Lanka, Brazil, and Madagascar. The mines in Malawi have produced the highest grades of yellow Tourmaline.

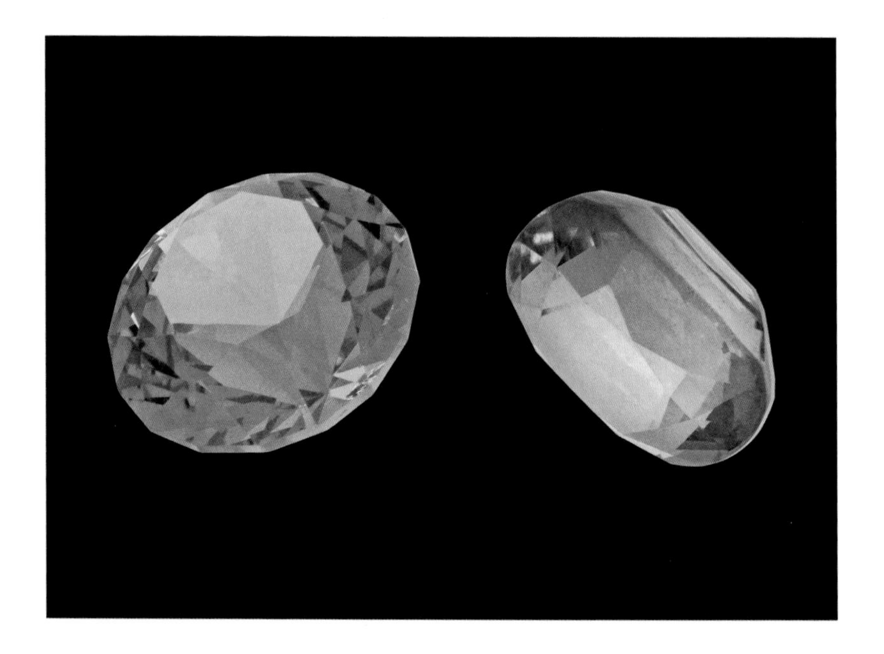

FIGURE 141. Faceted yellow Elbaite Tourmalines (canary) of 8.1 and 9.8 carats in weight from Madagascar (FMNH H2069, H2184).

MULTICOLOR ELBAITE TOURMALINE

SYSTEM Inorganic

CLASS Silicate

GROUP Tourmaline

SUBGROUP Alkali Tourmaline

SPECIES *Elbaite*

VARIETIES bicolor Elbaite; tricolor Elbaite; watermelon Elbaite

COMPOSITION complex Borosilicate $[Na(Li,Al)_3Al_6(BO_3)_3[Si_6O_{18}](OH)_4]$

TRACE ELEMENTS FOR COLOR

Iron (Fe), Manganese (Mn), and Titanium (Ti)

HARDNESS ON MOHS SCALE

7–7½

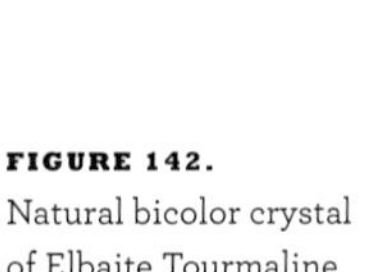

FIGURE 142. Natural bicolor crystal of Elbaite Tourmaline from Minas Gerais, Brazil. Crystal is 12 × 12 x × 62 mm and weighs 107.8 carats (FMNH H2340).

Multicolor Elbaite Tourmalines are most commonly combinations of pink and green. These include **POLAR BICOLOR** and **POLAR TRICOLOR** stones, which are different colors at each end of the crystal (figs. 142 and 148), and **RADIAL BICOLOR** stones, which are one color in the center surrounded by a "rind" of a different color (figs. 146 and 147). These gems are popular, unusual, and very scarce as large eye-clean stones.

Polar bicolor and tricolor Elbaite is usually pink on one end and red on the other, but some stones have a narrow colorless zone in between the two ends generally counted as a third "color." The "pink" end can also be more of a red or purple in some stones, and the "green" end can range to more of a blue-green or blue. Some bicolor stones exist that are green on one end and blue or transparent on the other, and rarely yellow on one end and blue on the other. Polar bicolor stones are frequently filled with inclusions, and eye-clean stones of size, such as the gem in fig. 143, are rare. There is a fine rose Gold bracelet with six faceted polar bicolor Elbaites from Mozambique in the Grainger Hall of Gems (fig. 144).

Polar bicolor in gems is rare, and Elbaite Tourmaline exhibits this more than

FIGURE 143.
Large, nearly flawless, faceted bicolor Elbaite Tourmaline from Minas Gerais, Brazil, in top and three-quarter view. This gem is 63.4 carats in weight and measures 32 × 17 × 11 mm (FMNH H2329).

FIGURE 144.
Left: Six faceted bicolor Elbaite Tourmaline gems from Mozambique, weighing from 5.3 to 6.9 carats each (36.4 carats total). *Bottom:* The same gems from above set in a 14-karat rose Gold and white Gold bracelet with 200 small ideal-cut Diamonds. Designed by Monica Lilak of Lester Lampert, Inc., in 2009 for the Grainger Hall of Gems and named "Mosaic" by the designer (FMNH H2591).

FIGURE 145.
Natural crystals of watermelon Tourmaline on Quartz and Albite matrix from Minas Gerais, Brazil. Largest crystal is 90 mm in length (FMNH H2324).

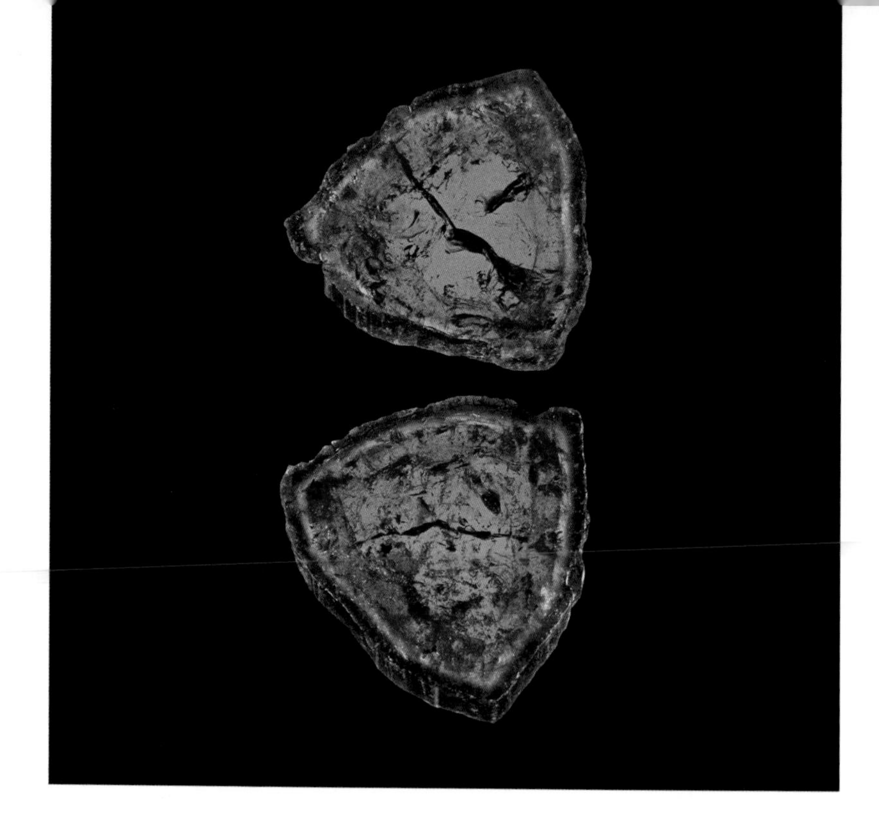

FIGURE 146.
Slices from a watermelon Tourmaline crystal from Minas Gerais, Brazil, showing cross section of color pattern. Slices are 21–25 mm in diameter and 37.3 and 44.2 carats in weight (FMNH H2360.1, H2360.2).

FIGURE 147.
Facing page: A 14-karat yellow Gold necklace with three polished watermelon Elbaite Tourmaline slices weighing 112.18 carats total. Set with 45 ideal-cut Diamonds of 1.27 carats total weight. Designed by Lester Lampert in 2009 for the Grainger Hall of Gems and named "Trilogy" by the designer (FMNH H2592).

any other gem. Polar bicolor is also found in ametrine Quartz (fig. 68), Liddicoatite Tourmaline (fig. 156), Oregon Feldspar, and sapphire (fig. 41), but it is best known in Elbaite Tourmaline. It is generally formed as the result of changing conditions in the HYDROTHERMAL SOLUTION during formation of the gemstone.

Radial bicolor Elbaite, or watermelon Tourmaline, is one of the most interesting and popular of all Tourmaline varieties. It gets its name from its appearance in cross section. A slice of a watermelon Tourmaline crystal normally has a watermelon pink center and a green "rind" (fig. 146). As with polar bicolor gems, the color zoning of radial bicolor gems is the product of changing conditions and chemistry during formation of the crystal. Although green and pink are the most common colors of watermelon Tourmaline, blue and red are also sometimes found.

Watermelon Tourmaline always contains inclusions. Although polar bicolor and tricolor stones occasionally exist as large eye-clean stones, true watermelon slices always contain visible flaws. Nevertheless, because of its natural beauty, good color watermelon Tourmaline can make beautiful pieces of fine jewelry (fig. 147). Watermelon Tourmaline is cut in slices perpendicular to the long axis, or C-AXIS, of the crystal rather than being faceted, to show the unique radial color-zoning beauty of the gemstone.

Multicolored Elbaite stones are mined primarily in Brazil, South Africa, East Africa, and California, although the cleanest polar-colored stones come from Brazil.

COLORLESS ELBAITE TOURMALINE (ACHROITE)

SYSTEM Inorganic

CLASS Silicate

GROUP Tourmaline

SUBGROUP Alkali Tourmaline

SPECIES *Elbaite*

VARIETY achroite

COMPOSITION complex Borosilicate $[Na(Li,Al)_3Al_6(BO_3)_3[Si_6O_{18}](OH)_4]$

TRACE ELEMENT FOR COLOR none

HARDNESS ON MOHS SCALE 7–7½

Colorless Elbaite, or achroite, is a very rare form of Tourmaline in nature. Natural crystals are often very clear with few or no inclusions, but they are usually very small. Cutting material for large natural achroite gems is generally unavailable. Very light pink Tourmaline can be made colorless by heat treatment. The name achroite is derived from the Greek *achros*, meaning "without color." Even though the colorless form is one of the rarest varieties of Elbaite, it is one of the least expensive as a faceted gem because it is not very popular for jewelry and the faceted stones are quite small. Among the many colorless species of gemstones, there are many that are more attractive and more readily available as large eye-clean stones than colorless Elbaite. One natural crystal in the Grainger Hall of Gems that contains a large section of clear achroite is a polar bicolor that is colorless on one end and green on the other end (fig. 148).

As with other varieties of Elbaite, achroite usually forms in pegmatites. It occurs in deposits in England, Afghanistan, Madagascar, and the United States (Maine).

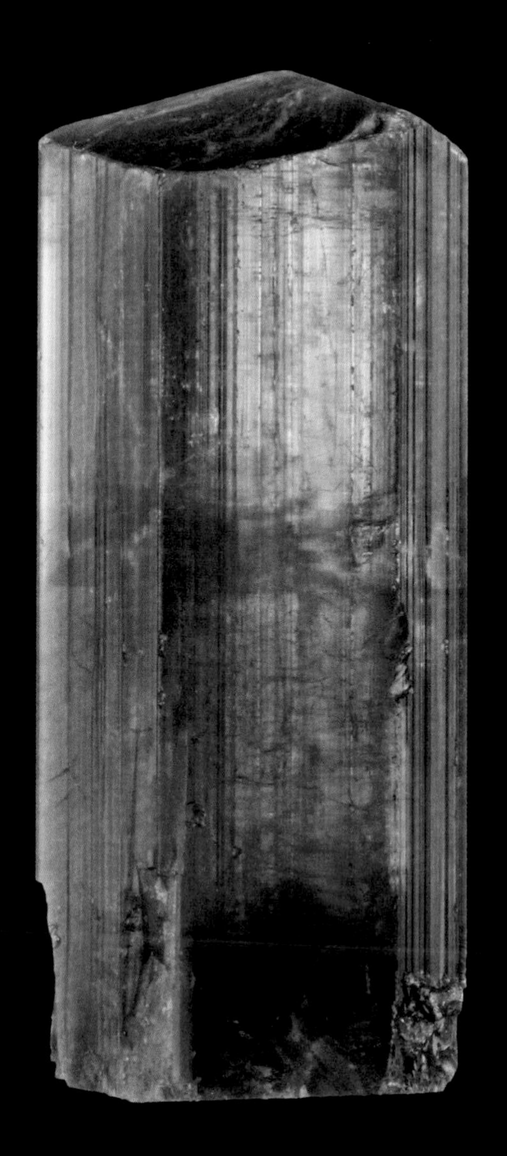

FIGURE 148.
Right: Natural crystal of bicolor Elbaite Tourmaline containing part colorless achroite and part green verdelite from Mesa Grande, San Diego County, California. Length of crystal is 38 mm (FMNH M11500).

FIGURE 149.
Below: Faceted colorless Elbaite Tourmaline gems (achroite) from Mount Mica, Oxford County, Maine, ranging in weight from 1.96 to 4.46 carats (FMNH H1872–H1877, H2475).

CUPRIAN ELBAITE TOURMALINE

SYSTEM Inorganic

CLASS Silicate

GROUP Tourmaline

SUBGROUP Alkali Tourmaline

SPECIES *Elbaite*

VARIETIES blue cuprian Elbaite; green cuprian Elbaite; purple cuprian Elbaite; pink cuprian Elbaite; bicolored cuprian Elbaite

COMPOSITION complex Borosilicate $[Na(Li,Al)_3(Al,Cu)_6(BO_3)_3[Si_6O_{18}](OH)_4]$

TRACE ELEMENTS FOR COLOR Copper (Cu) and Manganese (Mn)

HARDNESS ON MOHS SCALE 7–7½

The varieties of cuprian Elbaite, or cuprian Tourmaline, are rare and discussed together in this section. They differ from regular blue, green, and red Tourmaline by having Copper as part of their chemical structure and also in having an intensely bright "neon" look. First recognized in the 1980s, cuprian Elbaite Tourmaline (particularly the transparent turquoise-blue Tourmaline from Paraíba, Brazil) currently includes the rarest, most expensive varieties of Tourmaline. There is nothing comparable to the blue of a high-grade cuprian Elbaite. Its neon sky-blue hue seems to glow. Although cuprian Elbaite occurs in blue, green, purple, and pink, the neon blue color is the most desirable. Bicolored stones (blue and various shades of red) also exist. Some of the stones are heat-treated, which removes the pink and purplish colors to leave a pure turquoise blue. Good-quality turquoise-colored stones in sizes larger than a few carats are extremely rare, due primarily to two factors. First, true blue cuprian Elbaite is known only from very limited sources that now appear to be greatly diminished. It has so far been mined primarily in the state of Paraíba in Brazil and in the Alto Lingonha region of Mozambique. The best

FIGURE 150. Natural crystal of blue cuprian Elbaite Tourmaline in white Quartz matrix from Paraíba, Brazil. Specimen is 28 mm in height (FMNH H2190).

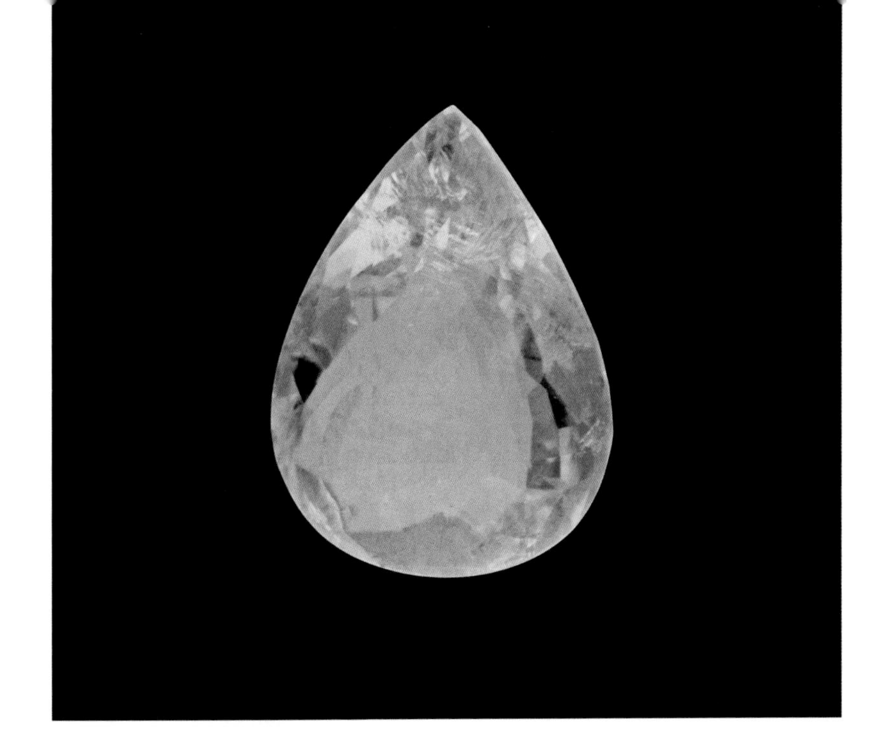

FIGURE 151.
Faceted blue cuprian Tourmaline stone of 3.05 carats from Paraíba, Brazil. Pear shape with Portuguese cut (FMNH H2503).

localities in the Paraíba region are said to have now been largely worked out, and the value of stones from those localities has risen dramatically in recent years. The deposits found in Africa rarely match the beauty of the best blue stones from Paraíba, and gems from outside of Brazil have been the subject of legal battles over the right to use the name paraiba. Does the name paraiba represent a VARIETY based on its composition and other physical properties, or does it simply represent a specific locality, based on geography of the source material? The current controversy is not one based on science or logic; it is instead one based on fierce market competition and politics. Here we use the term "cuprian" for the variety, since variety names should be tied to composition rather than locality, as with other gemstones.

The second reason that large cuprian Elbaite gems are so rare is that the Brazilian stones are usually found as crystals that are completely embedded within other hard rocks rather than in open pockets or vugs. Removing the stones from the enclosing rock usually results in breaking the gems into smaller fragments, thus making large natural crystals extremely hard to come by.

Highest-quality South American paraiba blue gems can today fetch more than $30,000 per carat, and the stones are becoming rarer on the market as time goes on. The African material, while not quite as valuable, is also priced above most other varieties of Tourmaline but remains threatened with legal issues surrounding the use of the name paraiba. Hopefully, the legal issues will be settled with time and additional deposits of this gem will be discovered. In the meantime, this gem will remain very rare in collections and very expensive in the marketplace.

Dravite Tourmaline

SYSTEM Inorganic

CLASS Silicate

GROUP Tourmaline

SUBGROUP Alkali Tourmaline

SPECIES *Dravite*

GEM VARIETIES brown Dravite (champagne Tourmaline); green Dravite (chrome Tourmaline); yellow Dravite; colorless Dravite

COMPOSITION complex Borosilicate $[NaMg_3Al_6(BO_3)_3[Si_6O_{18}](OH)_4]$

TRACE ELEMENTS FOR COLOR

Iron (Fe), Titanium (Ti), Chromium (Cr), and Vanadium (V)

HARDNESS ON MOHS SCALE

7–7½

FIGURE 152.
Top: Natural crystal of Dravite Tourmaline, from Western Australia, weighing 526 grams. Although not transparent, this specimen shows fine natural crystal faces and measures 85 × 65 × 60 mm (FMNH M19627).

FIGURE 153.
Bottom: Faceted Dravite gem from Sri Lanka weighing 58.2 carats and measuring 30 × 22 × 18 mm (FMNH H251).

Dravite, or Sodium Magnesium Tourmaline, is rare as a gem and is one of the least known varieties of Tourmaline. It is named after the Drava River of Europe, which runs near the district of Drava, Austria, where Dravite was first reported in the late nineteenth century. It is almost always a dark brown color, although the color can be lightened by HEAT TREATMENT. Lighter shades of brown can be orange-brown to dark greenish-yellow and occasionally show good transparency. Yellow, golden, and colorless varieties also occur. Dravite is strongly pleochroic. Transparent, light brown material is sometimes faceted for the gem trade, where it is sometimes referred to as champagne Tourmaline. Although opaque brown Dravite is fairly abundant, gem-quality transparent stones are not common. Inclusion-free brown stones make fine and attractive gemstones. Their earth-like coloration gives them an unusual hue for a gem.

The most valuable Dravite gems are the intense emerald-green stones containing traces of Chromium called chrome Tourmaline or chromdravite. Chrome Tourmaline also forms in Uvite (next section). Gem-quality red Dravite also exists, but is very rare. The only variety of gem-quality Dravite included in the Grainger Hall of Gems is the brown variety (fig. 153).

Dravite occurs in IGNEOUS ROCKS rich in Boron and in pegmatites. Large, double-terminated crystals occur in Western Australia (fig. 152). Other localities include Sri Lanka, Russia, Sweden, Austria, Tajikistan, Canada, the United States (New York and Maine), Mexico, Bolivia, and Brazil.

Schorl Tourmaline

SYSTEM Inorganic

CLASS Silicate

GROUP Tourmaline

SUBGROUP Alkali Tourmaline

SPECIES *Schorl*

GEM VARIETY black Schorl

COMPOSITION complex Borosilicate $[NaFe_3Al_6(BO_3)_3[Si_6O_{18}](OH)_4]$

TRACE ELEMENT FOR COLOR

none (color is from heavy Iron content of the basic structure)

HARDNESS ON MOHS SCALE

7–7½

FIGURE 154. Large natural crystal of black Schorl Tourmaline with Quartz and Albite from Pakistan. Specimen is 135 mm in height and weighs 1,221 grams (FMNH H2352).

Black Tourmaline, or Schorl, is the most common type of Tourmaline by far. It is estimated that it may account for more than 90 percent of all Tourmaline found in nature. It is also known as Iron Tourmaline after its Iron content. The name Schorl is derived from a fourteenth-century town in Saxony, Germany.

Schorl is opaque black and not used much in jewelry today; but during Victorian times in Britain, it was used in mourning jewelry. Schorl mourning jewelry is extremely rare today, and most black mourning jewelry that survived from the Victorian era is made from jet, a type of hard coal.

Schorl is a hard stone that is resistant to wear, so it is occasionally used for inlay work in modern jewelry. The most common use of it in modern jewelry is as free-form natural crystal pendant. Schorl is a relatively common occurrence in pegmatites and is widely found in localities that produce other forms of Tourmaline. It forms beautifully striated crystals, and natural mineral specimens of Schorl are very popular with mineral collectors (fig. 154). It is also found as long thin crystal inclusions in colorless Quartz. Quartz with such Tourmaline inclusions is called tourmalinated Quartz, a stone that is also occasionally used in jewelry.

Liddicoatite Tourmaline

SYSTEM Inorganic

CLASS Silicate

GROUP Tourmaline

SUBGROUP Calcic Tourmaline

SPECIES *Liddicoatite*

GEM VARIETIES red or pink Liddicoatite; green Liddicoatite; multicolor Liddicoatite

COMPOSITION complex Borosilicate $[CaLi_3Al_6(BO_3)_3[Si_6O_{18}](OH)_4]$

TRACE ELEMENTS FOR COLOR Iron (Fe), Manganese (Mn), and Titanium (Ti)

HARDNESS ON MOHS SCALE 7–7½

Liddicoatite Tourmaline, also called Calcium Lithium Tourmaline, is a relatively new gem species, recognized in 1977. It has rarely been used in jewelry so far, because quality material is much rarer than Elbaite and it contributes no color variety not already provided by Elbaite. Also, it is so new that no specific demand has built up for it as of yet. It is mainly popular as a collector mineral and is only recently becoming known in the gem trade. Liddicoatite often shows complicated multicolor zoning patterns and strain lines. Most gems cut from Liddicoatite are currently from multicolored stones. Multicolor stone is sometimes used in carvings or cut and polished as decorative gem "wings" for jewelry, such as the butterfly-winged 14-karat Gold fairy pendant in the Grainger Hall of Gems designed by Martha Richter (fig. 156).

Liddicoatite was named after Richard Liddicoat (1918–2002), a gemologist and former president of the Gemological Institute of America. The best material comes mainly from pegmatite deposits from Anjanabonoina, Madagascar. It is sometimes considered to be a form of Elbaite Tourmaline, as Liddicoatite-Elbaite, but it clearly has a different chemical composition than Elbaite, so we retain it as a different species in this book.

FIGURE 155. Natural crystal of Liddicoatite Tourmaline from Myanmar. Specimen is 56 × 42 mm and weighs 892.5 carats (FMNH H2492).

FIGURE 156.
Liddicoatite Tourmaline gems and jewelry. *Top:* Pendant of a fairy or an angel, with carved Liddicoatite wings from Madagascar, set within a solid 14-karat Gold carving. Pendant is 54 mm in width. Designed in 2004 by Martha Richter of Martha Richter Ltd., who calls the piece "Angelica" (FMNH H2485). *Bottom:* Faceted tricolor Liddicoatite gem from Madagascar weighing 3.1 carats and measuring 17 × 5 × 3.5 mm (FMNH H2466).

Uvite Tourmaline

SYSTEM Inorganic

CLASS Silicate

GROUP Tourmaline

SUBGROUP Calcic Tourmaline

SPECIES *Uvite*

GEM VARIETIES brown Uvite; red Uvite; brownish-yellow Uvite; green Uvite

COMPOSITION complex Borosilicate $[CaMg_3Al_6(BO_3)_3[Si_6O_{18}](OH,O)_4]$

TRACE ELEMENTS FOR COLOR

Iron (Fe) for the browns and reds; Vanadium (V) and Chromium (Cr) for green

HARDNESS ON MOHS SCALE

7–7½

Uvite Tourmaline, another of the Calcium Lithium Tourmaline species, is a relative newcomer to the gem market. It is named after the province of Uva in Sri Lanka, where it was first described in 1929 by a German mineralogist. Uvite was once thought to be very rare, but new finds and mining efforts have made Uvite gems much more widely available on the gem market. Most gem-quality material of red, green, or yellow has a brownish hue to it, but very fine red, green, and yellow gemstones also exist. The most valuable Uvite gem is the intense emerald-green variety containing traces of Vanadium and occasionally Chromium called "chrome" Tourmaline. Technically, use of the prefix "chrome" should be reserved for varieties that contain traces of Chromium as the main coloring agent, but the vernacular "chrome" is occasionally used more loosely in the gem trade to refer to any variety with a rich emerald-green hue. Traces of Vanadium in gemstones can sometimes produce hues similar to those resulting from Chromium.

Gem-quality red Uvite is also extremely rare. Black Uvite also exists and is one of the more common hues of the species. Black Uvite is occasionally mistaken for Schorl Tourmaline. Colorless Uvite is also known, but not used as a gemstone. Uvite is usually found in lightly metamorphic rocks that are rich in Calcium, and it forms stubby, prismatic to tabular crystals with pyramidal terminations (fig. 157, *top*). Uvite is often found as a solid solution (BLEND) with Dravite Tourmaline, generally as brownish stones.

Uvite is found in Sri Lanka, Myanmar, Brazil, Switzerland, and New York State. Recently, deposits of extremely fine natural crystals of red and green Uvite crystals in Magnesite matrix have been discovered in Brumado, Bahia, Brazil (fig. 157, *top*). There are also several very fine faceted brownish-yellow, red, and green Uvites in the Grainger Hall of Gems (fig. 157, *bottom*).

FIGURE 157.
Facing page, top: Natural Uvite Tourmaline crystals in matrix with colorless magnesite and Quartz crystals, from Brumado in Bahia, Brazil. Specimen with red Uvite crystals is 65 mm in height (FMNH H2595), and specimen with green Uvite crystals is 60 mm in height (FMNH H2597). *Facing page, bottom:* Faceted red, brownish-yellow, and green Uvite Tourmaline gems from Sri Lanka weighing from 1.1 to 9.5 carats each, with the largest stone measuring 13 × 10 × 9 mm (FMNH H261.1–261.3, H2401, H2419).

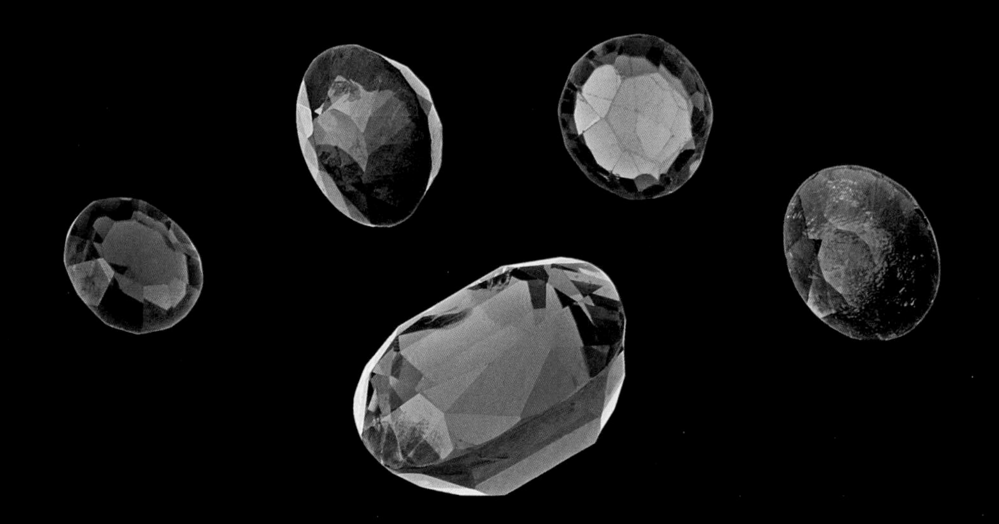

GARNET GROUP

SYSTEM Inorganic

CLASS Silicate

GROUP Garnet

PYRALSPITE GARNET SPECIES

Almandine Garnet; *Pyrope* Garnet; *Spessartine* Garnet

UGRANDITE GARNET SPECIES

Grossular Garnet; *Andradite* Garnet; *Uvarovite* Garnet

Garnets have been valued as gems for more than 5,000 years. Egyptians were using them in jewelry as early as 3100 BC. The name Garnet is derived from the Latin *granatum*, meaning "grain" or "seed." It is thought that the origin of this etymology is the red Garnet's resemblance to a pomegranate seed. The name Garnet (now applied to the entire Garnet group) was originally used only for red stones of the Pyrope-Almandine series. Garnet has a rich mythical and theological history ranging from Greek mythology to the Old Testament of the Bible. It has been associated with miraculous luminescence, protection from evil, and healing powers. On the practical side, Garnets were also used as bullets during the India rebellion of 1892. Sand made from some of the harder Garnet varieties, such as Almandine (Mohs 7–7½), is highly valued as an abrasive material.

Garnet, like Tourmaline, is an extremely complex group containing several species and many varieties of gemstones. It differs from many other gem groups in that different Garnet species readily combine with each other to form INTERMEDIATE varieties, or BLENDS, much like species of Feldspar and some species of Tourmaline do. For example, Almandine and Pyrope represent end members in a series of Almandine-Pyrope blends, with many intermediates. It is rare to find a Garnet that is of a pure "ideal" composition for Almandine or Pyrope. Most Pyrope Garnet is a blend of mostly Pyrope with a little Almandine and most Almandine Garnet is a blend of mostly Almandine with a little Pyrope. A 1:2 mix of Almandine and Pyrope is called rhodolite (a named intermediate type). The same is true for the Grossular-Andradite series, which contains the intermediate variety named grandite, and the Pyrope-Spessartine series, which contains an intermediate variety, umbalite. Sometimes there is also Spessartine within the Almandine-Pyrope mix. There is some controversy in regard to Garnet names among gemologists. Here we discuss the intermediate types with their closest series end members.

Garnets can be classified into two subgroups: the Pyralspites, or Calcium-free species, with Iron (Fe), Magnesium (Mg), or Manganese (Mn) in their basic composition; and the Ugrandites, or Calcium species, with Calcium (Ca) in their basic composition. The Calcium-free Garnet gem species tend to be mostly shades of red to orange, while the Calcium Garnets, particularly varieties of Grossular, come in a much wider variety of colors. There are several varieties in each subgroup that have been of great importance to the gem trade.

Like Tourmalines, Beryls, and sapphires, Garnets come in many colors, although most people are familiar mainly with the abundant red varieties. The only primary color lacking in Garnet is pure blue. Unlike Beryl, Corundum, and certain other species whose pure form is colorless and whose color variants are due to "impurities" or trace elements (ALLOCHROMATIC gems), the color variation of some Garnet species is due to the basic chemical structure of each

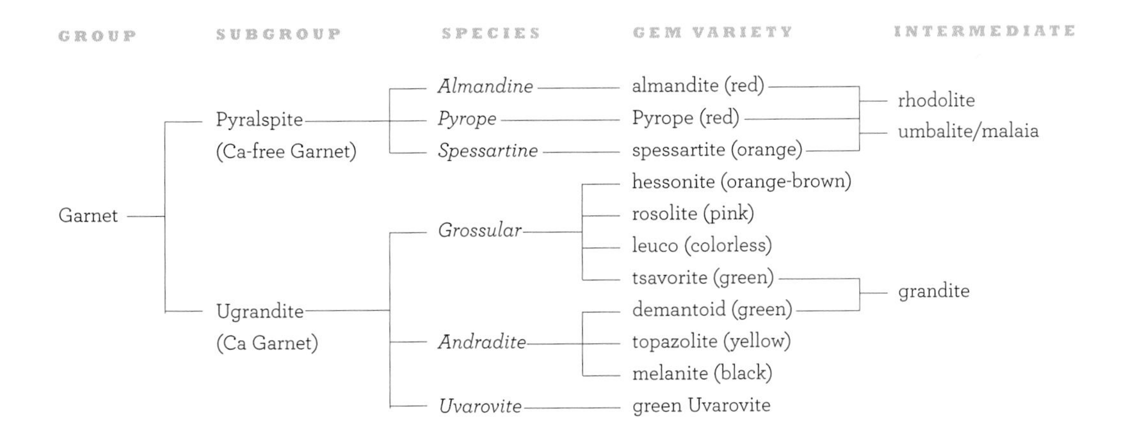

FIGURE 158. Classification of Garnet gemstone varieties discussed in this book. We list only six of the fifteen Garnet species because these six contain the varieties most significant as gemstones. Not shown on the diagram is the Spessartine-Almandine intermediate, kashmerine.

species or variety (they are IDIOCHROMATIC). Most pure Garnet species (other than Grossular) are not colorless, and the more than thirty color variations of Garnet are often due to blending of different species.

In addition to the diverse color variants, there are also star varieties of Almandine Garnets, and color-changing varieties (ALEXANDRITIC varieties) consisting of Pyrope and Spessartine blends that change from red or pink in daylight to blue or green in incandescent light.

Garnet is found in both igneous and metamorphic rocks, but most gem-quality stones appear to be derived from pegmatites (fig. 11), mica schists (fig. 159), or placer deposits sourced by those rocks (fig. 163). Garnet deposits, particularly of the Almandine-Pyrope series, are widely distributed around the world.

Currently, the most valuable Garnets are the green varieties tsavorite and demantoid, particularly in sizes over 3 carats. Mandarin Spessartine and sizable color-change Garnets are also increasing in value, although market conditions often change when new deposits are discovered. For further reading on Garnets, we recommend Jackson (2006) and Rouse (1986).

Almandine Garnet

SYSTEM Inorganic

CLASS Silicate

GROUP Garnet

SUBGROUP Pyralspite

SPECIES *Almandine*

GEM VARIETY

almandite (dark red)

COMPOSITION (IDEAL)

Iron Aluminum Silicate

$[Fe_3Al_2(SiO_4)_3]$—but most stones are intermediate between the ideals of Almandine and Pyrope

TRACE ELEMENT FOR COLOR

Color due to idiochromatic Iron (Fe)

HARDNESS ON MOHS SCALE

7–7½

Almandine is one of the two most common Garnet gemstones, the other being Pyrope. Occasionally it goes by the variety name of almandite. It is generally a dark violet-red, somewhat darker than Pyrope. It is often nearly opaque due to excessive depth of color. Sometimes darker stones are cut as a cabochon with the back hollowed out to let more light through and lighten its appearance. The best-quality transparent Almandine resembles ruby and red Spinel in color. Stones with a brownish tinge or poor transparency are much less valuable. Lighter, more transparent stones are the most valuable. Some crystals have inclusions of fine, hair-like needles of Rutile, or even needle-shaped void resulting in red star Garnets (fig. 161).

The name Almandine is derived from the city of Alabanda, Asia Minor (Turkey), where gems were traded more than 2,000 years ago; Almandine is still found there today. There is a 3,400-year-old Almandine necklace from ancient Egypt in the Grainger Hall of Gems (fig. 162). Almandine is usually

FIGURE 159. Natural crystals of Almandine Garnet in gray mica schist from Wrangell, Alaska. Specimen measures 76 × 203 mm (FMNH M2851).

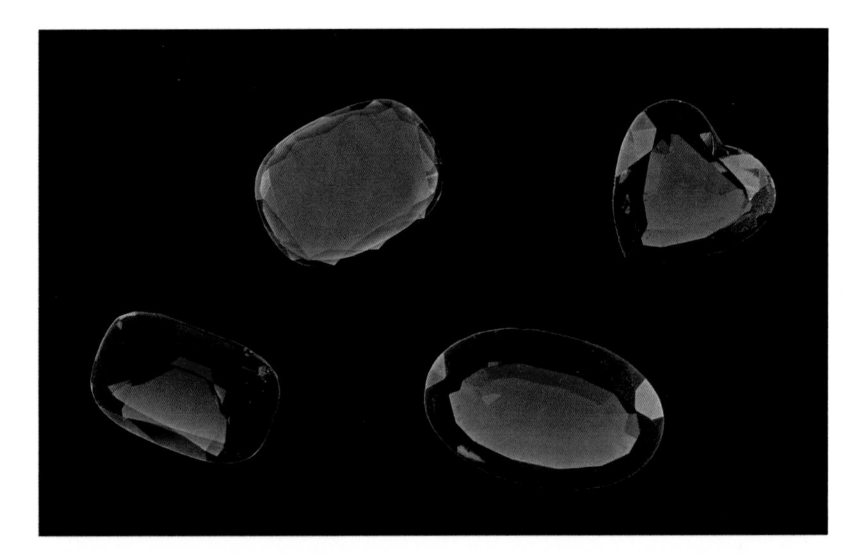

FIGURE 160.
Faceted Almandine Garnets from Bago (formerly Pegu), Myanmar, ranging in size from 12.8 to 36.0 carats (FMNH H252–H255, H257).

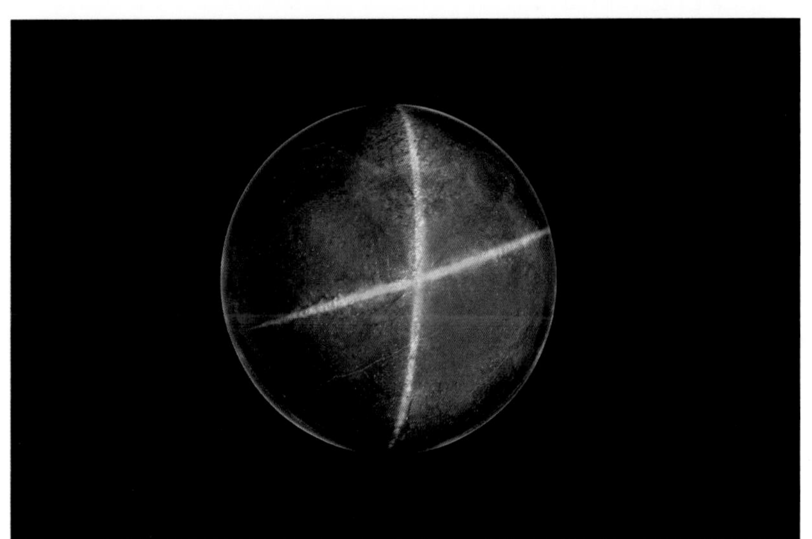

FIGURE 161.
Almandine star Garnet sphere of 1,060 carats from Bago (formerly Pegu), Myanmar. Piece is 44 mm in diameter (FMNH H277).

found in gray schists called Garnet schist (a metamorphic rock composed mostly of small pieces of mica; see fig. 159). Pure Almandine is rare in nature, and most Almandine is a mixture of Almandine and its close relative, Pyrope. Any combination with more than 50 percent Almandine is generally called Almandine. Unlike its cousin, Pyrope, good crystals of Almandine in matrix are fairly common.

There are many productive mining localities around the world for Almandine, most notably India, Sri Lanka, Greenland, Brazil, Myanmar, Australia, South Africa, Argentina, the United States, Scotland, Switzerland, and Tanzania. Natural Garnet crystals are often found with well-formed crystal faces and make attractive mineral specimens (figs. 11 and 159).

FIGURE 162.
Ancient necklace of Almandine Garnet from Egypt's New Kingdom (1400 BC). Necklace is about 640 mm in length (FMNH A31221).

Pyrope Garnet

SYSTEM Inorganic

CLASS Silicate

GROUP Garnet

SUBGROUP Pyralspite

SPECIES *Pyrope*

GEM VARIETIES blood-red Pyrope; rhodolite (raspberry-red intermediate between Pyrope and Almandine)

COMPOSITION (IDEAL)

Magnesium Aluminum Silicate [$Mg_3Al_2(SiO_4)_3$]—but most stones are intermediate between the ideals of Pyrope and Almandine

TRACE ELEMENTS FOR COLOR

Pure Pyrope is colorless, but it is almost never found that way. The red color is either due to idiochromatic Iron (Fe) due to blending with Almandine, or from traces of Chromium (Cr).

HARDNESS ON MOHS SCALE

7–7½

Pyrope is another of the red Garnets and is fairly common, although not as common as Almandine. It is blood-red, often showing a tinge of yellow. It generally lacks the violet tints present in Almandine, although color alone is not a sure diagnostic feature. Transparent stones are often lighter than Almandine in color, and generally the lighter the color, the more valuable it is. Pyrope sometimes goes by the name Bohemian Garnet. The name Pyrope is derived from the Greek *pyropus*, meaning "fiery." This Garnet is a commonly used gem in the jewelry industry (fig. 165).

As opposed to Almandine and most other varieties of Garnet, Pyrope rarely occurs as distinct crystals. Fine Pyrope crystals in matrix are rare, and unlike most other Garnet varieties, the most common origin of Pyrope is igneous rather than metamorphic rock.

Like Almandine, most Pyrope is a combination of Pyrope and Almandine. The stone is generally Pyrope if more than 50 percent of the mix is Pyrope, or, in other words, if the Magnesium component is greater than the Iron component. Pyrope is known primarily from Bohemia, South Africa, Tanzania, the United States, Mexico, Brazil, Argentina, Australia, and Sri Lanka.

FIGURE 163. Raw Pyrope Garnet faceting material from placer deposits in Meronitz, Czechoslovakia (FMNH M2779).

FIGURE 164.
Rhodolite Garnet cushion-cut gem from the Umba Valley of Tanzania weighing 31.9 carats. Dimensions are 21 × 18 × 16 mm (FMNH H2343).

FIGURE 165.
Facing page: Pyrope Garnet necklace with five strands containing several hundred faceted Garnets and a Gold clasp with small Pearls (FMNH H1405).

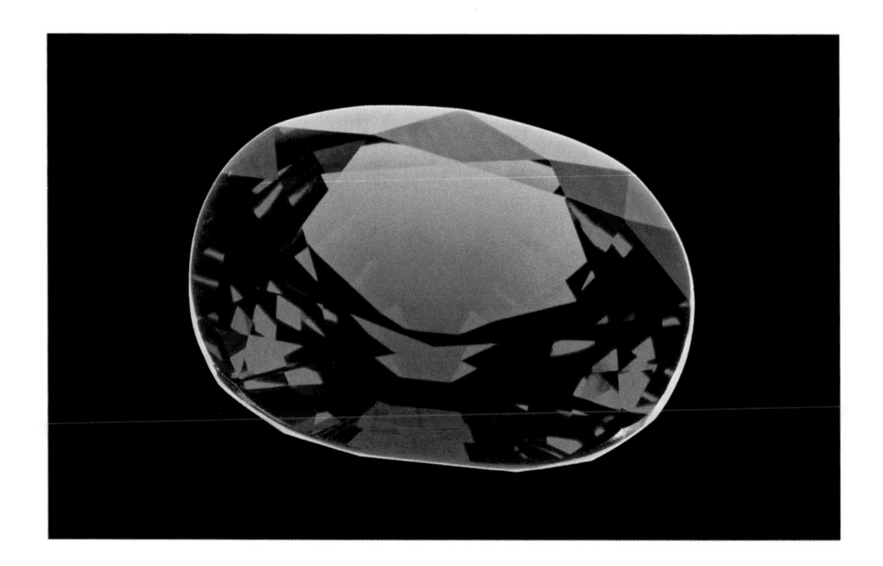

A special variety of intermediate Pyrope-Almandine BLEND called rhodolite is recognized by many gemologists. Rhodolite is compositionally a mix of roughly two parts Pyrope to one part Almandine (composition $2[Mg_3Al_2(SiO_4)_3] + Fe_3Al_2[SiO_4]_3$). Not all mineralogists recognize this as a distinct variety of Garnet. Rhodolite is a deep pinkish-red or pale purplish-red color, and large transparent stones are among the most valuable in the Almandine-Pyrope series. The name rhodolite is derived from the Greek *rhodon,* meaning "rose," and *lithos,* meaning "stone." Rhodolite crystals can be very large. Rhodolite is found in North Carolina, Sri Lanka, Tanzania, and Zimbabwe, and is rarer than either Almandine or Pyrope. Here we include it in our discussion of Pyrope because it contains more Pyrope than Almandine, and thus falls into the Pyrope range. There is a fine faceted rhodolite gem of 31.9 carats in the Grainger Hall of Gems (fig. 164).

Spessartine Garnet

SYSTEM Inorganic

CLASS Silicate

GROUP Garnet

SUBGROUP Pyralspite

SPECIES *Spessartine*

GEM VARIETIES spessartite (orange); mandarin (orange-red); kashmerine (reddish-orange Spessartine-Almandine intermediate); malaia (pinkish-orange Spessartine-Pyrope intermediate); color-change malaia (alexandritic Spessartine-Pyrope intermediate); umbalite (pinkish-purple Spessartine-Pyrope intermediate)

COMPOSITION (IDEAL)

Manganese Aluminum Silicate [$Mn_3Al_2(SiO_4)_3$]

HARDNESS ON MOHS SCALE

7–7½

Spessartine Garnet is named after the Spessart region of western Germany, where it was first discovered in the mid-1800s. The common varieties are sometimes referred to as spessartite. Pure Spessartine is yellow-orange in color. One of the most valuable varieties of Spessartine is a recently discovered bright fiery orange-red variety called mandarin Garnet. Mandarin Garnets were first discovered in 1991 in Namibia, Africa. They are unusually fine, intensely orange-colored gems, and they greatly increased the demand for Spessartine Garnets in the gem trade. The original mines for these stones are now largely exhausted, although new deposits have been discovered in Nigeria.

Like other species of Garnet, Spessartine is often found as an intermediate blend with other Garnet species. A Spessartine-Almandine blend (adding Iron to the Spessartine composition) makes the stone darker reddish-orange or brownish-orange. The reddish-orange Almandine-Spessartine blend is sometimes referred to as kashmerine. A Spessartine-Pyrope blend results in varieties such as malaia (a red-orange to pink-orange variety from Tanzania and

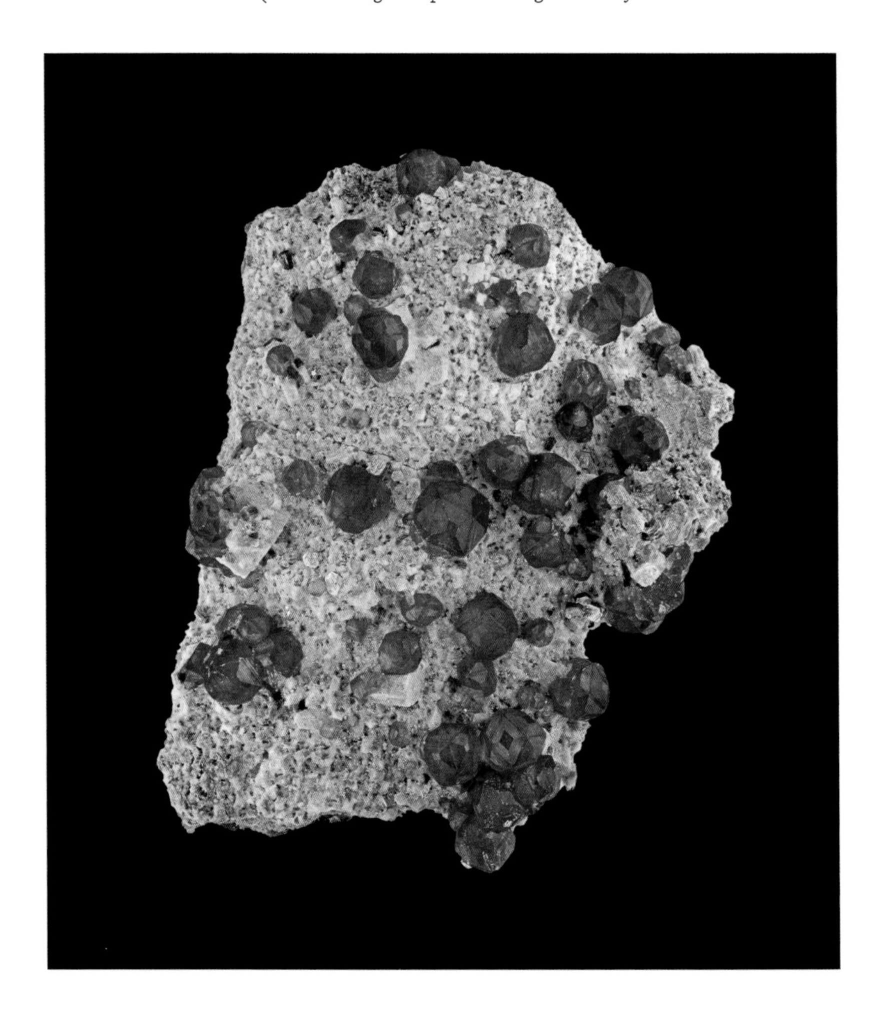

FIGURE 166.
Natural crystals of Spessartine Garnet on matrix from the Tongbei area of Fujian Province, China. Specimen is 80 mm in height (FMNH H2442).

FIGURE 167.
A flowing, free-form necklace with an overlapping floral centerpiece made to enhance the strong faceted color of 17 orange Spessartine Garnets from the Amelia Courthouse site in Virginia. The Spessartine gems weigh a total of 27 carats. Set in 18-karat Gold, this piece also contains 282 ideal-cut Diamonds weighing a total of 2.54 carats. The necklace was designed by Pavel Myagkov of Lester Lampert, Inc., in 2009 for the Grainger Hall of Gems and is named "Tangerine" (FMNH H2585).

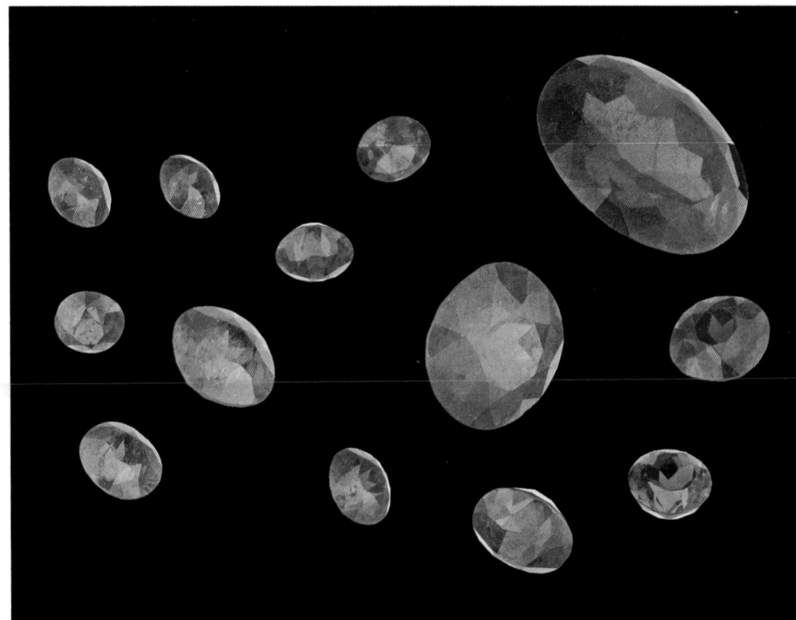

FIGURE 168.
Right: Thirteen faceted Spessartine Garnets from Amelia Courthouse, Virginia, ranging from .82 to 24.7 carats in weight (FMNH H286.1, H286.3–H286.5, H286.8, H286.10–H286.14, H1826, H1827, H2484).

FIGURE 169.
Above: A natural crystal of orange Spessartine Garnet from Tanzania weighing 85 carats and measuring 25 × 23 × 16 mm (FMNH H2441).

Kenya) or umbalite (a pinkish-purple variety from the Umba Valley of Tanzania, sometimes not distinguished from malaia). A rare subvariety of malaia is the color-change Garnet, which can change from red to green, showing a strong ALEXANDRITE EFFECT. A blend of spessartite and Grossular Garnet can produce a variety that resembles hessonite. Spessartite has lace-like or feather-like inclusions that help distinguish it from hessonite.

Spessartine Garnet was once extremely rare as a gem, but major deposits developed in Africa in the 1990s have increased the supply of spessartite gems considerably. These deposits were mined for much of the decade until about 2001. Material from these deposits provided a large boost to the popularity of Spessartine in the gem market. Pure mandarin Spessartine is still one of the most valuable of the Garnet varieties. The reddish- or brownish-orange stones are somewhat less valuable, but also more plentiful. Spessartine gems of 10 carats or more in size, particularly mandarin Garnets, are extremely rare and valuable.

Spessartine is usually found in pegmatites and placer deposits. Localities where it is mined include Nigeria, Tanzania, Pakistan, Sri Lanka, Brazil, Madagascar, Mozambique, Myanmar, the United States (California, Virginia, Colorado, Nevada), China, Sweden, and Australia. There is a fine Spessartine necklace in the Grainger Hall of Gems designed by Pavel Myagkov (fig. 167).

Grossular Garnet

SYSTEM Inorganic

CLASS Silicate

GROUP Garnet

SUBGROUP Ugrandite

SPECIES *Grossular*

GEM VARIETIES hessonite (orange-brown); tsavorite (emerald green); rosolite (pink); leuco (colorless)

Grossular Garnet has more colored gem varieties than any other Garnet species. Some of the variety names are sometimes referred to as grossularite. Unlike most Garnet species, whose color variants are mostly IDIOCHROMATIC (the result of basic chemical structure), the color varieties of Grossular are mostly ALLOCHROMATIC (the result of trace elements). Here we focus on the four main varieties of Grossular: hessonite, tsavorite, rosolite, and leuco.

The name Grossular is derived from the Latin *Grossularia,* which is the name for gooseberries. This is because the pale green varieties of Grossular Garnet were thought to resemble gooseberries (*Ribes grossularia*). Grossular is most often found in metamorphosed limestone. Unlike the Pyralspite Garnets that are Aluminum-based, Grossular is a Calcium-based Ugrandite Garnet. Gem-quality Grossular is sometimes found as a blend with other Garnet species (Spessartine or Andradite). The most valuable of the Grossular gem varieties is emerald-colored tsavorite, in which the color is the result of Chromium and Vanadium. Pure Grossular is colorless.

There are also deposits of massive (nontransparent) Grossular that produce large pieces of opaque or translucent green Grossular. Occasionally, this variety is incorrectly called "African jade," "Garnet jade," or "Transvaal jade," although it is Garnet, not jade. This form is more correctly called hydrogrossular Garnet. Hydroglossular Garnet is a MICROCRYSTALLINE variety not generally used for fine jewelry, but it is occasionally used for carving small statues and figurines.

GOLDEN-ORANGE GROSSULAR GARNET (HESSONITE)

SYSTEM Inorganic

CLASS Silicate

GROUP Garnet

SUBGROUP Ugrandite

SPECIES *Grossular*

VARIETY hessonite (orange-brown)

COMPOSITION (IDEAL) Calcium Aluminum Silicate [$Ca_3Al_2(SiO_4)_3$]

TRACE ELEMENTS FOR COLOR Iron (Fe) and Manganese (Mn)

HARDNESS ON MOHS SCALE 7–7½

Hessonite—also known as essonite or cinnamon stone—comes in hues ranging from golden to orange-brown to bright red. The most desirable color is a bright golden-orange resembling honey with an orange tint. Some hues resemble spessartite Garnet, but spessartite has a different REFRACTIVE INDEX and usually has peculiar lace-like or feather-like inclusions, while hessonite inclusions are usually granular.

The name hessonite is derived from the Greek *esson,* meaning "inferior," probably referring to the fact that hessonite was thought to be softer than other Garnets known in ancient times. Hessonite has been popular for thousands of years in Greek and Roman cultures, where it is used in jewelry and cameos. VEDIC astrologists believed hessonite to be a powerful TALISMAN to increase longevity and happiness. Ancient Hindus believed hessonite to be the fingernails of a great demon.

The vast majority of gem-quality hessonite comes from placer deposits in Sri Lanka, but it is also found in Madagascar, Brazil, Canada, Russia (Siberia), Maine, California, Washington, and New Hampshire. There are two exceptional

Golden-Orange Grossular Garnet (Hessonite)

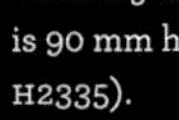

FIGURE 170.
Natural crystals of hessonite Garnet on matrix from Vesper Peak, Washington. Specimen is 90 mm high (FMNH H2335).

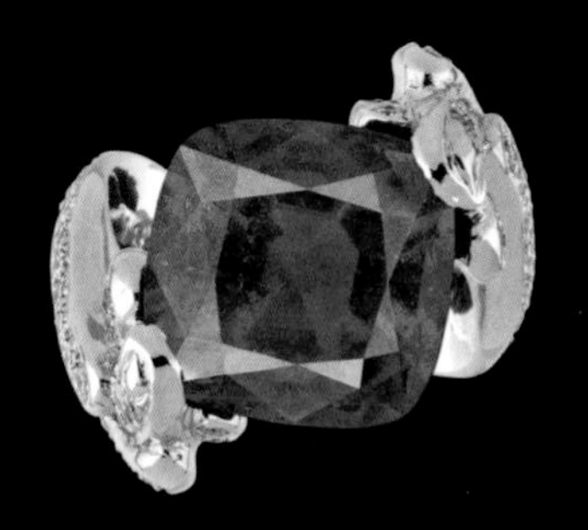

FIGURE 171.
Ring with 13.94-carat hessonite Garnet from Sri Lanka mounted in 18-karat white Gold, with 75 round ideal-cut brilliant Diamonds. Top, side, and front views. Named "Flight" by designer David Lampert of Lester Lampert, Inc. Setting designed for the Grainger Hall of Gems in 2008. Ring is 30 mm in height (FMNH H2534).

FIGURE 172.
Brooch called "Bloom" with 5.65-carat hessonite Garnet from Sri Lanka, set in 18-karat white and 14-karat rose Gold, with 102 ideal-cut brilliant accent Diamonds. Designed by David Lampert of Lester Lampert, Inc., in 2008 for the Grainger Hall of Gems (FMNH H2542).

pieces of hessonite in the Grainger Hall of Gems, including a ring with a 13.94-carat gem mounted in 18-karat white Gold, named "Flight" by designer David Lampert (fig. 171).

GREEN GROSSULAR GARNET (TSAVORITE)

SYSTEM Inorganic

CLASS Silicate

GROUP Garnet

SUBGROUP Ugrandite

SPECIES *Grossular*

VARIETY tsavorite (green)

COMPOSITION (IDEAL)

Calcium Aluminum Silicate $[Ca_3Al_2(SiO_4)_3]$

TRACE ELEMENTS FOR COLOR

Chromium (Cr) and/or Vanadium (V)

HARDNESS ON MOHS SCALE

7–7½

Green Grossular is generally referred to as tsavorite, although occasionally the very light green stones are called gooseberry Garnet. The best, most expensive tsavorite has an emerald-green hue. The name tsavorite was first proposed by Tiffany & Co. president Henry Platt in honor of Tsavo National Park in Kenya, where it was first discovered in 1967. Tsavorite is also found in neighboring Tanzania. Although it is a relative newcomer to the gem trade, tsavorite has quickly become a precious gem in high demand. Tsavorite's beautiful green hues can be virtually indistinguishable from that of the finest emeralds. It is as rare and beautiful as the other green Garnet gem, demantoid, but is more durable because of its greater hardness (Mohs 7–7½ vs. Mohs 6½–7 for demantoid). Tsavorite's popularity began to surge in 1974, when Tiffany & Co. staged a special campaign to promote it in the United States. Since then it has had large worldwide demand. There are currently very few mines where tsavorite is produced in commercial quantities. Most stones are less than 2 carats in size, and fine faceted stones over 5 carats are extremely rare.

Much "Transvaal jade" is an opaque microcrystalline form of tsavorite and not jade at all. Green Grossular Garnet also can form a blend with demantoid Ugrandite Garnet to form mali Garnets, also called grandite. We discuss these in the section on demantoid Garnets.

FIGURE 173.
Natural crystals of tsavorite Garnet from the Umba Valley, Tanzania. Specimen on matrix is 35 mm high (FMNH H2444) and isolated crystal measures 37 × 30 × 23 mm (FMNH H2443).

FIGURE 174.
Facing page: Tsavorite ring with 2.3-carat round brilliant-cut tsavorite from the Umba Valley, Tanzania. Stone is set in 18-karat yellow Gold with 47 round brilliant-cut and 24 baguette Diamonds. Top, side, and front views. (FMNH H2445).

PINK GROSSULAR GARNET (ROSOLITE)

SYSTEM Inorganic

CLASS Silicate

GROUP Garnet

SUBGROUP Ugrandite

SPECIES *Grossular*

VARIETY rosolite (rose pink)

COMPOSITION (IDEAL)

Calcium Aluminum Silicate [$Ca_3Al_2(SiO_4)_3$]

TRACE ELEMENT FOR COLOR

Iron (Fe) or Manganese (Mn)

HARDNESS ON MOHS SCALE

7–7½

FIGURE 175.
Natural crystals of pink rosolite Garnet on matrix from Morelos, Mexico. Specimen is 68 mm high (FMNH H290).

FIGURE 176.
Faceted pink rosolite Garnet from Mexico weighing 2 carats (FMNH H2474).

Rosolite is a pink variety of Grossular Garnet also known as landerite or xalostocite. The name rosolite refers to the stone's rose-pink color. It was first discovered in the nineteenth century in limestone deposits of Xalostoc, in the state of Morelos, Mexico, where fine, transparent rose-pink crystals still occur (fig. 175), some of which is gem quality. Rosolite makes a very beautiful faceted stone, and only the extreme rarity of eye-clean faceting material has prevented it from being more popular. Most stones from this locality are densely filled with inclusions. Similar crystals also occur in Chihuahua, Mexico, but they are even more heavily filled with inclusions. Very light pinkish rosolite crystals are also found at the Jeffrey Mine in Quebec, Canada, but these lack the rich color of the Xalostoc stones. Rosolite is an uncommon variety of Grossular, and true pink transparent stones without inclusions are very rare. Large eye-clean rosolite of good color are among the rarest of all gems.

Rosolite also occurs in South Africa as an opaque microcrystalline form known as "Transvaal jade" (which also occurs in green, yellow, and white). This massive form of Grossular (not true jade) is used for carving but not in fine jewelry.

COLORLESS GROSSULAR GARNET (LEUCO)

SYSTEM Inorganic

CLASS Silicate

GROUP Garnet

SUBGROUP Ugrandite

SPECIES *Grossular*

VARIETY leuco (colorless)

COMPOSITION (IDEAL)

Calcium Aluminum Silicate [$Ca_3Al_2(SiO_4)_3$]

TRACE ELEMENT FOR COLOR

none

HARDNESS ON MOHS SCALE

7–7½

Leuco Grossular is the colorless form of the species. It is a relatively rare variety of the species and is usually found in association with colored varieties of Grossular. Colorless Garnet is rarely faceted or set into jewelry, although there is available material (fig. 177). Of the numerous colorless gems, leuco Garnet is not in demand for fine jewelry, probably due to its inferior optical properties compared to other colorless gems such as colorless Diamond, leuco sapphire, and goshenite. Colorless Grossular from East Africa can be irradiated to produce a yellow-green Garnet, but the color fades relatively quickly in daylight. Colorless Grossular can be found in Tanzania, Sri Lanka, Quebec, California, Myanmar, and China.

FIGURE 177. Faceted colorless Grossular (leuco) Garnets from the Umba Valley, Tanzania, totaling 2 carats in weight (FMNH H2440.1–H2440.4).

Andradite Garnet

SYSTEM Inorganic
CLASS Silicate
GROUP Garnet
SUBGROUP Ugrandite
SPECIES *Andradite*
GEM VARIETIES demantoid (green); topazolite (yellow); grandite or mali (yellowish-green Andradite-Grossularite intermediate); melanite (black)

Andradite Garnet is named after the Brazilian mineralogist Jose Bonifácio d'Andrada e Silva. It is a Calcium-Iron Garnet with three main varieties: green demantoid, yellow topazolite, and black melanite. The crystals have good luster, and they are found in metamorphosed limestone and occasionally in certain igneous rocks. Transparent demantoid stones of good color over 5 carats in size are rarities. Like Grossular Garnet, color varieties are strongly influenced by trace elements. Andradite is one of the softest of all Garnet species (Mohs 6½–7) and is usually even softer than Quartz (Mohs 7). No colorless Andradite is known. Andradite also blends with other species. Mali Garnet, or grandite, is a BLEND of Andradite and Grossular Garnet. Andradite and Grossular are the end members of a compositional series of blends, much like Almandine and Pyrope, mentioned in previous sections. Andradite is also is known to occasionally blend with Uvarovite.

GREEN ANDRADITE GARNET (DEMANTOID)

SYSTEM Inorganic
CLASS Silicate
GROUP Garnet
SUBGROUP Ugrandite
SPECIES *Andradite*
VARIETY demantoid (green)
COMPOSITION (IDEAL)
Calcium Iron Silicate [$Ca_3Fe_2(SiO_4)_3$]
TRACE ELEMENT FOR COLOR
Chromium (Cr) (enhances green hue)
HARDNESS ON MOHS SCALE
6½–7

Demantoid Garnet is one of the rarest and most valuable varieties of Garnet. It formerly had the nickname "Uralian emerald," although it is clearly Garnet, not Beryl. Like the Grossular variety tsavorite, it is a vivid green color, but it also has exceptional Diamond-like luster. The name demantoid is derived from the German *demant*, meaning "Diamond-like." It can further be distinguished from tsavorite in that it is softer. Tsavorite will scratch demantoid, but demantoid will not scratch tsavorite. Also, demantoid often has a diagnostic inclusion called horsetail, which is a bunch of fine, hair-like asbestos inclusions resembling the tail of a horse.

Demantoid Garnet was discovered in Russia's Ural Mountains in 1868. Productive mining of this gem lasted only a few decades there, and today the deposits are largely exhausted. Small-scale mining for demantoid continues today, but most demantoids currently come from localities outside Russia. Demantoid was particularly popular in the nineteenth century, but as the Russian deposits became exhausted, the stone's availability decreased so much that it also decreased the stone's popularity. Its softness is also thought to have had a negative impact on its popularity.

The best-colored demantoid is that with trace components of Chromium, sometimes called chrome demantoid. These are the richest-colored stones and the ones that most closely approach emerald green. These stones are much more valuable than the pale green, olive green, or brownish-green stones. The brown and blackish-brown varieties of demantoid are not generally used as gemstones.

Because of its softness (Mohs 6½–7) and its structural weakness caused by the asbestos "horsetail" inclusions, the facet edges of demantoid are easily

FIGURE 178.
Far left: Natural crystals of demantoid Garnet on matrix from Lanzada, Italy. Specimen is 165 mm high (FMNH M13373).

FIGURE 179.
Three views of a striking demantoid Garnet ring, with 3.07-carat center stone from Poldnevaya, Ural Mountains, Russia. Set in 22-karat yellow Gold and 18-karat white Gold with 79 small brilliant-cut Diamonds and 12 small demantoids. Pavé Diamond petals support the cushion-shaped demantoid. Top, side, and front views. Design of setting, named "Garnet of Eden," was created by David Lampert of Lester Lampert, Inc., in 2008 for the Grainger Hall of Gems (FMNH H2535).

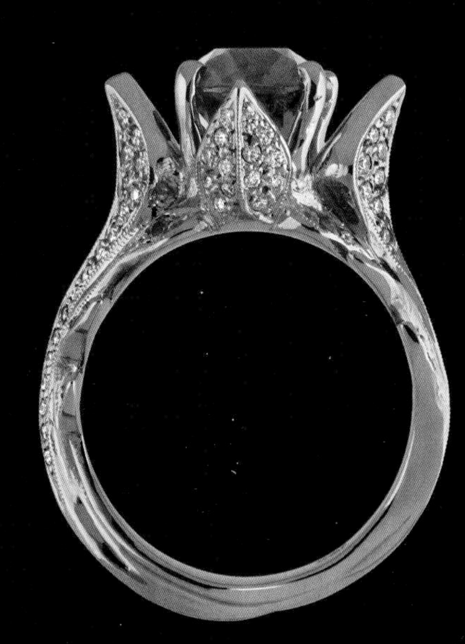

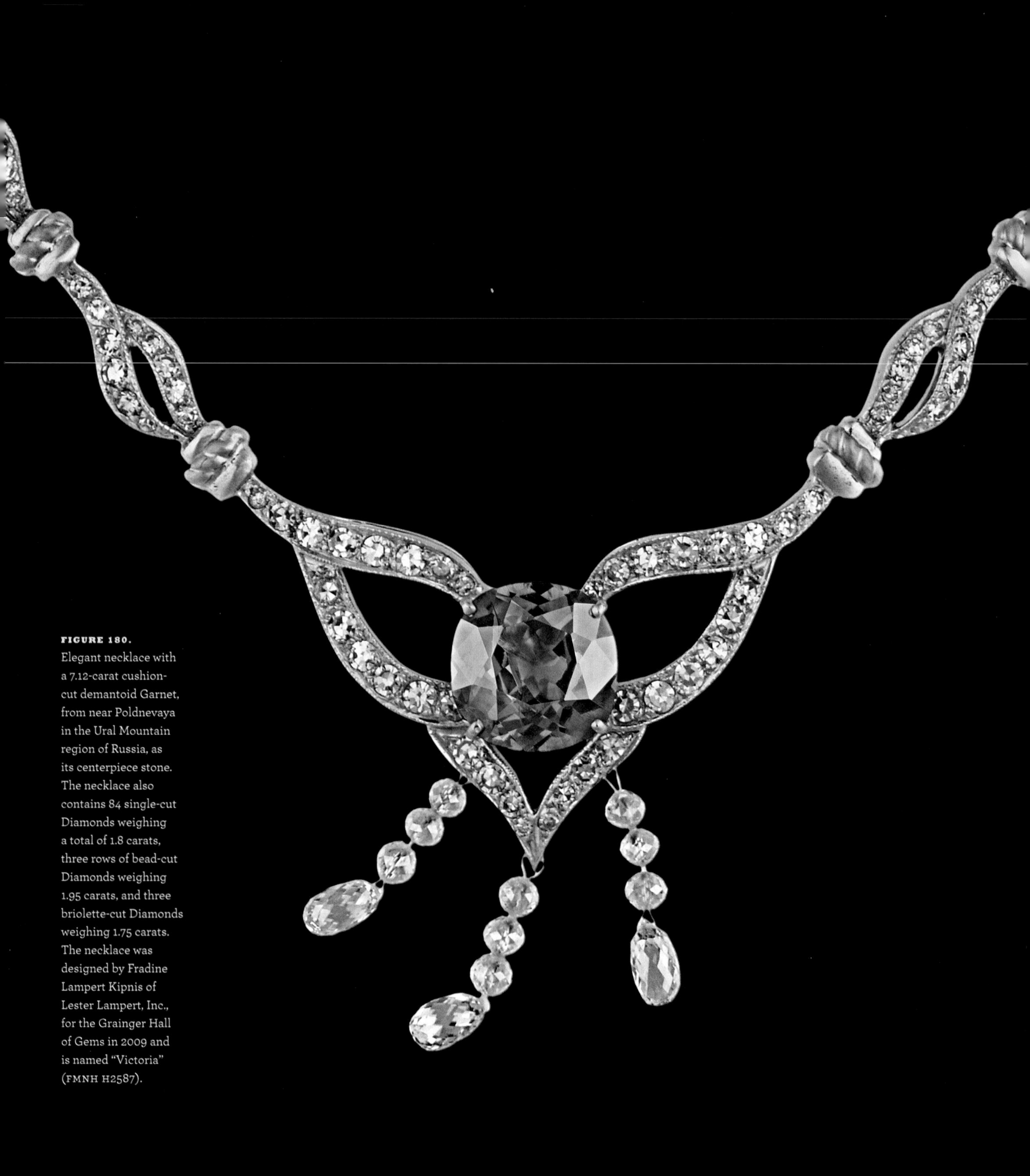

FIGURE 180.
Elegant necklace with a 7.12-carat cushion-cut demantoid Garnet, from near Poldnevaya in the Ural Mountain region of Russia, as its centerpiece stone. The necklace also contains 84 single-cut Diamonds weighing a total of 1.8 carats, three rows of bead-cut Diamonds weighing 1.95 carats, and three briolette-cut Diamonds weighing 1.75 carats. The necklace was designed by Fradine Lampert Kipnis of Lester Lampert, Inc., for the Grainger Hall of Gems in 2009 and is named "Victoria" (FMNH H2587).

damaged or worn down with repeated use. Because of the lack of durability and the high value of demantoid stones, this is not a good choice for jewelry that gets hard daily use.

The best known and largest demantoid stones still are the ones from the Ural Mountain mines in Russia. Other localities include Zaire, Kenya, Namibia, and northern Italy. Gems of over a few carats are extremely rare. There are three extremely fine pieces of demantoid jewelry in the Grainger Hall of Gems. One is a 3.07-carat Russian demantoid set in an exquisite Gold ring designed by David Lampert (fig. 179). The other is a 7.12-carat Russian demantoid set in a spectacular necklace illustrated in figure 180. Also see figure 83.

YELLOW AND YELLOWISH-GREEN ANDRADITE GARNET (TOPAZOLITE AND GRANDITE)

SYSTEM Inorganic

CLASS Silicate

GROUP Garnet

SUBGROUP Ugrandite

SPECIES *Andradite*

VARIETIES topazolite (yellow to orangish-yellow); grandite (yellowish-green intermediate between Andradite and Grossular)

COMPOSITION (IDEAL)

Calcium Iron Silicate [$Ca_3Fe_2(SiO_4)_3$]

TRACE ELEMENT FOR COLOR

Titanium (Ti) for orange color; yellow has no Titanium

HARDNESS ON MOHS SCALE

6½–7

Topazolite is the yellow form of Andradite Garnet. Its hue ranges from pale yellow to dark yellow, but generally it is a honey yellow. It is similar in transparency and color to yellow Brazilian Topaz, which is responsible for the name. Topazolite is an extremely rare gemstone, even rarer than demantoid. Only small stones are available, and gems over 3 carats are extremely rare. Most of the largest stones are also currently in the hands of private collectors. Although much rarer than demantoid, it is less valuable due to lack of demand. In general, green varieties of most gem species, particularly those colored with traces of Chromium, are much more popular than yellow varieties.

Topazolite is found in metamorphic rocks of the Swiss and Italian Alps, and in California. Like demantoid, topazolite is a relatively soft gem (Mohs 6½–7). There is a specimen of natural topazolite crystals on matrix in the Grainger Hall of Gems (fig. 181).

Another yellowish-green Garnet, discovered in 1994 in Mali, Africa, is a BLEND of Andradite and Grossular. It is called grandite or mali Garnet (fig. 182). Huge stockpiles of this material are said to be waiting to enter the market, although most of it lacks the rich green hues of tsavorite and the best demantoid.

FIGURE 181.
Natural crystals of yellow topazolite Garnet on matrix from Piedmont, Italy. Specimen is 115 mm in height (FMNH M3213).

FIGURE 182.
Faceted grandite Garnet, a mix of Andradite and demantoid Garnet, from Mali. Tear-drop-shaped 4.1-carat gem, measuring 11 × 9 × 6 mm (FMNH H2476).

BLACK ANDRADITE GARNET (MELANITE)

SYSTEM Inorganic

CLASS Silicate

GROUP Garnet

SUBGROUP Ugrandite

SPECIES *Andradite*

VARIETY melanite (black)

COMPOSITION (IDEAL)

Calcium Iron Silicate [$Ca_3Fe_2(SiO_4)_3$]

TRACE ELEMENTS FOR COLOR

Color primarily due to idiochromatic presence of Iron, but Titanium (Ti) is also a major component of the black color.

HARDNESS ON MOHS SCALE

6½–7

Melanite is the darkest form of Andradite. It is usually glassy black, but can also be dark brown or blackish-red. The name is derived from the Greek *melas*, which means "black." Melanite crystals can be quite large (fig. 183) and contain varying amounts of Titanium. Titanium usually only occurs as a trace element, but when it is occasionally present in more substantial amounts, the stone is sometimes referred to as Titanium Andradite. It has a **VITREOUS** to **METALLIC LUSTER**, and it has been reported to have been occasionally used in mourning jewelry during the Victorian era. Melanite mourning jewelry is extremely rare today, and most existing black mourning jewelry from Victorian times is made from jet, a type of hard coal.

Melanite occurs in volcanic and metamorphic rock, and is found in France, Germany, Italy, Norway, Colorado, Arkansas, and California.

FIGURE 183. Natural crystals of black melanite Garnet. *Top:* A large single crystal from Mali, Africa, weighing 2,167.5 carats and measuring 75 × 58 × 48 mm (FMNH H2439). *Bottom:* Numerous crystals in black matrix from the Czech Republic. Specimen is 50 mm wide (FMNH H2598).

Uvarovite Garnet

SYSTEM Inorganic

CLASS Silicate

GROUP Garnet

SUBGROUP Ugrandite

SPECIES *Uvarovite*

VARIETY drusy Uvarovite (emerald green)

COMPOSITION

Calcium Chromium Silicate [$Ca_3Cr_2(SiO_4)_3$]

TRACE ELEMENT FOR COLOR

none

HARDNESS ON MOHS SCALE

7–7½

Uvarovite Garnet is the rarest Garnet species and also one of the hardest. It is a hard Garnet with an extremely attractive emerald-green color, unsurpassed for beauty. Seemingly, this should be a very valuable gem variety of Garnet except for one thing: it is only known as very small crystals, and the only known faceted gems are tiny stones that are a fraction of a carat in weight. Larger crystals when found are either opaque or too heavily included to make attractive faceted stones. Uvarovite is also brittle and generally fragile. Eye-clean stones of good colored Uvarovite weighing a half-carat or more are simply unavailable. The green color and overall appearance of the tiny crystals in matrix is so striking that occasionally jewelry is made with small free-form pieces of tiny DRUSE-encrusted matrix (fig. 184).

Uvarovite was discovered in 1832 by Henri Hess, who named it after Count Sergei Semenovitch Uvarov (1786–1855), a Russian statesman and amateur mineral collector who was once the president of the St. Petersburg Academy. Uvarovite develops on metamorphic rocks such as serpentine and occurs mainly as thin crusts of crystals on matrix. It is highly lustrous and matches the finest emerald in color. It is sometimes found as a blend with Andradite and/or Grossular.

The best Uvarovite is found in serpentine rocks of the Ural Mountains of Russia, lining rock fissures and cavities. It is also known to occur in Finland, Norway, Turkey, and Italy.

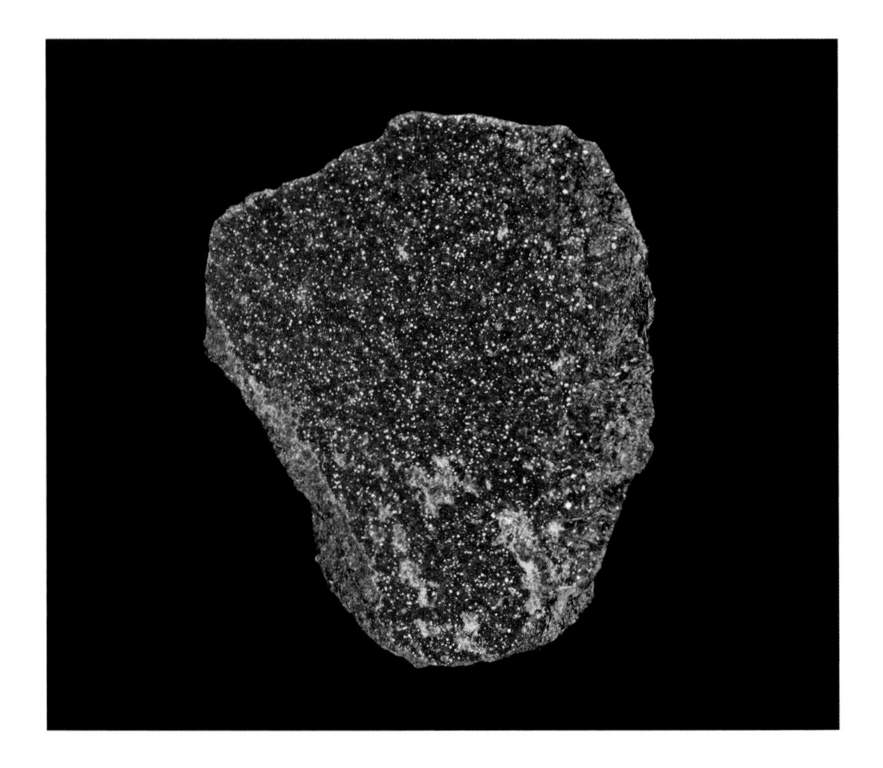

FIGURE 184. Natural crystals of Uvarovite Garnet on matrix from the Ural Mountain region of Russia, measuring 58 × 46 × 6 mm (FMNH H2314).

Zircon

SYSTEM Inorganic

CLASS Silicate

GROUP Zircon

SPECIES *Zircon*

GEM VARIETIES hyacinth, or jacinth (brownish-red, red, or brown); jargon (yellow, yellowish-green); starlight (blue); matara (colorless)

Zircon is one of the most misunderstood gemstones. On the negative side, Zircon is somewhat brittle and is vulnerable to chipping when knocked against hard objects; but this is also true of the much more popular gem Topaz. And to its credit, Zircon is a hard, scratch-resistant stone that makes one of the most beautiful of gems. So why isn't Zircon more popular as a gemstone? It has suffered in popularity for two main reasons: (1) the name has been confused with that of the man-made Diamond SIMULANT cubic zirconia (CZ); and (2) for years colorless natural Zircon was used as a Diamond simulant, eventually making the name Zircon synonymous with "cheap imitation" in many people's minds. Both perceptions are unfair to such a beautiful and hard gemstone. Zircon is a natural gemstone with the composition $ZrSiO_4$ and a Mohs hardness of 7½. It is *not* the same mineral as CUBIC ZIRCONIA, a man-made Diamond simulant that has the composition ZrO_2 and a Mohs hardness of 8¼.

Also, there was a good reason that natural colorless Zircon was used as a Diamond simulant: it is one of the only colorless gems that matches the brilliant fire of Diamond. Its beauty should in theory be an asset and not a liability to its popularity. Unfortunately, popularity does not always go hand in hand with logic.

Zircon is also interesting because of its remarkably great age. In fact, nothing on earth is older, as far as we know. The oldest known piece of the earth is a Zircon crystal from Western Australia, which is dated at 4.4 billion years before present. This Zircon is in the collection of the natural history museum in Perth, Western Australia.

Zircon is doubly refractive; that is, light splits into two rays as it passes through the gem. This results in the back facets appearing as double images, giving optical depth to the stone. This factor, together with its Diamond-like ADAMANTINE LUSTER, makes it sparkle much like a Diamond. Zircon is also one of the heaviest gemstones, with a smaller-per-carat size than many other gems. Zircon also often contains traces of radioactive elements such as Uranium or Thorium, although not enough to pose danger to the wearer.

The name Zircon is derived from the Persian *zarqun*, meaning "golden color." Gem-quality Zircons have been mined for more than 2,000 years in Sri Lanka, and Zircon jewelry first became fashionable in the West in the 1920s. Zircon is widely distributed, found in Cambodia, Myanmar, Ratnapura in Sri Lanka, Thailand, France, Norway, Tanzania, Victoria (Australia), Norway, Ontario, Germany, Japan, Maine, Colorado, New Jersey, and New York. Zircons have even been found on the moon.

Common names for Zircon varieties include starlight for blue Zircon, and matara for colorless Zircon. The names for the other colors are more confusing and somewhat controversial. Hyacinth (or jacinth) is an old term that has been used for reddish-brown, yellowish-red, orange, and yellow Zircons. Jargon is an old term that has been used for yellow, straw-yellow, and near-colorless Zircon. The overlap of color range between the hyacinth and jargon causes confusion and contradictions. For the purpose of simplification, we follow those authors who restrict yellow and yellowish-green Zircons to the variety name jargon, and use the name hyacinth for Zircons of red, reddish-brown, orange, or brown hues. For further reading on Zircon, we recommend Bauer (1968) and O'Donoghue (2006).

RED TO BROWN ZIRCON (HYACINTH)

SYSTEM Inorganic

CLASS Silicate

GROUP Zircon

SPECIES *Zircon*

VARIETY hyacinth (jacinth)

COMPOSITION

Zirconium Silicate ($ZrSiO_4$)

TRACE ELEMENT FOR COLOR

Uranium (U)

HARDNESS ON MOHS SCALE

7½

Hyacinth is sometimes also referred to as jacinth. Both names are thought to be different forms of the Greek *huakinthos*, a name of uncertain derivation. Hyacinth is mentioned in the Bible as a colored stone, but it is not certain if the name was referring to Zircon or to some other gemstone. The use of the name hyacinth has varied throughout history. For example, in the Middle Ages, the name was used for all yellow stones of East Indian origin, including yellow Zircon, orange-yellow or yellow-brown Topaz, yellow Garnet, yellow Corundum, citrine Quartz, and other mineral species. Eventually the use of the name was restricted to the species we think of today as Zircon (defined by the chemical composition $ZrSiO_4$), and it has been used for nearly all colors of Zircon at one time or another. Here we follow a more restricted definition for this variety, including only reddish-brown, brown, red, and orange Zircon as hyacinth, and excluding yellow, blue, green, and colorless hues.

Brown and reddish-brown are the two most common hues of hyacinth (fig. 187). Eye-clean reddish-brown to brownish-orange stones are faceted for fine jewelry, and their fire and sparkle make them attractive gems. Of the brownish stones, those with orange or red overtones are the most in demand for jewelry. Plain brown stones are not highly prized as gemstones because brown is not currently a popular color for jewelry. Both brown and reddish-brown Zircons can be heat-treated to produce colorless Zircons (matara) and blue Zircons (starlight). Brown Zircon comes mainly from Myanmar, Nigeria, Vietnam, Cambodia, Sri Lanka, Russia, and Norway.

Red and pink hues are the scarcest colors of hyacinth and are the most valuable. The red-colored stones can sometimes resemble red Spinel, and just like red Spinel, fine red Zircon is scarce. Red to pink Zircons come mainly from Cambodia, Nigeria, Tanzania, Thailand, and Sri Lanka.

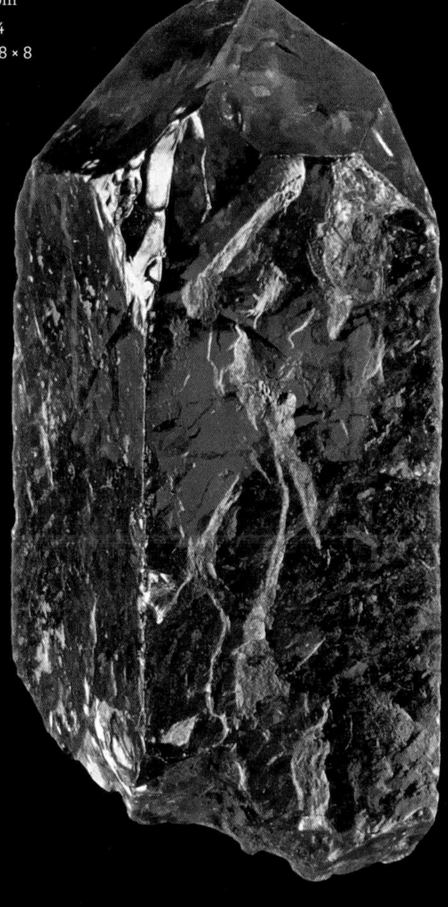

FIGURE 185.
Left: Natural crystals of brownish hyacinth Zircon on matrix from the Ural Mountain region of Russia, measuring 72 mm in height (FMNH M2883).

FIGURE 186.
Right: Natural crystal of red hyacinth Zircon from Tanzania weighing 25.4 carats, measuring 22 × 8 × 8 mm (FMNH H2486).

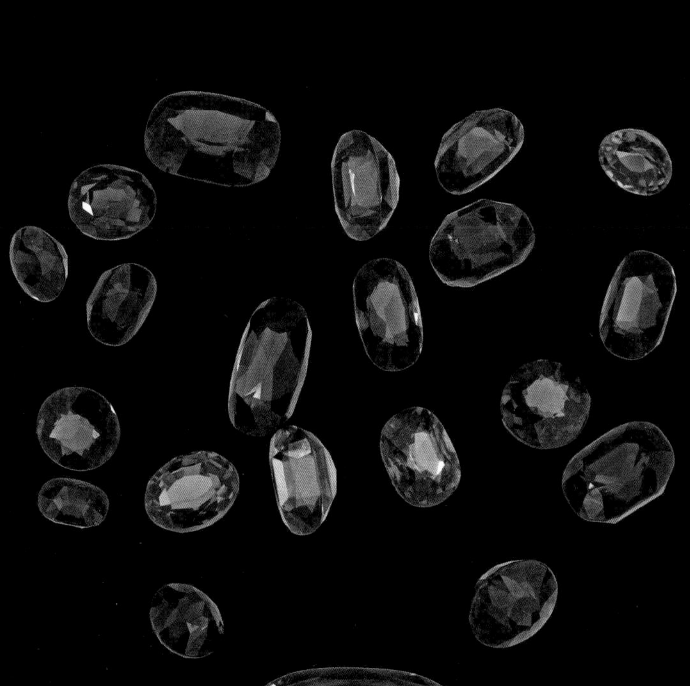

FIGURE 187.
Faceted hyacinth Zircons from Sri Lanka showing a range of color from brownish- to pinkish-red. Gems range from 1.7 to 47.9 carats in weight (FMNH H228.1–H228.4, H228.6–H228.17, H234, H2222, H2424–H2426).

YELLOW TO YELLOWISH-GREEN ZIRCON (JARGON)

SYSTEM Inorganic

CLASS Silicate

GROUP Zircon

SPECIES *Zircon*

VARIETY jargon

COMPOSITION

Zirconium Silicate ($ZrSiO_4$)

TRACE ELEMENT FOR COLOR

Uranium (U)

HARDNESS ON MOHS SCALE

7½

Jargon (also known as jargoon) is derived from the Persian word *zarqun*, meaning "golden color." This name also eventually evolved into the word Zircon, the name for the entire mineral species. We include all hues of yellow in our usage of the name. It can be pale yellow, canary yellow, golden yellow, brownish-yellow, and greenish-yellow (fig. 189).

Yellow stones occur naturally, but some are also produced by heat-treating hyacinth. Golden yellow and pure green (without brown) are the two scarcest color hues of jargon. Many authors include golden yellow within their definition of hyacinth. We find this to be confusing, and follow previous authors who group all yellow hues of Zircon together in the variety jargon.

Yellow jargons come mainly from Sri Lanka, Cambodia, Thailand, and Vietnam. Sri Lanka produces many naturally colored stones, but most stones from elsewhere are heat-treated brown stones. As with other colored Zircons, the value of jargon today is not as great as many other colored stones due to lack of popularity. In the classical period of Rome and in the Middle Ages, yellow Zircon was very popular.

FIGURE 188.
Facing page: Hyacinth Zircon ring with faceted Tanzanian Zircon of 6.57 carats set in 18-karat Platinum with 89 small ideal-cut Diamonds weighing 1.01 carats and 18 single-cut Diamonds weighing .31 carats set upside down with points facing up. Top, side, and front views. Designed by David Lampert, who named the piece "Tanzania Twist" (FMNH H2357).

FIGURE 189.
Right: Faceted jargon Zircons from Sri Lanka, ranging from 1.2 to 10.2 carats each, showing a range of color from yellow or brownish-yellow to green (FMNH H228.5, H228.18, H228.19, H241, H1860, H1862, H2221, H1369).

BLUE ZIRCON (STARLIGHT)

SYSTEM Inorganic

CLASS Silicate

GROUP Zircon

SPECIES *Zircon*

VARIETY starlight

COMPOSITION

Zirconium Silicate ($ZrSiO_4$)

TRACE ELEMENT FOR COLOR

Uranium (U), but virtually all blue Zircon is colored by heat treatment

HARDNESS ON MOHS SCALE

7½

Starlight Zircon is another of the beautiful blue varieties of gemstone, along with aquamarine and blue Topaz. This is currently the most popular hue of Zircon for jewelry. Blue Zircon is exceedingly rare in nature, but brown hyacinth Zircon can be converted to blue through heat treatment. Because brown hues of gemstones are typically not as popular for jewelry as other colors, much of the eye-clean brown hyacinth that is mined is converted to blue to meet popular demand. Since this is simply a process of color-shifting natural Zircon, heat-treated stones are acceptable as natural gems for use in jewelry (fig. 190). The process involves heating the brown stones to temperatures of around 1,800°F (1,000°C) in a reducing (Oxygen-free) atmosphere until they turn blue. Yellow jargon Zircons can also be heat-treated to make starlight. Heat-treated stones can sometimes fade after repeated exposure to direct sunlight over a period of years, but faded stones can be treated again to rejuvenate the blue color. Although nearly all blue Zircon has been artificially heat-treated to obtain its blue color, it is currently the most popular of Zircon varieties in the gem trade.

The highest-grade color for starlight is a light electric blue. Starlight has a higher degree of PLEOCHROISM than other Zircon varieties, and this can make the stone look blue in one direction and greenish in another. Blue Zircons (almost exclusively heat-treated stones) come from Vietnam, Cambodia, Thailand, and Sri Lanka. The Grainger Hall of Gems features an antique blue Zircon ring with several starlight gems set in 20-karat Gold that came from Thailand in the mid-twentieth century (fig. 191).

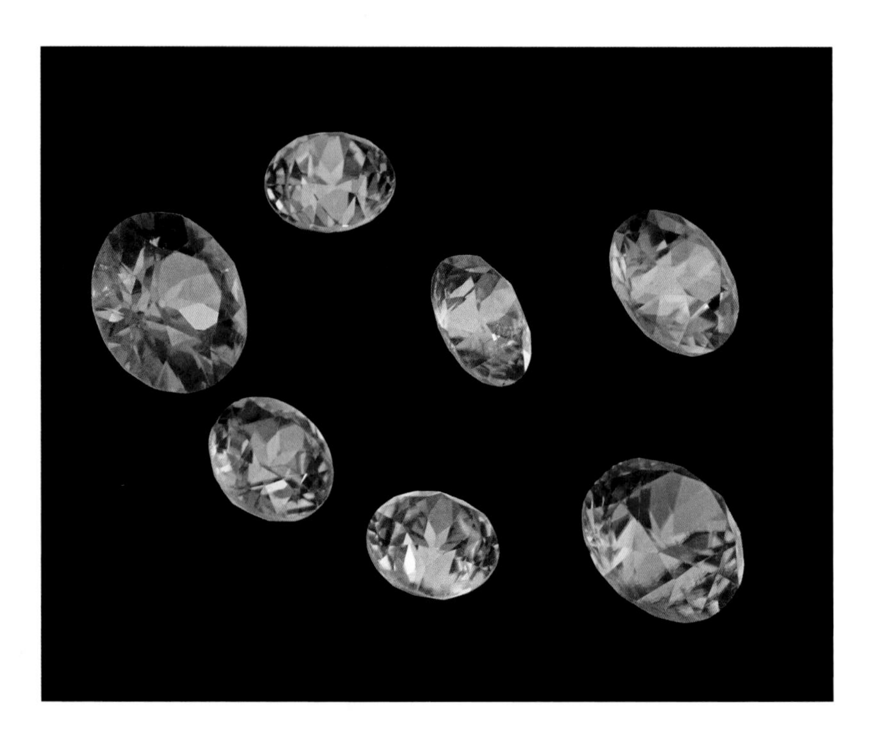

FIGURE 190. Faceted blue starlight Zircons from Sri Lanka processed in Thailand, ranging from 2.9 to 7.8 carats each (FMNH H237, H238, H186–H1866, H2423).

FIGURE 191. *Facing page:* Blue Zircon ring from the early twentieth century, with three faceted starlight gems set in a 20-karat Gold ring with filigree design, made in Bangkok, Thailand. Top, side, and front views (FMNH H1385).

COLORLESS ZIRCON (MATARA)

SYSTEM Inorganic

CLASS Silicate

GROUP Zircon

SPECIES *Zircon*

VARIETY matara

COMPOSITION

Zirconium Silicate ($ZrSiO_4$)

TRACE ELEMENT FOR COLOR

none

HARDNESS ON MOHS SCALE

7½

Matara, or colorless Zircon, comes closer to resembling Diamond than any other natural gem. In fact, in the gem trade this variety is frequently referred to as "matara Diamond" or "Ceylon Diamond." It has the adamantine luster and fiery sparkle of Diamond and is accented further by the marked BIREFRINGENCE of Zircon, giving double reflections of light from the back facets. Matara's sheer beauty and close resemblance to Diamond has worked against it in some respects. The fact that it was used as a cheaper Diamond substitute or "imitation" for years has cheapened its overall image in the eyes of the gem-buying public. Now that it is no longer used as a Diamond substitute, demand for matara has dropped considerably because the jewelry industry has not embraced this variety for its own merits. Ironically, matara itself, which was once used mainly to imitate Diamond, is now imitated by cheaper synthetic simulants, such as synthetic Spinel, YAG, and cubic zirconia (all three of which are occasionally marketed under the name of "Zircon"). As we said before, none of these—not even cubic zirconia—is true Zircon.

Natural colorless Zircon is scarce, and most of it is produced by heat-treating brownish stones. In fact, natural colorless Zircon is rarer even than Diamond, the stone it used to imitate. And while the best-quality colorless Diamonds sell for tens of thousands of dollars per carat, the best-quality colorless Zircons sell for little over a hundred of dollars per carat.

Matara is best known as coming from Sri Lanka and in fact is named after the city of Matara at the southern tip of the island. Matara also comes from Thailand and Vietnam. There is an exceptional piece of colorless Zircon jewelry in the Grainger Hall of Gems that was designed by Marc Scherer (fig. 193).

FIGURE 192.
Faceted colorless Zircon gems from Sri Lanka, ranging from 1.1 to 6.9 carats (FMNH H239, H242, H2547.1–H2547.3).

FIGURE 193.
Facing page: Oval faceted colorless Zircon weighing 15.26 carats with small Diamonds mounted as a pendant in 14-karat white Gold. Designed by Marc Scherer of Marc and Co., created in 2008 for the Grainger Hall of Gems, and named "Light's Labyrinth" (FMNH H2589).

PYROXENE GROUP

SYSTEM Inorganic

CLASS Silicate

GROUP Pyroxene

SPECIES *Spodumene; Jadeite*

Two important gemstone species, Spodumene and Jadeite, are in the Pyroxene group of minerals and are also classified together in a subgroup of Pyroxenes called the Clinopyroxenes. This group is characterized by a particular atomic and crystalline structure that is shared by these two species, and it unites them on our tree of gem species. Jadeite has an extremely long history of value to human culture, dating back thousands of years to prehistoric times. Spodumene, in contrast, is a relatively recent entry into the gem market, recognized only in the late nineteenth century.

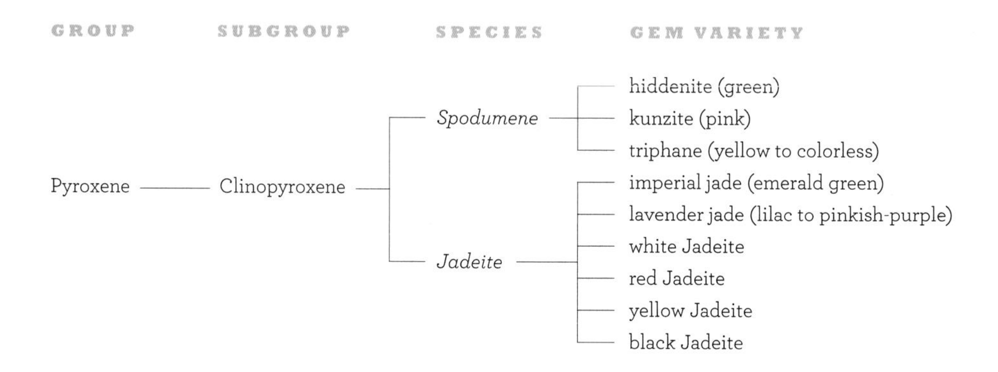

FIGURE 194. Classification of Pyroxene gemstone varieties discussed in this book.

Spodumene

SYSTEM Inorganic

CLASS Silicate

GROUP Pyroxene

SPECIES *Spodumene*

GEM VARIETIES kunzite (pink to lilac); hiddenite (green); triphane (yellow to colorless)

The transparent gem varieties of Spodumene are relatively recent discoveries of the gem world. Hiddenite (green Spodumene) gems were first identified in the late 1870s, and kunzite (pink Spodumene) was first described in the early 1900s. Triphane (yellow to colorless Spodumene), first discovered in 1877, is not widely used in the gem trade today, and some authors include triphane within their definition of hiddenite. We follow the older definition of the yellow to colorless form to provide a more precise and useful categorization. A blue Spodumene weighing 35 carats has been reported (O'Donoghue 2006), but this is exceedingly rare and little is known about it. Most Spodumene is an opaque ash-gray crystal that is unsuitable for gems. This non-gem form is probably the source of the name Spodumene, which is derived from the Greek *spodumenos*, meaning "burned to ashes." The opaque crystals can be quite large, reaching lengths of 50 feet or more and weighing up to 90 tons. This material is a major ore of the metal Lithium. Gem-quality crystals have a characteristic flattened and striated prismatic shape, and are generally small, particularly hiddenite. For further reading on Spodumene, we recommend Sinkankas (1959).

PINK OR LILAC SPODUMENE (KUNZITE)

SYSTEM Inorganic

CLASS Silicate

GROUP Pyroxene

SPECIES *Spodumene*

VARIETY kunzite

COMPOSITION

Lithium Aluminum Silicate [$LiAl(SiO_3)_2$]

TRACE ELEMENT FOR COLOR

Manganese (Mn)

HARDNESS ON MOHS SCALE

6½–7

Kunzite is a relative newcomer to the gemstone world. It was first described in 1902, a couple of decades after the discoveries of the other Spodumene gem varieties, triphane and hiddenite. Of the three Spodumene gem varieties, kunzite is the most popular and best known by far. It was first found in significant quantities in the Pala district of California in the late nineteenth and early twentieth centuries, and was named after the noted gem expert George F. Kunz, who was vice president of Tiffany & Co. of New York at the time. Kunzite is usually a delicate pale pink to lilac-colored gemstone that is frequently found as clean, nearly inclusion-free crystals. It is distinguishable from pink sapphire and morganite in that the pink hue often contains at least a hint of lilac overtones, and it is softer than those two gem varieties. Most kunzite gems are pale or very light in color, and the more intense the color, the more valuable these are as gems.

Kunzite is strongly PLEOCHROIC, and the deepest pink coloration is much stronger when viewed in a particular direction. In fact, some kunzite crystals are deep pink when viewed from the end (or principal crystal axis), and much lighter when viewed from the side. For this reason, gem cutters must carefully orient the stones with the TABLE facet across the principal crystal axis in order to produce gems of good color. Color on some light-colored stones can fade over time with repeated exposure to bright sunlight. The pink hue of faded stones and stones with brownish tints can be improved by heat-treating the stones to about 300°F (150°C). Some stones are also irradiated to improve the

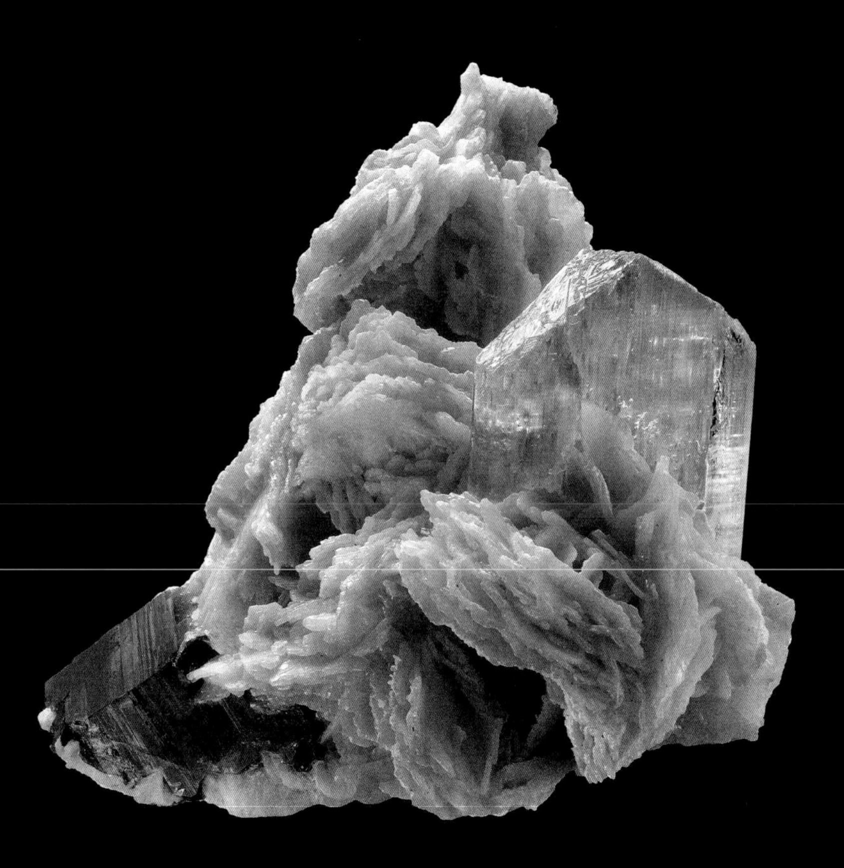

FIGURE 195.
Right: Natural crystal of pink kunzite together with smoky Quartz crystal on Feldspar matrix from Pakistan. Specimen is 140 mm in height (FMNH H2353).

FIGURE 196.
Above left: Large faceted kunzite gems from Minas Gerais, Brazil, ranging from 21.7 to 505.4 carats. Largest gem measures 62 × 35 × 30 mm (FMNH H1548, H1552, H1573, H1576, H1717, H1733, H2465).

FIGURE 197.
Right: Pendant with 16-carat kunzite set in 14-karat white Gold surrounded by small round brilliant-cut Diamonds. Pendant is 30 mm in height. Designed by Ashish Jhalani of MySolitaire.com. Setting © MySolitaire.com (FMNH H2499).

color. Gem-quality kunzite crystals are often sizable, and faceted stones in excess of 10 carats are not uncommon, with much larger ones also available. In the Grainger Hall of Gems, there are a number of faceted stones, including one that is 505.4 carats (fig. 196). There is also a large piece of clear faceting-grade kunzite of good color weighing 4,908.5 carats in the Grainger Hall of Gems.

California remains an excellent source of kunzite, but good quantities of quality material are now also known from Brazil, Madagascar, Maine, and Afghanistan. Occasionally, pockets of intensely purple kunzite crystals called amethystine kunzite have been found in California and Brazil. This hue is rarer than the pink to lilac-pink hues of kunzite but usually fades to pink over time with exposure to light.

GREEN SPODUMENE (HIDDENITE)

SYSTEM Inorganic

CLASS Silicate

GROUP Pyroxene

SPECIES *Spodumene*

VARIETY hiddenite

COMPOSITION

Lithium Aluminum Silicate [$LiAl(SiO_3)_2$]

TRACE ELEMENT FOR COLOR

Chromium (Cr) (for the emerald-green material)

HARDNESS ON MOHS SCALE

6½–7

Hiddenite, or green Spodumene, is the rarest of the major Spodumene gems. It can be an exceptionally beautiful bright emerald color, due to occasional traces of the element Chromium. Such stones are referred to as chrome hiddenite. Like so many other top-tier green gems such as emerald, chrome Tourmaline, demantoid Garnet, and tsavorite Garnet, the presence of Chromium seems to be the key to hiddenite's allure. Chrome hiddenite is known primarily from the locality where hiddenite was first discovered: Alexander County, North Carolina. It is sometimes called "lithia emerald," but it is clearly not emerald, which is a variety of Beryl. It is, in fact, much rarer than emerald. Chrome hiddenite is not known to occur in large crystals. Faceted gems of eye-clean material weighing a carat or more are extremely rare. We know of only a single stone larger than 10 carats, which currently resides in a private collection.

Some kunzite, when first mined, is a greenish color that turns pink after exposure to light. Other colorless or pink material can sometimes be turned a pale green color through heat or radiation treatment. Such stones also fade with exposure to sunlight. The first author purchased a piece of green Brazilian Spodumene that turned pink and then nearly colorless after only two weeks on a window ledge exposed to full sunlight. The color of true chrome hiddenite, however, is stable and does not fade with exposure to light.

The first gem-quality hiddenite gemstones were discovered in Alexander County, North Carolina, in 1879. The green Spodumene variety was named after William E. Hidden, a geologist who had been commissioned by Thomas Edison to search for sources of Platinum in North Carolina. He failed to find Platinum, but was one of a chain of people who brought hiddenite to the attention of the world.

Hiddenite crystals are generally found in pegmatitic pockets of igneous rock. In addition to North Carolina, pockets of fine-quality hiddenite have

FIGURE 198.
Above: Natural 12-carat crystal and two faceted gems of hiddenite from Alexander County, North Carolina. Crystal measures 17 × 9 × 3 mm (FMNH M8945), and gems are .56 and .67 carats (FMNH H314, H1904).

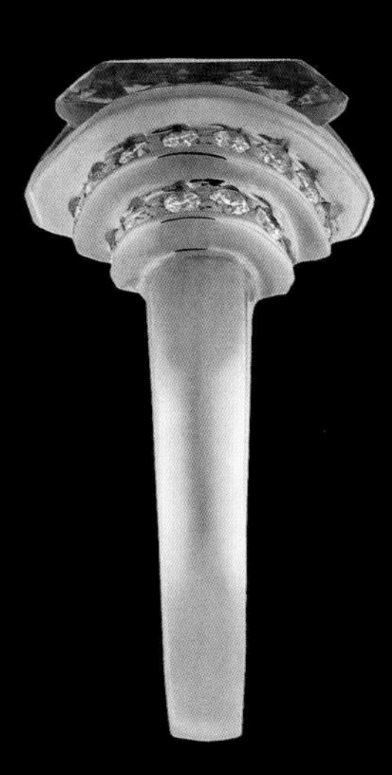

FIGURE 199.
Right: Fine faceted 2.1-carat hiddenite from Nuristan, Afghanistan, set in an 18-karat yellow Gold ring with 28 ideal-cut Diamonds weighing .39 carats total. Designed by Dennis and Lester Lampert and named "Legacy" by Lester Lampert (FMNH H2594).

also been found in Madagascar and Afghanistan. Stones that are pale green to yellowish-green are much less valuable than the chrome (emerald-green) hiddenite. There is a fine natural crystal and two faceted stones of hiddenite from North Carolina in the Grainger Hall of Gems (fig. 198). The exhibit also includes a 2.1-carat faceted eye-clean gem with fine color from Nuristan, Afghanistan, set in an 18-karat Gold ring with Diamonds (fig. 199).

YELLOW TO COLORLESS SPODUMENE (TRIPHANE)

SYSTEM Inorganic

CLASS Silicate

GROUP Pyroxene

SPECIES *Spodumene*

VARIETY triphane

COMPOSITION

Lithium Aluminum Silicate $[LiAl(SiO_3)_2]$

TRACE ELEMENT FOR COLOR

Iron (Fe)

HARDNESS ON MOHS SCALE

6½–7

Triphane, the yellow to colorless variety of Spodumene, was the earliest discovered gem variety of Spodumene. First reported from Brazil in 1877, the name triphane is derived from a Greek word meaning "three aspects," referring to the strong TRICHROISM, which makes this gem appear in three different color hues that change when viewed from different directions. The name triphane is used by some authors as a synonym for Spodumene, including hiddenite and kunzite. We follow earlier usage of the variety distinctions because we find it more useful in classifying the gem varieties.

Crystals of triphane, like those of other Spodumene, often have a characteristic "Roman sword" shape (fig. 200). Of the three gem varieties of Spodumene discussed here, triphane is the least popular and least expensive. As is often the case, pale yellow is not a popular color in the gem trade. Completely colorless triphane of faceting grade is not common, but it nevertheless has little or no use in the fine-gem trade because its aesthetic properties are inferior to several other varieties of colorless gemstones (e.g., colorless Zircon, colorless sapphire, and colorless Diamond). Yellow triphane gems from Brazil occasionally find their way into the gem markets mistakenly identified as "Chrysoberyl" or "Brazilian Chrysoberyl." This is partly due to the similar color and partly due to the fact that triphane is occasionally found in the same place as true Chrysoberyl.

Fine gem-quality natural triphane crystals can reach large size. The Pala Mine of San Diego County, California, has produced fine crystals reported to be up to 300 mm in length. Gem-quality triphane is known from deposits in California, Brazil, Afghanistan, and Madagascar. There is a beautiful, fine natural 1,528-carat gem-quality crystal of triphane in the Grainger Hall of Gems (fig. 200), as well as a fine faceted 163.3-carat gem (fig. 201).

FIGURE 200.
Left: Natural 1,528-carat gem-quality triphane crystal from Nuristan, Afghanistan, measuring 183 × 29 × 23 mm (FMNH H2491).

FIGURE 201.
Below: Faceted triphane gem from Minas Gerais, Brazil, weighing 163.3 carats (FMNH H1680).

Jadeite

SYSTEM Inorganic

CLASS Silicate

GROUP Pyroxene

SPECIES *Jadeite*

GEM VARIETIES green Jadeite; other colored Jadeite

The use of jade for adornment goes back thousands of years in human history, and it has been long revered, from North America to South America to Asia. But it was not until the mid-nineteenth century that the French mineralogist Alexis Damour discovered that what had previously been called "jade" was in fact two completely different mineral species. One species, for which he proposed the name Jadeite, is a Sodium Aluminum Silicate measuring 6½–7 on the Mohs hardness scale. The other "jade" species, called nephrite, is a Calcium Magnesium Iron Silicate that measures 6½ on the Mohs scale. In some respects, the chemical and crystalline structure of Jadeite more closely resembles that of Spodumene gemstones such as kunzite and hiddenite than nephrite jade. Jadeite superficially resembles nephrite in that it is a tough, opaque to translucent aggregate, often green in color, that is cut and polished into CABOCHONS, beads, or carvings. It also entered the deeply entrenched Chinese jade culture in the eighteenth century, originally imported as a particularly fine variety of "jade." But Jadeite is much rarer than nephrite and has a number of physical properties that today make it much more valued as a gemstone.

The term "Jadeite," derived from "jade," stems from the time of the Spanish conquest in Mesoamerica. The word "jade" acquires its name from the Spanish *piedra de ijada*, "stone of the flank," referring to the area of the kidney. Jade was alleged to have therapeutic powers when laid next to an affected area, or when worn as jewelry, or when ground into powder and drunk with water. On the other side of the world in China, the name *yu* was used to refer to nephrite and Jadeite. In Chinese *yu* means "the most beautiful stone" or "royal gem," but this term has not been generally accepted in the Western world.

Jade has particularly deep roots in two ancient civilizations: Mesoamerica and Asia. In the Mesoamerican culture, Jadeite was known very early, with carvings going back at least as far as 1200 BC to the Olmec civilization. To the ancient civilizations of Mesoamerica—including the Olmec, Aztec, and Mayan—Jadeite was valued much more highly than Gold. The Aztec culture restricted Jadeite to persons of high social rank. Jadeite production in Mesoamerica continued up to the coming of the Spanish conquest in the early sixteenth century, after which it came to a virtual halt. The industry saw a revival in the mid-1970s with the discovery of new deposits.

In China, as in Mesoamerica, jade was valued much more highly than Gold. Early "jade" production, going back thousands of years, consisted of nephrite rather than Jadeite. Good-quality Jadeite was not available in China until the mid-eighteenth century, when it was imported from Burma (Myanmar). The Burmese deposits were discovered in the thirteenth century, but commercial export to China did not start until several centuries later.

Although Jadeite is not extremely hard (Mohs 6½–7), it is extremely tough due to its granular, closely interlocking crystals. It is tougher than most other minerals (other than nephrite), so it is extremely resistant to FRACTURE and breakage, and can be finely carved (figs. 204, 207). Semi-transparent, emerald-green imperial jade is the most valuable variety of Jadeite, but lavender Jadeite, pure translucent white Jadeite, and "moss-in-snow" (green-mottled white) Jadeite are also rare and valuable if they have not been artificially colored or treated.

The amount of artificial treatment is also a factor in determining the value of Jadeite, with treatment "grades" or "types" ranging from A to C. Grade A Jadeite is natural Jadeite without any artificial treatment. Grade B Jadeite is natural Jadeite that has been bleached with acid and impregnated with polymer to improve color and transparency. Grade C Jadeite is natural Jadeite that has been dyed to improve color. Some treated jadeite is Grade B + C, which is bleached, treated with polymer, and dyed. Grade A Jadeite is the most durable and stable type of jade, and is also the rarest and most valuable by far.

Jadeite originates in metamorphic rocks that form at high pressure but relatively low temperatures, generally forming as nodular or lens-shaped pieces in serpentine. But few pieces of Jadeite are found IN SITU. They are instead usually found as nodules to small boulders that have weathered out of the source rock. Nodules from rivers produce some of the best gems. Localities for Jadeite include Myanmar, Guatemala, Russia, Kazakhstan, California, and Japan. There are some reports of Jadeite from China, but it is not of high quality. Gem-quality Jadeite used in China has always been imported from Myanmar (formerly Burma), the source of highest-quality Jadeite. The best sources of gem-quality Jadeite today are the mines near the town of Hpakan, in north-central Myanmar.

For further reading about Jadeite, we recommend Sinkankas (1959) and Hughes (1999).

GREEN JADEITE

SYSTEM Inorganic

CLASS Silicate

GROUP Pyroxene

SPECIES *Jadeite*

VARIETIES emerald-green Jadeite (imperial jade); pale green or opaque green Jadeite; moss-in-snow Jadeite; glass jade; opaque blue-green (Olmec blue)

COMPOSITION

Sodium Aluminum Silicate ($NaAlSi_2O_6$), but Mg or Fe may also be incorporated

TRACE ELEMENTS FOR COLOR

Chromium (Cr) for imperial Jadeite, Iron (Fe) for duller green colors

HARDNESS ON MOHS SCALE

6½–7

One of the most highly valued gemstones in the world is a variety of green Jadeite called imperial jade or gem jade. This variety is a rich emerald-green color with a high level of translucency, and, weight for weight, it can be more valuable than high-quality Diamond. The intense green comes from the trace element Chromium, just as in emerald, tsavorite, chrome hiddenite, and chrome Tourmaline. Jadeite is particularly revered in Asia. Although the imperial jade variety of Jadeite is often associated with China, this is a relatively recent historical development. The main commercial source for Chinese Jadeite has always been Myanmar (formerly Burma). Jade has been an important ornamental stone in Chinese culture for about 10,000 years; but prior to the development of trade with Burma in the mid-eighteenth century, the only type of jade available in China was nephrite, also known as Ming jade.

There are several factors that are used in evaluating the quality of imperial jade. The main criteria are the degree of translucency or transparency, luster, and even saturation of emerald-green color without brown or gray tints. Jadeite that approaches complete transparency is sometimes called glass jade or crystal jade, and can be extremely valuable even with uneven or low color saturation (fig. 206). A rare variety of Jadeite called moss-in-snow is a white-colored Jadeite with swirls and splotches of emerald green. This variety can rank close to imperial Jadeite in value.

Jadeite for jewelry is not faceted, even when semi-transparent. It is either finished as a small cabochon, or as beads, small rings, or carvings. The record price for imperial Jadeite was a necklace known as "Doubly Fortunate" consisting of 27 beads ranging from 15.09 to 15.84 millimeters of fine emerald-green color that sold at Christie's in 1997 for $9.3 million.

Pale green, opaque green, and dark green Jadeite with dark mottling or heavy inclusions is less valuable than other types of Jadeite. This material is used less often in jewelry, and more often in carvings of small statues and ceremonial tools, like nephrite. Occasionally white Jadeite is dyed to resemble imperial jade, but this is easy to detect under magnification because the dye color is concentrated in the minute fractures and cracks rather than being evenly contained within individual crystals. The color of dyed Jadeite is also less stable than that of natural (Grade A) Jadeite.

A deep blue-green variety of jadeite from Central America called "Olmec blue" is extremely rare as translucent gem-quality material, but more abundant as opaque carving stone. Olmec blue was used by the Olmec peoples of Central America for making sculptures, ceremonial tools, and other carvings until about 400 BC. Vast deposits of Olmec blue have recently been discovered in Guatemala.

There are many fine pieces of green Jadeite in the Grainger Hall of Gems and even more in The Field Museum's Malott Hall of Jades.

FIGURE 202.
Above: Green Jadeite boulder from Myanmar, with small window cut into it to reveal the good quality of the olive-green Jadeite inside. Specimen weighs 972 grams and measures 100 × 65 × 60 mm (FMNH M21516).

FIGURE 203.
Right: Green imperial Jadeite cabochon of 41 carats set in a Platinum ring surrounded with 48 Diamonds totaling 4.1 carats. Jadeite cabochon measures 25 × 18 × 10 mm (FMNH H2308).

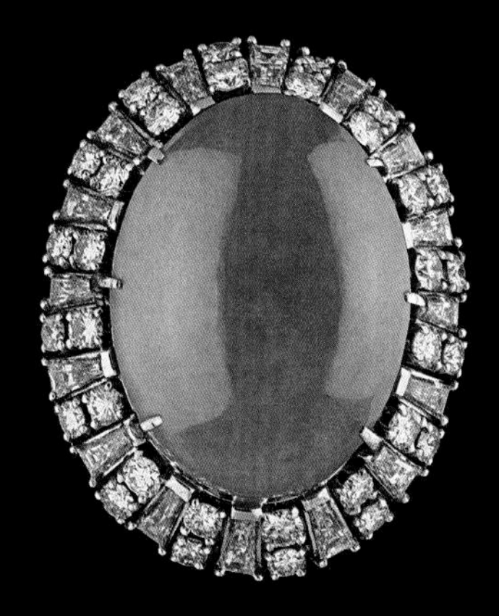

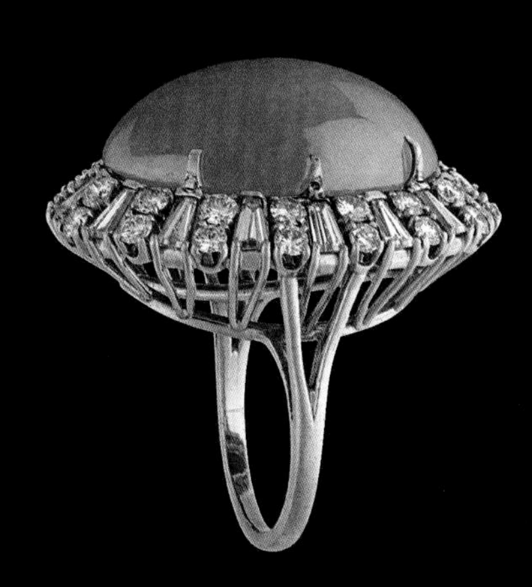

FIGURE 204.
Left: Green imperial Jadeite pin with dragon carving set in 14-karat Gold pin from Hong Kong. Piece is 61 × 21 × 12 mm (FMNH H2367).

FIGURE 205.
Below: Apple-green Jadeite necklace with 115 beads ranging in size from 8.4 to 9.3 mm in diameter (FMNH H2367).

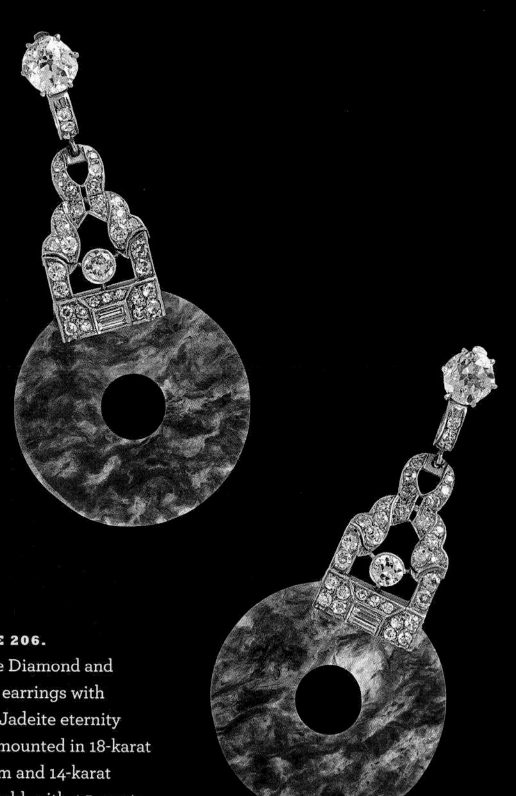

FIGURE 206.
Antique Diamond and Jadeite earrings with "glass" Jadeite eternity circles mounted in 18-karat Platinum and 14-karat white Gold, with 4.5 carats of brilliant-cut and old mine-cut Diamonds. Art deco style from the early twentieth century. Each Jadeite disc measures 26 mm in diameter (FMNH H2382). Jadeite with such transparency and fine color is both rare and extremely valuable.

OTHER COLORED JADEITE (OTHER THAN GREEN)

SYSTEM Inorganic

CLASS Silicate

GROUP Pyroxene

SPECIES *Jadeite*

VARIETIES white Jadeite (crystal jade or ice jade); lavender Jadeite (lavender jade); red Jadeite (konpi); yellow Jadeite (chicken Jadeite); black Jadeite

COMPOSITION

Sodium Aluminum Silicate ($NaAlSi_2O_6$), but Mg or Fe may also be incorporated

TRACE ELEMENTS FOR COLOR

Iron (Fe) for red and yellow, Manganese (Mn) for lavender

HARDNESS ON MOHS SCALE

6½–7

FIGURE 207.
Facing page: White Jadeite bodhisattva statue from early twentieth-century China. The piece is 300 mm in height (FMNH A232696).

There is a wide range of colors of Jadeite other than green, and some of them are highly valued. As with the green Jadeite, the most valuable source of other colored Jadeite is Myanmar. Although the most valuable color of Jadeite is green, Chinese culture also highly values translucent white Jadeite. White Jadeite tends to be the purest form, with little or no trace elements adding color.

Another rare, highly desirable variety of Jadeite is lavender Jadeite, also called lavender jade. It ranges from pastel blue-violet to pink-violet. The lavender color comes mostly from traces of Manganese. Lavender Jadeite comes primarily from Myanmar and can be very expensive.

Red Jadeite is commercially mined only in Myanmar, although it is also known in Guatemala. The red color comes from Iron Oxide and is often a product of the Iron-rich soil it is found in. The red color in red Jadeite is rarely evenly saturated throughout a stone, and the intensity of color is variable. This variety is significantly less valuable than imperial, moss-in-snow, lavender, translucent white, or "glass" Jadeite, but the most valuable red stones are translucent with rich, evenly saturated color, resembling fine carnelian.

Yellow Jadeite is not a popular color of Jadeite for gems and jewelry manufacture. Sometimes referred to as chicken Jadeite due to its color, it is occasionally made into beads and combined with other varieties of Jadeite to make multicolor necklaces. Similarly, black Jadeite is not often used in the fine-gem trade. It has often been confused with black nephrite because of its similar appearance. Unlike black nephrite, some black Jadeite can be polished to exhibit star-like flashes of reflected light due to its particular crystalline structure. Some black Jadeite is actually extremely dark green, opaque Jadeite. Some multicolored Jadeite is very valuable when several colors are present in a single finished stone. For example, gems containing red, green, lavender, yellow, and white in the same stone are very valuable. The Chinese name for this Jadeite variety is "Wufu linmen."

There are many fine pieces of white Jadeite on exhibit in The Field Museum's Malott Hall of Jades (fig. 207), and the Grainger Hall of Gems shows fine examples of lavender, red, orange, and black Jadeite (figs. 208, 209). Jadeite is a beautiful stone for jewelry and fine carvings, but its greatest popularity by far is in Asia.

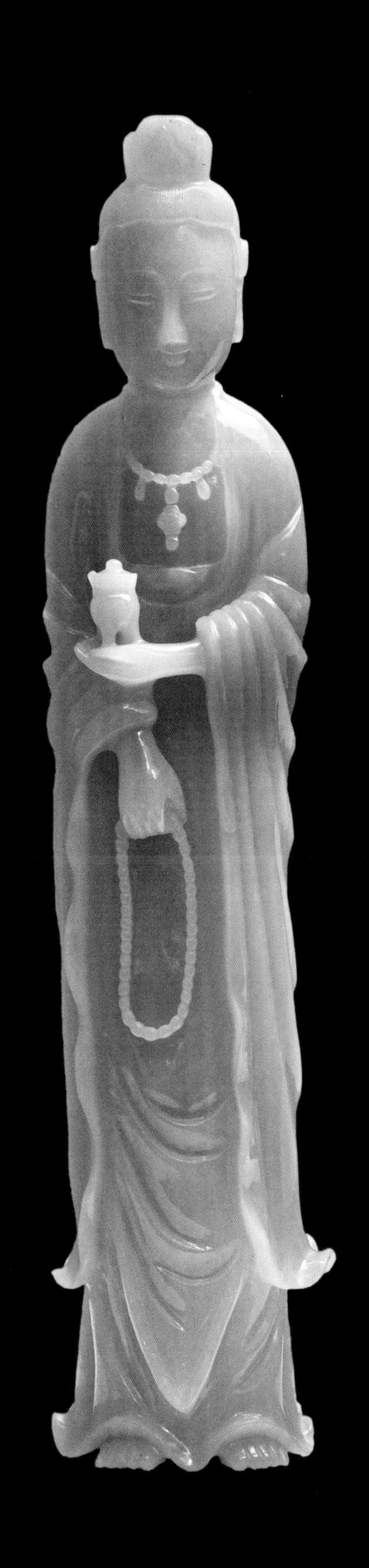

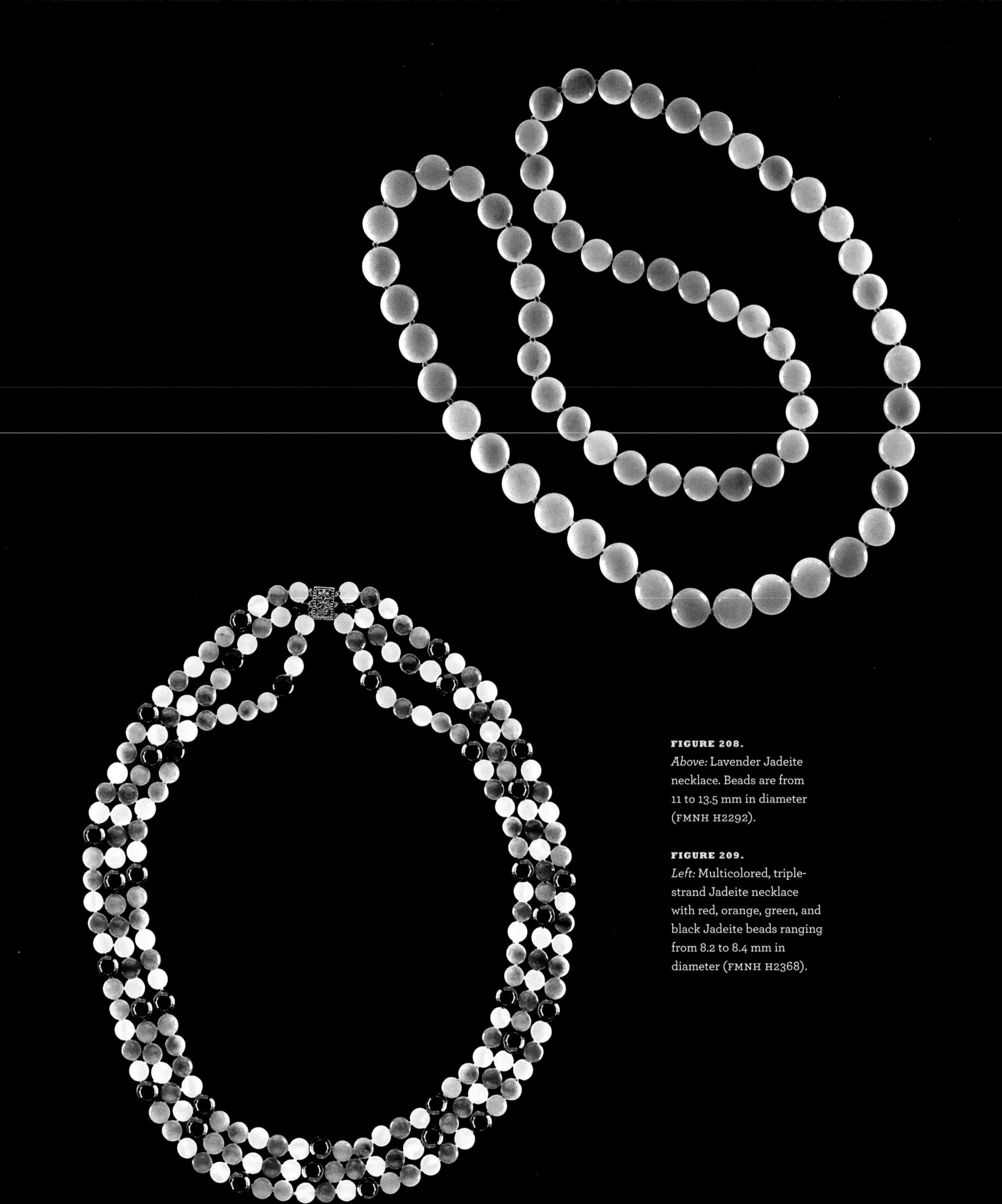

FIGURE 208.
Above: Lavender Jadeite necklace. Beads are from 11 to 13.5 mm in diameter (FMNH H2292).

FIGURE 209.
Left: Multicolored, triple-strand Jadeite necklace with red, orange, green, and black Jadeite beads ranging from 8.2 to 8.4 mm in diameter (FMNH H2368).

Actinolite (or Ferro-Actinolite-Tremolite)

SYSTEM Inorganic

CLASS Silicate

GROUP Amphibole

SPECIES *Actinolite*

GEM VARIETIES nephrite; cat's-eye Actinolite

The peculiar molecular structure of Actinolite causes it to form long, thin crystals, sometimes so thin that they give the stone a fibrous texture. In nephrite, these fibrous crystals interlock making it tougher and stronger than steel. In cat's-eye Actinolite, these fibers cause CHATOYANCY, or a CAT'S-EYE EFFECT. In other forms of Actinolite where the fibers are unconsolidated, the fibers can even form asbestos, which can cause lung damage. There is no such health danger from nephrite, cat's-eye Actinolite, or transparent Actinolite.

The most popular and historically significant varieties of Actinolite in the gem industry are the varieties of nephrite, sometimes known as nephrite jade. Although Actinolite also occurs in other forms, such as transparent Actinolite and cat's-eye Actinolite, neither approach the great popularity of nephrite. The name Actinolite is derived from the Greek *aktinos*, meaning "rays," alluding to the occurrence of Actinolite in nature as fibrous aggregates of crystals radiating from a common central point in the host rock. Actinolite [$Ca_2(Mg,Fe)_5Si_8O_{22}(OH)_2$] is a mineral species that consists of a mix between two other mineral species: ferro-Actinolite [$Ca_2Fe_5Si_8O_{22}(OH)_2$] and tremolite [$Ca_2Mg_5Si_8O_{22}(OH)_2$]. The various blends of these two minerals are collectively referred to as Actinolite. It should logically be referred to as a "super-species" or "series" because it contains two species itself (Back and Mandarino 2008), but Actinolite is an old and entrenched name, and is accepted as a valid mineral species for reasons of tradition and by consensus of metamorphic petrologists. Tradition sometimes wins over logic, even in science.

NEPHRITE

SYSTEM Inorganic

CLASS Silicate

GROUP Amphibole

SPECIES *Actinolite*

GEM VARIETIES green nephrite; white nephrite; yellow nephrite; black nephrite

COMPOSITION Calcium Magnesium Iron Silicate [$Ca_2(Mg,Fe)_5Si_8O_{22}(OH)_2$]

TRACE ELEMENTS FOR COLOR Chromium (Cr) and Iron (Fe)

HARDNESS ON MOHS SCALE 6–6½

Nephrite is steeped in ancient history. Once a gem of nobility, it is today the "affordable" jade and much more common than Jadeite jade. In the twentieth century, immense nephrite deposits were developed in Canada, where nephrite boulders weighing several tons can be found. As a result, the market became fairly saturated with high-quality nephrite, and to this day its value remains relatively low. Although less valuable than Jadeite, nephrite's historical value is no less rich, especially in China, where nephrite was the only type of jade known for thousands of years. Nephrite, like the later imported Jadeite, was known as *yu*, or "most beautiful stone," in ancient Chinese culture. Confucius praised nephrite as a symbol of righteousness and knowledge. The less colorful boulders of nephrite also had great practical uses in Chinese culture due to the supreme toughness of the stone; it was used to make hammers, axheads, and anvils. Although nephrite is generally softer than Jadeite (Mohs 6–6½ vs. Mohs 6½–7), it is tougher and is in fact the toughest known natural mineral. It forms in igneous and metamorphic rocks, where its tiny fibrous crystals interlock, making it even stronger (tougher) than steel. Nephrite is so tough

Nephrite

FIGURE 210.
Natural green botryoidal nephrite boulder from Santa Cruz, California. Piece weighs 1,863 grams and measures 160 × 90 × 70 mm (FMNH H1736).

that it cannot be chiseled, but must instead be ground using abrasive sand and points of Corundum or Diamond. Nephrite jade was mined in the Kunlun Mountains of northwestern China at least as far back as 5000 BC. Historically, green nephrite has sometimes been referred to as greenstone, particularly in early Mesoamerican cultures. (Some Central American Jadeite has historically also been referred to as greenstone.)

Less than 1/20th of 1 percent of nephrite is gem-quality material, and most of that is green. The most valuable nephrite for jewelry is the apple-green nephrite jade with a rich color resembling that of imperial jade due to the presence of Chromium. Nephrite is somewhat softer and lacks the glassy VITREOUS LUSTER of Jadeite. As with the much more expensive Jadeite, the best-quality green nephrite has even color saturation and is at least semi-translucent, although nephrite is not as transparent as the best-quality Jadeite. Nephrite, like Jadeite, makes a fine carving material. Because of its toughness, fine details of nephrite statues and other carvings are highly resistant to breakage and fracture (e.g., figs. 4, 211–213). Other varieties of green nephrite include spinach jade, green and white nephrite, and dark green nephrite.

Green jade used for jewelry and other ornamental purposes in China was exclusively nephrite beginning around 6000 to 5000 BC until the mid-eighteenth century, when Jadeite began to be imported from Burma (now Myanmar). The immense deposits of high-quality green nephrite developed in British Columbia during the last century have kept the supply of gem-grade material at an inexpensive price, with high-quality apple-green material selling for $100 to $200 per pound. Green nephrite is the official gemstone of British Columbia. Green nephrite may have even color saturation or may be blotchy or banded.

In China milk-white nephrite is the most highly valued nephrite. It is used

FIGURE 211.
Top: Green nephrite dragon boat carving from China, probably early to mid-nineteenth century, measuring 236 mm in length (FMNH A82613).

FIGURE 212.
Bottom left: Ceremonial knife from eighteenth-century Jaipur, India, with jeweled green nephrite handle including Pearls, emeralds, and rubies. Knife measures 359 mm in length (FMNH A259380).

FIGURE 213.
Middle right: Carved white nephrite cup from the Ming dynasty (fourteenth to seventeenth century), China. Cup is 102 mm high (FMNH A183337).

primarily for fine carvings, has a greasy-looking luster, and is sometimes known as "mutton fat" jade, a name that is not used by the Western gem industry for marketing reasons. White nephrite with swirls of emerald green is also a highly valued form of nephrite, although not nearly as valuable as moss-in-snow Jadeite, which has a similar color and pattern.

Black nephrite is mined primarily in Australia, New Zealand, and western North America. It is often pure black but occasionally has some light mottling. It sometimes shows slight greenish undertones under bright light. It is used primarily for inlay and cabochons, and usually sells for under $10 per pound. Grayish white jade has little value as a gem or ornamental stone, and yellow nephrite is similarly of little value.

Nephrite is found in China, British Columbia, Russia, New Zealand, Australia, Wyoming, Mexico, Brazil, Taiwan, Zimbabwe, Italy, Poland, Switzerland, and Germany. There are many fine pieces of carved nephrite in the Grainger Hall of Gems and in The Field Museum's Malott Hall of Jades.

For further reading on nephrite, we recommend Sinkankas (1959) and O'Donoghue (2006).

CAT'S-EYE ACTINOLITE

SYSTEM Inorganic

CLASS Silicate

GROUP Amphibole

SPECIES *Actinolite*

GEM VARIETIES green cat's-eye Actinolite; yellow cat's-eye Actinolite; black cat's-eye Actinolite

COMPOSITION Calcium Magnesium Iron Silicate $[Ca_2(Mg,Fe)_5Si_8O_{22}(OH)_2]$

TRACE ELEMENTS FOR COLOR Iron (Fe), Chromium (Cr)

HARDNESS ON MOHS SCALE 6

Other than nephrite, there are very few gemstone varieties of Actinolite. Transparent Actinolite, which is most often green, is extremely rare and a collector's stone, but it is not normally used in jewelry and not generally considered to be a gemstone. Cat's-eye Actinolite is occasionally used in jewelry (also known as cat's-eye jade or cat's-eye nephrite), where it is cut as a cabochon. It occurs in green, yellow, brown, and black, but it is the green variety that is most used in jewelry. Cat's-eye Actinolite is an inexpensive gem, with fine-quality gems selling for under $12 per carat wholesale. Its relatively low value is probably due to its softness (Mohs 6), making it less than optimal for use in rings or other jewelry subject to normal wear.

Cat's-eye Actinolite is most frequently mined in Russia, but it is also known from Taiwan, Canada, Tanzania, Madagascar, and the United States.

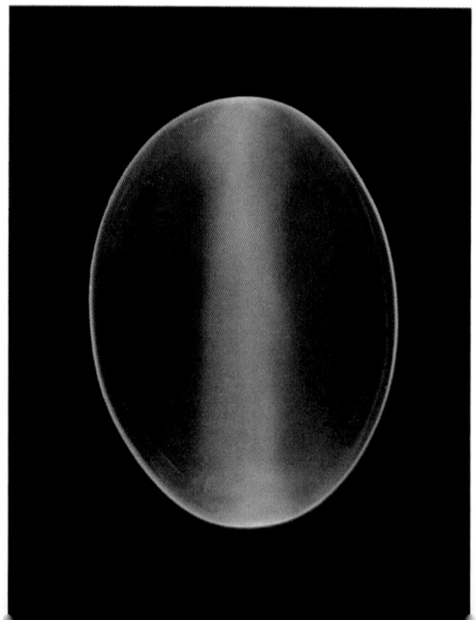

FIGURE 214. Cat's-eye Actinolite cabochon from Russia. Gem is 16 carats in weight and measures 20 × 14 × 5 mm (FMNH H2543).

Zoisite

SYSTEM Inorganic

CLASS Silicate

GROUP Epidote

SPECIES *Zoisite*

GEM VARIETIES tanzanite (blue-violet); thulite (pink); anyolite (green)

The mineral Zoisite was first discovered in 1805, in the Sau-Alp Mountains of Austria. It was originally called saualpite, after the locality in which it was discovered, but eventually it became known as Zoisite, after the scientist who first reported it, Sigmund Zois. There are three varieties of Zoisite gems: tanzanite, thulite, and anyolite. Of these, only tanzanite regularly occurs as a transparent stone that is of faceting grade, and only tanzanite has significant value in the fine-gem market. Thulite and anyolite are usually opaque, irregular-shaped rocks that are used for making carvings or cabochons for inexpensive jewelry. Transparent Zoisite also occurs rarely as colorless, yellow, brown, or green, but these varieties are not used as gemstones because of scarcity, small crystal size, or for aesthetic reasons.

Transparent Zoisite is also strongly pleochroic. Some tanzanite, for example, can appear blue in one direction, red in another, and green in a third. More commonly it is blue in one direction and purplish-blue in other directions.

BLUE-VIOLET ZOISITE (TANZANITE)

SYSTEM Inorganic

CLASS Silicate

GROUP Epidote

SPECIES *Zoisite*

VARIETY tanzanite

COMPOSITION Calcium Aluminum Hydroxysilicate [$Ca_2(Al,OH)Al_2(SiO_4)_3$]

TRACE ELEMENT FOR COLOR Vanadium (V)

HARDNESS ON MOHS SCALE 6–6½

Tanzanite is another relative newcomer to the fine-gem market. Discovered in 1967 in the foothills of Mount Kilimanjaro, it is named after its country of origin, Tanzania. According to local sources, the blue gem was first found by Masai cattle herders and was kicked up by the feet of moving cattle. Sadly, the days of finding tanzanite close to the earth's surface are long gone as it only occurs in a few localities, which have been mined extensively.

Few stones match the beauty of the highest-grade tanzanite, which comes partly from its TRICHROIC optical properties: it radiates three different colors, each depending on the angle of view. This provides an optical mix of blue, violet, and burgundy in the best-quality stones. There is also a slight color change between natural and incandescent light. Most tanzanite gems appear more amethyst-violet under incandescent lights, and ultramarine to sapphire blue in natural light. Natural crystals of tanzanite sometimes have regions or streaks of brown or yellowish tints. These less desirable tints are sometimes removed through heat treatment of 750°–930°F (400°–500°C). This treatment also deepens the blue color of the stone, and the color enhancement is usually permanent. Transparent tanzanite is faceted for use in jewelry, but there is also a rare cat's-eye form of tanzanite that is made into cabochons.

Tanzanite is known only from small deposits in Tanzania and neighboring Kenya, and it is more than a thousand times rarer than Diamond. Its great beauty and rarity make it an expensive gem, although its relative softness (Mohs 6–6½) compared to other precious gems keeps the price of fine tanzanite far below that of fine Diamond. Tanzanite, like tsavorite Garnet discussed above, was first promoted by Tiffany & Co. Much like tsavorite, Tiffany coined the name

FIGURE 215.
Right: Natural crystal of tanzanite from the Umba Valley, Tanzania, weighing 1,263.8 carats and measuring 95 × 55 × 30 mm (FMNH H2339).

FIGURE 216.
Above: Two faceted tanzanite gems weighing 8.4 and 33.3 carats from the Umba Valley, Tanzania (FMNH H2344, H1664).

FIGURE 217.
Facing page: Tanzanite necklace with a 37-carat rounded square brilliant-cut tanzanite from the Umba Valley, Tanzania, surrounded by Diamonds and set in 18-karat white Gold. Pendant measures 27 mm in diameter. Designed by C.D. Peacock (FMNH H2358).

for tanzanite because they thought it sounded better than blue Zoisite. It was thought that the name Zoisite sounded too much like the word "suicide."

Tanzanite mining today is becoming more and more difficult. The limited commercial mining regions are requiring deeper and deeper (and more dangerous) mining. The supply of good-quality stones, especially large stones, is dwindling. The price of fine tanzanite has been climbing as a result of both the dwindling supply and controlled market conditions. Currently about 50–60% of all tanzanite reserves are owned by TanzaniteOne, a company that is taking a lesson from the De Beers marketing plan for Diamonds: tight control of supply and distribution. The current price of fine tanzanite can range from hundreds to thousands of dollars per carat, depending on quality and size of stone.

The world's largest known tanzanite crystal is a raw specimen weighing 16,839 carats, found in 2005. In the Grainger Hall of Gems, there is an exceptionally fine 37-carat faceted tanzanite set in an 18-karat white Gold pendant designed by C.D. Peacock company in 2006 (fig. 217). There is also a natural crystal of 1,263.8 carats in the hall (fig. 215).

PINK ZOISITE (THULITE)

SYSTEM Inorganic

CLASS Silicate

GROUP Epidote

SPECIES *Zoisite*

VARIETY thulite

COMPOSITION Calcium Aluminum Hydroxysilicate $[Ca_2(Al,OH)Al_2(SiO_4)_3]$

TRACE ELEMENT FOR COLOR

Manganese (Mn)

HARDNESS ON MOHS SCALE

6–6½

FIGURE 218.
Natural piece of thulite rough from Leksvik, Norway, weighing 376 grams and measuring 101 × 51 × 51 mm (FMNH M3105).

FIGURE 219.
Facing page: Eight polished cabochons of thulite (from unknown locality) set in Silver necklace made in Bangkok, Thailand. Each cabochon measures approximately 10 × 14 mm (FMNH H2514).

Based on outward appearance, it is hard to believe that thulite has much kinship with the gem tanzanite, but they are in fact two varieties of the same species: Zoisite. Unlike tanzanite, thulite (also known as rosaline or unionite) is usually an opaque massive pink rock, often mottled with white. It is used for small carvings, ornamental inlay, cabochons, and beads, and is very inexpensive. There are extremely rare examples of transparent faceting-grade pink thulite, but these are exceedingly rare, even rarer than tanzanite. One such faceted stone exists in the Smithsonian Institution.

Thulite was first described from Telemark, Norway, in 1820. It is often found as veins and fracture fillings in other types of rock. It is named after Ultima Thule, the ancient Greek name for Norway, and it is the national stone of Norway. Thulite is also found in Western Australia, Namibia, Italy, Austria, and the United States (Washington and North Carolina).

GREEN ZOISITE (ANYOLITE)

SYSTEM Inorganic

CLASS Silicate

GROUP Epidote

SPECIES *Zoisite*

VARIETY anyolite

COMPOSITION Calcium Aluminum Hydroxysilicate $[Ca_2(Al,OH)Al_2(SiO_4)_3]$

TRACE ELEMENT FOR COLOR Chromium (Cr)

HARDNESS ON SCALE 6–6½

FIGURE 220.
Natural piece of anyolite from Tanzania weighing 997 grams, measuring 162 × 118 mm. The green part of the stone is Zoisite, and the red inclusions are rubies (FMNH M18931).

FIGURE 221.
Facing page: Two cabochons and carving of two frogs made from Tanzanian anyolite. Cabochons are 11.94 and 12.3 carats, and the carving is 267.5 carats and measures 68 × 38 × 25 mm. The green is Zoisite, and the red inclusions are rubies. (Cabochons are FMNH H2511, H2512; carving is FMNH H2509.)

Green Zoisite is rarely found in a pure form. It is instead most often found as the principal component of an opaque, multicolored metamorphic rock called anyolite. Anyolite is a green Zoisite intergrown with ruby and black hornblend (Tschemakite). The mix of green, red, and black makes an attractive stone for carving (figs. 220, 221).

It is also used to make cabochons and beads for inexpensive jewelry. Diamond-tipped tools must be used to carve this material because of the presence of ruby, which makes it extremely hard. The Longido area of Tanzania supplies a large quantity of this material to world markets, so it is currently relatively inexpensive. It was first discovered in 1954 in Tanzania, and its name is derived from the Masai word for "green." Today anyolite is found in both Tanzania and Kenya.

Forsterite (Olivine)

SYSTEM Inorganic

CLASS Silicate

GROUP Olivine

SPECIES *Forsterite*

GEM VARIETIES

peridot (green Olivine); chrysolite (yellow to yellowish-green Olivine)

Gems from outer space! Although this sounds like some sort of tabloid headline, faceting-grade Olivine has been recovered from certain types of meteorites (fig. 225). These meteorites are classified in a meteorite family called Pallasites, named after the nineteenth-century scientist who described them, Peter Simon Pallas. Faceted stones of Olivine from meteorites are extremely rare and usually less than 1 carat in weight, but have been known since the beginning of the twentieth century.

Olivine is not a specific mineral species; it is instead a group of minerals that are combinations of two different mineral species: Fayalite (Fe_2SiO_4) and Forsterite (Mg_2SiO_4), and by convention, combinations more than half Forsterite are called Forsterite and more than half Fayalite are called Fayalite. The gem varieties, peridot and chrysolite, are both mostly Forsterite and are therefore classified within the species Forsterite. The name Olivine is derived from the olive-green color that is common for the peridot variety of this species. First described in the late eighteenth century, Olivine is one of the most common minerals on earth, but large gem-quality stones are very rare. Historically, there have been mixed and confusing usages of the names Olivine, peridot, and chrysolite. Some authors use these three interchangeably, while others have used them to distinguish a color variety. We are promoting a less ambiguous, more practical use of the available names, so here we recognize "peridot" as the green variety and "chrysolite" as the greenish-yellow variety of Olivine. This convention follows earliest usage, where the name chrysolite meant "golden stone" (Greek *chryso*, "gold," + *lithos*, "stone").

Mineralogically, the name Olivine refers to a large group of minerals that also includes many non-gem varieties, but in this book we focus primarily on gemstones. Rich green peridot is currently more costly than yellow chrysolite, even though true yellow chrysolite without the greenish hue is much scarcer than green peridot, and even most chrysolite is a greenish-yellow rather than pure yellow. Olivine occurs mainly in IGNEOUS rocks such as basalt and gabbro, although a Magnesium-rich variety is also found in meteorites. For further discussion of Olivine, we recommend Sinkankas (1959) and O'Donoghue (2006).

GREEN OLIVINE (PERIDOT)

SYSTEM Inorganic

CLASS Silicate

GROUP Olivine

SPECIES *Forsterite*

VARIETY peridot

COMPOSITION Magnesium Iron Silicate (Mg_2SiO_4) with more Manganese than Iron

AGENT FOR COLOR Color due to idiochromatic Iron (Fe)

HARDNESS ON MOHS SCALE 6½–7

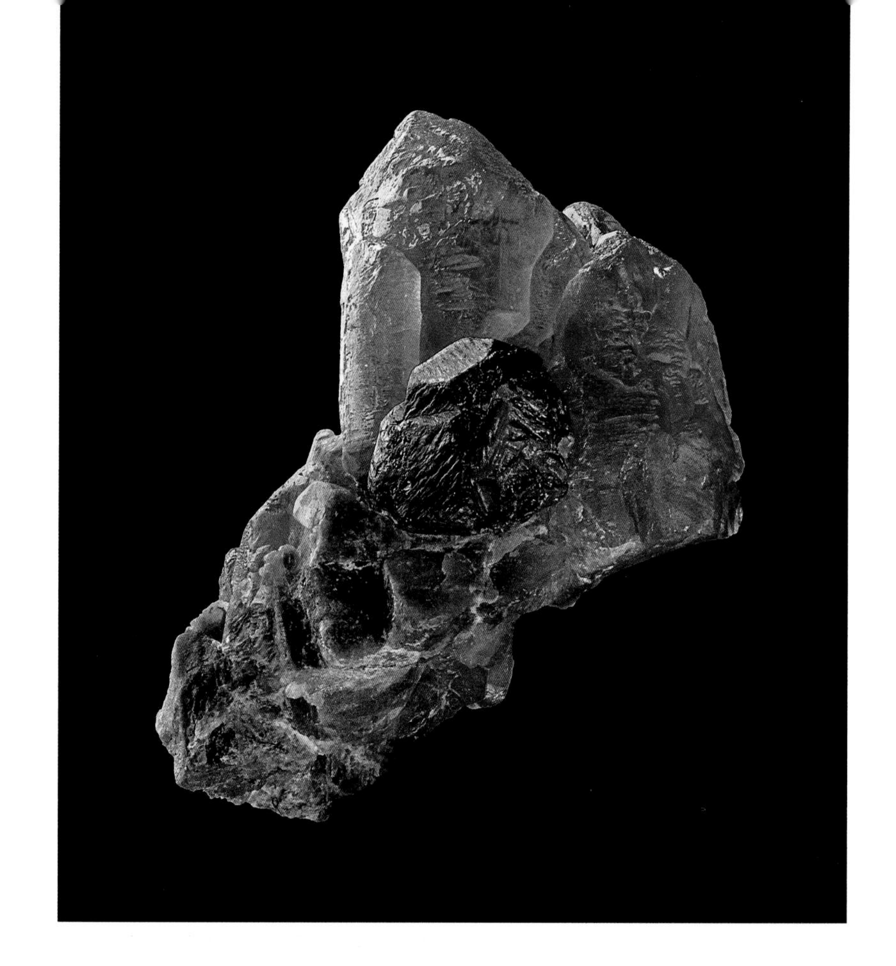

FIGURE 222.
Natural crystals of peridot from Pakistan. Specimen weighs 49.3 grams and measures 45 × 34 × 32 mm (FMNH H2436).

Most gem-quality peridot is terrestrial in origin, occurring in volcanic or intrusive igneous rocks. The word "peridot" is derived from the French *peritot*, "gold," because Olivine can vary from green to gold (although we classify true gold or yellow Olivine here as chrysolite). Peridot and chrysolite have been mined from St. John's Island in the Red Sea for over 3,500 years. The Crusaders brought peridot back to Europe in the Middle Ages from this locality to decorate churches and vestments. Peridot is still mined there today, producing some of the finest known gem-quality pieces. Large fine-quality stones of over 20 carats are rare. In recent times, peridot has become somewhat of a victim of fashion, being much less highly prized today than it was in the past. Ancient Egyptian royalty greatly prized peridot, and some of Cleopatra's "emeralds" were said to be peridots; occasionally fine peridot has been referred to as "evening emerald." Among green gemstones today, the emerald-green hues tend to be much more in demand than the olive-green hues, but most any large eye-clean faceted peridot of 20 carats or more is a rare and beautiful gem. The finest peridot is a rich emerald- to olive-green color with few or no inclusions.

One of the world's finest peridot gems is a flawless 154-carat pear-shaped gem from the classic peridot mines of St. John's Island (fig. 224). It is part of

FIGURE 223.
Faceted peridot gems from St. John's Island, Egypt (large stone), and New Mexico (small stones), weighing from 1.2 to 127.0 carats each (FMNH H1844, H311.1–H311.13, H1837, H1839, H1340).

FIGURE 224.
A marvelous pendant, "Green Goddess," with a flawless, pear-shaped peridot weighing 154 carats. This large faceted gem is surrounded by 3.24 carats of light yellow Diamonds and set in 18-karat yellow Gold, with a hand-carved Gold goddess on the back visible through the stone. Peridot is from St. John's Island, Egypt (Zagbargad). Designed by Lester Lampert in 2008 for the Grainger Hall of Gems (FMNH H2533).

The Field Museum's founding gem collection dating to the 1893 World Columbian Exposition. The stone was set in 2008 by Chicago jewelry designer Lester Lampert in a piece called "Green Goddess," highlighted by 78 light yellow Diamonds in an 18-karat Gold setting. A remarkable carving in 18-karat Gold on the back of the pendant is visible through the peridot.

In addition to St. John's Island, gem-quality peridot is also found in Australia, Brazil, China, Myanmar, Pakistan, South Africa, Norway, and the United States (Arizona, Colorado, Hawaii, and New Mexico). In Hawaii, peridot is found on the beaches in sand.

YELLOW OR GOLDEN OLIVINE (CHRYSOLITE)

SYSTEM Inorganic
CLASS Silicate
GROUP Olivine
SPECIES *Forsterite*
VARIETY chrysolite
COMPOSITION Magnesium Iron Silicate (Mg_2SiO_4) with more Manganese than Iron
TRACE ELEMENT FOR COLOR Color due to idiochromatic Iron (Fe)
HARDNESS ON MOHS SCALE 6½–7

Chrysolite is the earliest known Olivine gem, although there is some confusion about the early history of chrysolite and its sister variety, peridot. Pliny the Elder and other ancients referred to a gem he called "topazos," which is the derivation of the word "Topaz." "Topazos" is a term that in Pliny's day applied to nearly all known yellow gemstones. Today the term "Topaz" is used only for varieties of Aluminum Fluoro-Hydroxy-Silicate (see earlier discussion of Topaz). In retrospect, it appears that the "topazos" that Pliny was referring to may have been chrysolite, because the stones were from St. John's Island (formally called the Island of Topazos). St. John's Island has been one of the world's top producers of gem-grade chrysolite (and peridot) for thousands of years.

The name chrysolite is derived from Greek (*chryso*, "gold," + *lithos*, "stone"). Like peridot, it has an "oily" or GREASY LUSTER and is occasionally found in Pallasite meteorites. It is also found in most or all localities that produce gem-quality peridots, although virtually all stones show a greenish tinge. Occasionally, cat's-eye chrysolite as well as cat's-eye peridot are found, but fine stones are relatively rare. Some of the Olivine crystals found in Pallasite meteorites fall into the chrysolite category. A polished slice from one of the chrysolite-bearing meteorites is in the Grainger Hall of Gems (fig. 225).

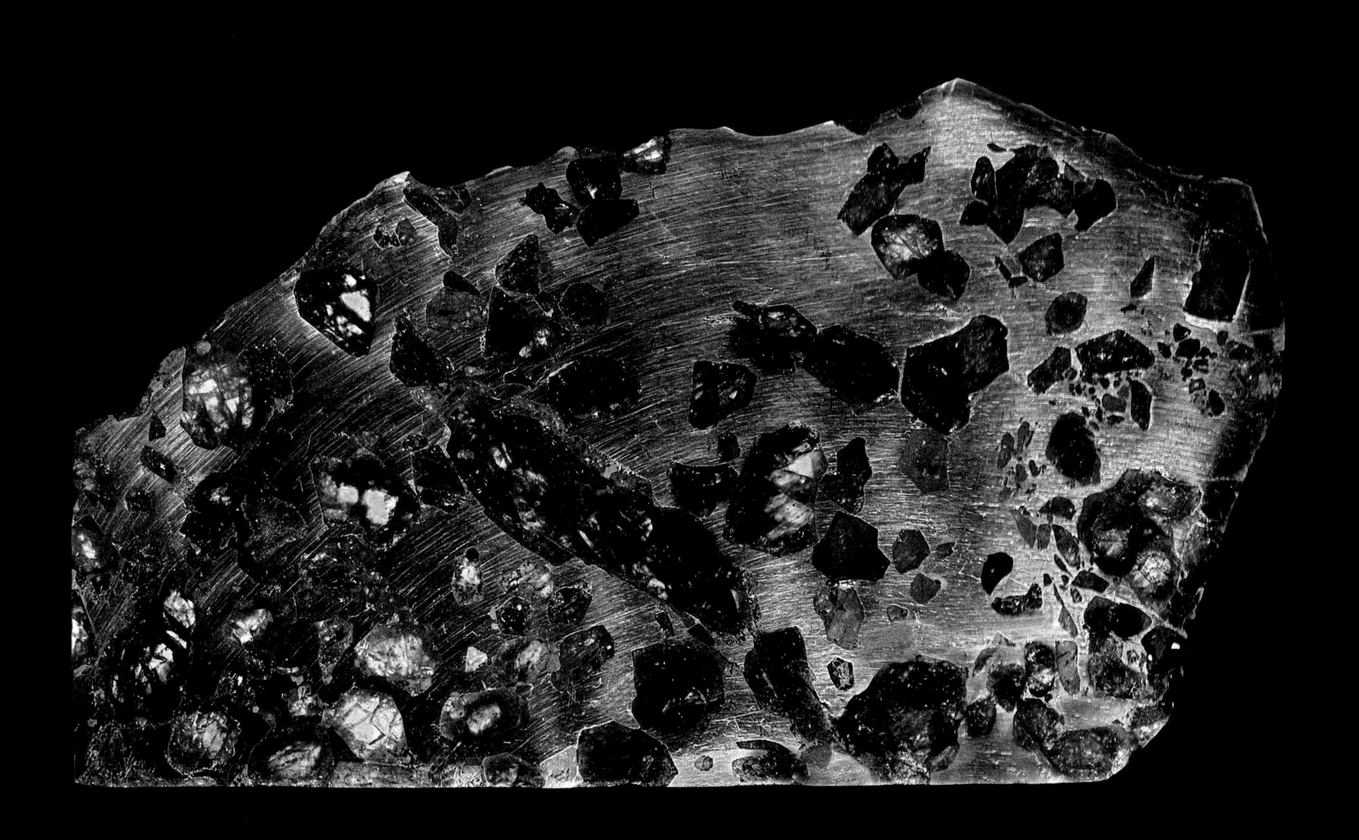

FIGURE 225.
Above: Slice of the Pallasite meteorite named Esquel, which fell in Chubut, Argentina, showing chrysolite inclusions. Slice measures 250 × 113 mm and weighs 363 grams (FMNH Me 4079).

FIGURE 226.
Left: Faceted chrysolite of 5.7 carats (FMNH H1370).

FELDSPAR GROUP

SYSTEM Inorganic

CLASS Silicate

GROUP Feldspar

K-SPAR SPECIES

Orthoclase Feldspar

PLAGIOCLASE SPECIES

Albite Feldspar; *Albite-Anorthite* Feldspar

Feldspar, formerly called "felspar," is the most abundant of all mineral groups and makes up about 60 percent of the earth's crust. There are two subgroups of Feldspars containing gem varieties. One is the K-Spar subgroup (Potassium Feldspars) containing gem varieties within the mineral species Orthoclase. The other is the Plagioclase subgroup (Sodium-Calcium Feldspars) containing gem varieties composed of intermediate combinations of two mineral species that readily combine with each other: Albite ($NaAlSi_3O_8$) and Anorthite ($CaAl_2Si_2O_8$). Consequently, the intermediate Feldspars chemically differ from each other in the proportion of Sodium (Na) to Calcium (Ca), while the K-Spar varieties generally all have the same composition other than trace elements.

The crystal structure of all Feldspars is a blocky one, because the well-defined CLEAVAGE planes occur nearly at right angles. There are many varieties of Feldspar, but here we describe only the few used as gemstones. For industry, Feldspars are used for such things as the production of ceramics and mild abrasive cleaners. Gem varieties of Feldspars are mainly moonstone (one variety of Orthoclase and one of an Orthoclase/Albite mix), sunstone (a variety of oligoclase), and Oregon sunstone and rainbow moonstone (varieties of labradorite). Most Feldspar gemstones share a special opalescent sheen called a SCHILLER EFFECT. Adularia moonstone is currently the most valuable of all Feldspar gems. For further reading on the Feldspar group, we recommend O'Donoghue (2006).

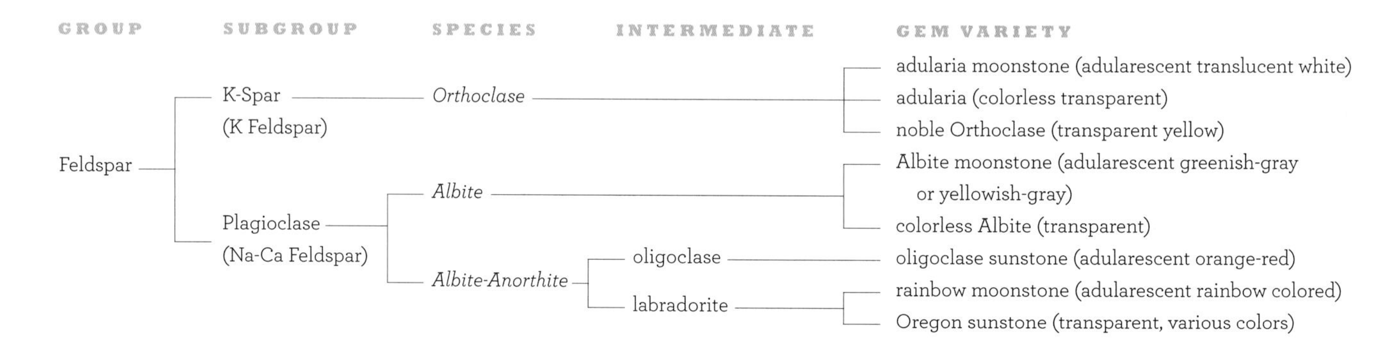

FIGURE 227. Classification of Feldspar gemstones discussed in this book.

Orthoclase Feldspar

SYSTEM Inorganic

CLASS Silicate

GROUP Feldspar

SUBGROUP K-Spar

SPECIES *Orthoclase*

GEM VARIETIES adularia (transparent colorless); adularia moonstone (adularescent translucent white); noble Orthoclase (transparent yellow)

COMPOSITION Potassium Aluminum Silicate ($KAlSi_3O_8$)

TRACE ELEMENTS FOR COLOR None for adularia, Iron (Fe) for noble Orthoclase. Schiller effect of adularia moonstone caused by traces of Albite intergrowths.

HARDNESS ON MOHS SCALE 6

There are three main gem varieties of Orthoclase, the most prized of which is called adularia moonstone. All three varieties exist within the K-Spar subgroup (fig. 227). This variety ranges from translucent to nearly transparent with slightly milky overtones and a beautiful light-blue Schiller effect, or **ADULARESCENCE**, suggestive of moonlight. This light effect is the source of the name moonstone. In India, these gems are called "dream stones" and are claimed to bring the wearer beautiful visions at night. The adularescence is a reflection off of a planar surface inside the gemstone, so care must be taken to cut the stone so the base of the cabochon is oriented parallel to the plane of adularescence. The more transparent stones tend to have tiny inclusions, giving them a slightly turbid appearance. The most valuable stones are those that are near-transparent with minimal inclusions and a clear blue adularescence. Large stones of this quality are extremely rare. Adularia moonstone was extremely popular in art nouveau jewelry of the early twentieth century. It is cut as cabochons to highlight the showy adularescent sheen. There are several high-quality Sri Lankan adularia moonstone cabochons in the Grainger Hall of Gems (fig. 230).

There is also a variety of adularia that is colorless and lacks adularescence (fig. 229). This variety is occasionally faceted but has little value because there are several fine colorless gem varieties that are much harder (e.g., colorless Zircon, colorless Diamond, colorless sapphire, and goshenite).

Adularia gets its name from the Adula Mountains of Switzerland, where this variety was first described and where many fine crystals are found to this day (fig. 228). Adularia and adularia moonstone are also found in Sri Lanka, India, Myanmar, Australia, Madagascar, Tanzania, Brazil, and the United States (Colorado, New Mexico, and Idaho).

Another scarce variety of Orthoclase is the transparent yellow noble Orthoclase. This variety resembles heliodor or light citrine and can be faceted into fine gemstones with **VITREOUS LUSTER**. Most stones are faceted with a simple step-cut because they can be brittle and difficult to cut. Noble Orthoclase comes mainly from Madagascar, where it is found in pegmatites. A rare cat's-eye yellow Orthoclase is also known from Myanmar.

The name Orthoclase derives from the Greek (*orthos*, "straight," + *klasis*, "broken"), referring to the blocky-looking crystals. This characteristic structure is the result of strong cleavage planes that are nearly perpendicular to each other. Although Orthoclase is a common mineral in most igneous and metamorphic rocks, gemstone varieties are relatively rare.

FIGURE 228.
Left: Natural crystals of adularia and Quartz from Tyrol, Switzerland. Specimen is 128 mm in height (FMNH M6816).

FIGURE 229.
Below: Faceted colorless adularia gem from Gotthard, Switzerland, weighing 1.1 carats and measuring 8 × 8 × 4 mm (FMNH H496).

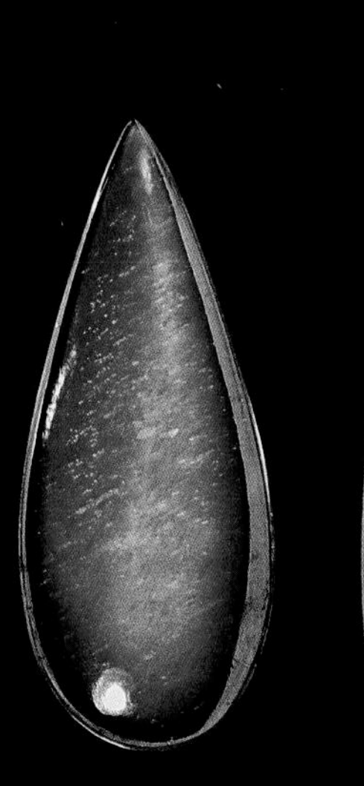

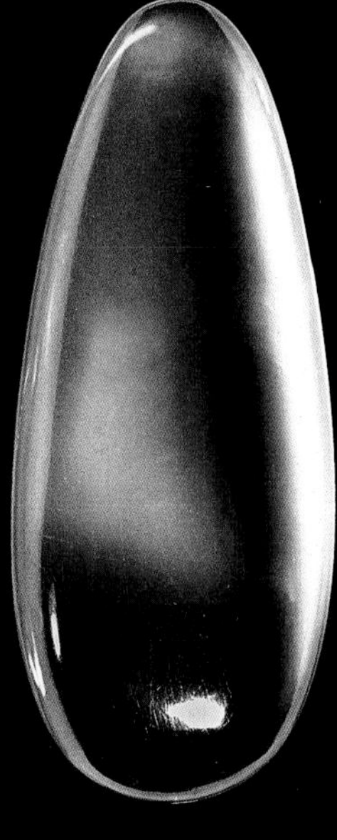

FIGURE 230.
Polished cabochons of high-quality adularia moonstone from Sri Lanka, ranging from 16.5 carats to 54.7 carats in weight. Largest gem measures 47 × 20 × 7 mm (FMNH H493, H2079, H2080).

FIGURE 231.
Left: Natural crystal of faceting-grade noble Orthoclase from Fianarantsoa Province, Madagascar. Specimen is 302.6 carats and measures 44 × 44 × 20 mm (FMNH H2493).

FIGURE 232.
Above: Faceted stones of noble Orthoclase from Madagascar, ranging in weight from 5.6 carats to 19.7 carats (FMNH H497, H1577, H2500–H2502).

Albite Feldspar

SYSTEM Inorganic

CLASS Silicate

GROUP Feldspar

SUBGROUP Plagioclase

SPECIES *Albite*

GEM VARIETIES Albite moonstone (adularescent pale greenish-gray or yellowish-gray); peristerite (cream or pinkish with bluish adularescence); colorless Albite (transparent)

COMPOSITION If pure, Sodium Aluminosilicate ($NaAlSi_3O_8$), but it may contain up to 10% Anorthite (Calcium Aluminosilicate) by definition. May also combine with Orthoclase in some moonstone varieties.

TRACE ELEMENT FOR COLOR

Iron (Fe)

HARDNESS ON MOHS SCALE

6–6½

Albite is usually white in color and occasionally found as colorless faceting-grade material; however, there is virtually no demand for colorless Albite in the gem trade. But Albite does have a variety that is valuable and in high demand: Albite moonstone. One particular variety, peristerite, is a white, cream, or brownish-pink body with a bluish adularescent flash, which comes primarily from Ontario and Quebec. Albite moonstone can also contain a percentage of Orthoclase in its composition, sometimes resulting in a greenish or bluish body color (fig. 233). Albite moonstone is very rare, and it is often simply lumped into a generic moonstone category with adularia moonstone, but despite the often similar appearance of the two types of adularescent gems, they belong to different mineral species, and we maintain that classification here. Albite moonstone is found in Canada, Kenya, Sri Lanka, and India. One very significant piece of jewelry in the Grainger Hall of Gems is a 408.6-carat cabochon of fine Albite moonstone set in an 18-karat yellow Gold and Platinum bracelet named "Lightning," designed by Paloma Picasso for Tiffany & Co. (fig. 234). Paloma Picasso is a famous jewelry designer in her own right and the daughter of the famous artist Pablo Picasso.

The most common form of Albite is a white or colorless stone. It is often the base matrix associated with other types of crystals in pegmatites because it is often the last mineral species to crystallize in a pegmatitic group of crystals (see fig. 11). The SCHILLER EFFECT in adularia moonstone is caused by traces of Albite intergrowths within the Orthoclase crystal.

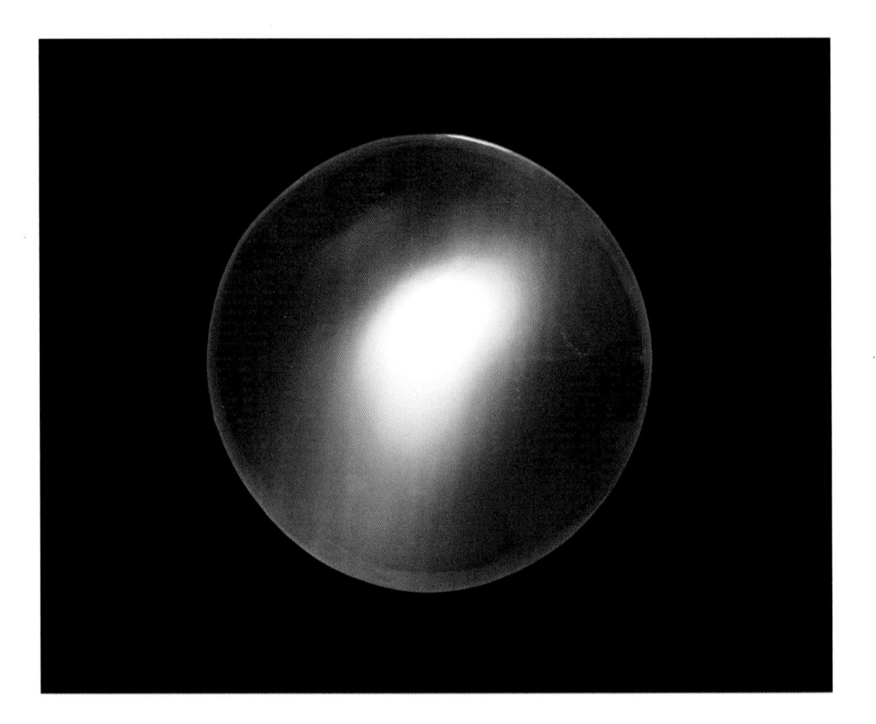

FIGURE 233.
Right: Albite moonstone cabochon of 303.6 carats from India. Cabochon measures 37 × 37 × 27 mm (FMNH H1716).

FIGURE 234.
Facing page: A 408.6-carat Albite moonstone from India set in 18-karat yellow Gold and Platinum. "Lightning" bracelet designed by Paloma Picasso for Tiffany & Co. The moonstone measures 50 × 34 × 34 mm (FMNH H2247).

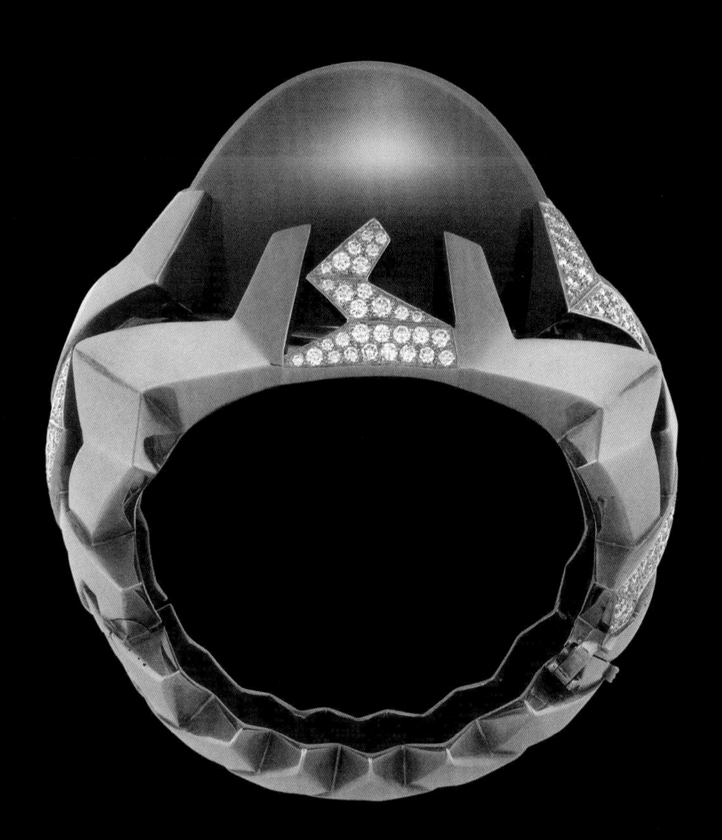

Albite-Anorthite Feldspar

SYSTEM Inorganic

CLASS Silicate

GROUP Feldspar

SUBGROUP Plagioclase

SPECIES *Albite-Anorthite*

GEM VARIETIES oligoclase sunstone (adularescent orange-red oligoclase); rainbow moonstone (labradorite); Oregon sunstone (transparent labradorite)

The Plagioclase gem varieties include Albite (discussed in the previous section) and a number of additional gem varieties that are simply different proportional combinations of the species Albite and Anorthite. These two species chemically combine in nature forming a continuous series of mineral types (intermediates) ranging from pure Albite to pure Anorthite (see fig. 235). Albite can contain up to 10 percent Anorthite and still be considered Albite just as Anorthite can contain up to 10 percent Albite and still be considered Anorthite. The combinations between these two end components have other named intermediates, including two that include gemstones discussed here: oligoclase and labradorite.

Oligoclase includes oligoclase sunstone (an adularescent orange-red variety of Orthoclase); and labradorite includes rainbow moonstone (classical opaque multicolored labradorite) and Oregon sunstone (many varieties of transparent labradorite). Although colorless andesine and bytownite are also occasionally faceted as collector stones, they are not generally used in the gem trade. Some Oregon sunstone occasionally falls into the andesine or bytownite category, but most Oregon sunstone is transparent laboradite.

Mineral Type			*Composition*	
SUBGROUP	SPECIES	INTERMEDIATE	*% Albite*	*% Anorthite*
Plagioclase	*Albite*		100–90%	0–10%
	Albite-Anorthite	oligoclase*	90–70%	10–30%
	Albite-Anorthite	andesine	70–50%	30–50%
	Albite-Anorthite	labradorite*	50–30%	50–70%
	Albite-Anorthite	bytownite	30–10%	70–90%
	Anorthite		10–0%	90–100%

*Contains major gemstone varieties.

FIGURE 235.
Classification of Plagioclase mineral subspecies and their compositions. Plagioclase intermediates fall within a series of Albite-Anorthite blends. Of the six Plagioclase Feldspar species and intermediates listed here, three contain the principal Plagioclase gemstones: Albite, oligoclase, and labradorite (although there are some rare instances of Oregon sunstone falling into the andesine or bytownite category). For details on the gem varieties within these three subspecies, see figure 227.

OLIGOCLASE SUNSTONE

SYSTEM Inorganic

CLASS Silicate

GROUP Feldspar

SUBGROUP Plagioclase

SPECIES *Albite-Anorthite*

GEM VARIETY

oligoclase sunstone

COMPOSITION Sodium Calcium Aluminosilicate [$NaAlSi_3O_8$]-[$CaAl_2Si_2O_8$] with Albite-Anorthite proportions ranging from 7:3 to 9:1. The Sodium comes from Albite and the Calcium from Anorthite.

TRACE ELEMENTS OR COMPOUNDS FOR COLOR

Iron (Fe) or Hematite (F_2O_3) "scales" that reflect orange light close to cleavage plane

HARDNESS ON MOHS SCALE

6–6½

Oligoclase gemstones are found as two main types: colorless faceting-grade material and an adularescent orange-red to yellowish-red opaque to semi-transparent stone called sunstone, Indian sunstone, or aventurine Feldspar. The colorless oligoclase is not generally used for jewelry because, like colorless Orthoclase, it lacks the hardness and finer optical properties of other colorless stones such as colorless Diamond, colorless Zircon, goshenite, and colorless sapphire.

Sunstone, on the other hand, is commonly used in the gem trade. It includes tiny flakes of Hematite or goethite disseminated in parallel orientations throughout the near-colorless body of the stone. This results in a strong Schiller effect. It is generally made into cabochons because it is usually opaque or semi-translucent, and proper orientation of the stone's cabochon dome can maximize the Schiller effect. As with all stones highlighting the adularescence or Schiller effect, the cabochon must be cut so that the back is parallel to the plane of Hematite orientation. Sunstone occurs in igneous and metamorphic rocks of Norway, India, Russia, and Canada.

Oligoclase sunstone, generally referred to simply as "sunstone," is a variety of oligoclase and should not be confused with "Oregon sunstone," which is a variety of labradorite.

RAINBOW MOONSTONE (LABRADORITE)

SYSTEM Inorganic

CLASS Silicate

GROUP Feldspar

SUBGROUP Plagioclase

SPECIES *Albite-Anorthite*

VARIETY rainbow moonstone

COMPOSITION Calcium Sodium Aluminosilicate $[NaAlSi_3O_8]$-$[CaAl_2Si_2O_8]$, with Albite-Anorthite proportions ranging from 3:7 to 5:5. The Sodium is from Albite and the Calcium is from Anorthite.

TRACE ELEMENT FOR COLOR

Iron (Fe), mostly in the form of Iron Oxide Magnetite platelets

HARDNESS ON MOHS SCALE

6–6½

FIGURE 236.
Facing page: Oligoclase sunstone; three raw slabs and a finished cabochon in a Silver pendent. Slabs are from Tvedestrand, Norway, weighing a total of 155.5 carats (FMNH H489), and cabochon is from India, weighing 12 carats (FMNH H2508).

FIGURE 237.
Above right: A natural rough piece of rainbow moonstone ("classic labradorite") from Madagascar. Specimen weighs 3,142.5 carats and measures 123 × 95 × 35 mm (FMNH H2506).

Rainbow moonstone, also called black moonstone, was the first gem variety of labradorite discovered, and it is still the variety most commonly associated with the name labradorite. Its opaque multicolor patterns usually have a predominance of blues, greens, and yellows, but red and orange are also often present. Its rainbow PLAY OF COLOR is unique in its metallic appearance, overlaid by a special type of iridescence called LABRADORESCENCE. It resembles the iridescence of the wings of tropical butterflies or peacock feathers. Rainbow moonstone rarely occurs in well-defined crystals. It instead is generally found as massive blocky pieces in metamorphic and igneous rock.

Rainbow moonstone was first discovered off the coast of Labrador (mainland Newfoundland) and first described by Moravian missionaries working among the Inuit Eskimos, who referred to the stone as "fire rock" or "firestone." Some of the Inuit folklore explains the colorful labradorescence of this gemstone as coming from the northern lights, or stars trapped in the rock. In the late eighteenth century, the Moravians sent quantities of rainbow moonstone to Europe. It became a popular gemstone in Europe during the late eighteenth and early nineteenth centuries and was used for pins, brooches, and other jewelry. Rainbow moonstone is still mined today in Newfoundland, but it is also found in other parts of Canada as well as the Ukraine, parts of the United States, Madagascar, and in Finland, where it is called spectrolite.

OREGON SUNSTONE (TRANSPARENT LABRADORITE)

SYSTEM Inorganic

CLASS Silicate

GROUP Feldspar

SUBGROUP Plagioclase

SPECIES *Albite-Anorthite*

VARIETIES yellow to champagne Oregon sunstone; red Oregon sunstone; green Oregon sunstone; bicolor Oregon sunstone; color-change Oregon sunstone; colorless Oregon sunstone

COMPOSITION Calcium Sodium Aluminosilicate $[NaAlSi_3O_8]$-$[CaAl_2Si_2O_8]$, with Albite-Anorthite proportions ranging from 3:7 to 5:5. The Sodium is from Albite and the Calcium is from Anorthite.

TRACE ELEMENT FOR COLOR Copper (Cu) for Oregon varieties

HARDNESS ON MOHS SCALE 6–6½

FIGURE 238. *Facing page:* Polished cabochons and necklace of rainbow moonstone from Labrador, Canada. Cabochons are 67.4 and 24.8 carats, measuring 57 and 28 mm in height. Necklace cabochons are each approximately 8 mm in height and are set in Silver (FMNH H2074, H2075, H2507).

Oregon sunstone is the complete name for the transparent varieties of labradorite. It comes almost exclusively from only a handful of mine fields in southern Oregon. It is confusing to use the name "sunstone" for transparent labradorite, because this name is already used for a different gemstone that is a variety of oligoclase rather than laboradite (fig. 227). "Oregon" generally precedes the name "sunstone" when referring to transparent labradorite varieties. We use this designation here to reflect current practice in the gem trade (e.g., Drucker 2008).

Oregon sunstone is much more than a single variety of gemstone. It is instead a complex of many varieties, most of which are too rare to yet be clearly promoted commercially. The vast majority of transparent Oregon sunstone gems are yellow, champagne, or orangish to pinkish in color, and some of this ranges to nearly colorless. Beyond this, there are rare varieties, such as red, green, bicolor, tricolor, and even ALEXANDRITIC (color-change varieties that shift from green to red). If Oregon sunstone were ever to be as successfully marketed on a broader level, these rare varieties would probably become much less available and much more valuable than they are today.

Oregon sunstone is a relative newcomer to the gem trade, first appearing in the market in the early twentieth century under the name "Plush diamonds," after the area around Plush, Oregon, where they were first mined. These gems were almost immediately popular in California, and Tiffany & Co. bought up the Oregon claims, intending to develop a broader commercial market for this gem. In the end, they were unable to develop the gem commercially, partly due to insufficient quantities of the most desirable varieties such as red and green. In order to initiate a broad and relatively self-sustaining market demand for new gem varieties, there is a minimum amount of supply required; otherwise there is little chance of sufficient economic return on the cost of mining and promoting a new gem variety. As with certain other gem varieties—such as bixbite, chrome hiddenite, and benitoite—a new gem's extreme rarity can sometimes hurt its chances of being optimally and broadly promoted, which keeps the gem's value from reflecting its scarcity.

Tiffany eventually abandoned its Oregon claims in the mid-twentieth century, and mining continued on a very small scale. Years later there was a renewed interest in development of Oregon sunstone. It was named the official state gemstone of Oregon in 1987, and in 1988 new claims were opened in nearby Burns, Oregon.

The complexity of Oregon Feldspar gems goes beyond the large number of color varieties. Some variants do not even appear to be labradorite, because some have been reported as having more Sodium (Na) and less Calcium (Ca) in their composition, placing them in the mineralogical category of andesine. Some have also been rarely reported with a composition indicating bytownite

FIGURE 239.
Left: Natural crystals of Oregon sunstone from the Dust Devil Mine, southern Oregon. Crystals weigh 25.9, 47.9, and 77.8 carats, and measure 18 × 31, 22 × 29, and 28 × 33 mm (FMNH H2515–H2517).

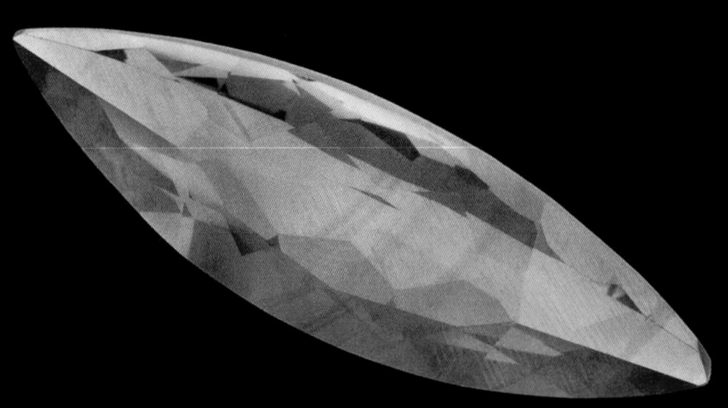

FIGURE 240.
Above: Faceted 3.35-carat red Oregon sunstone from the Dust Devil Mine in southern Oregon. Gem is 27 × 7 × 5 mm (FMNH H2518).

FIGURE 241.
Bottom: Exceptionally fine, faceted 10-carat green-and-red bicolor Oregon sunstone from the Dust Devil Mine of southern Oregon. Gem measures 18 × 13 × 8 mm (FMNH H2518).

FIGURE 242.
Champagne-colored AAA pink Schiller-grade Oregon sunstone from the Dust Devil Mine in southern Oregon set in 14-karat Gold pendant with three small Diamonds. Designed by Karla Proud of Exotic Gems, LLC, in 2007 and named "Schiller Elegance" by the designer (FMNH H2519).

(fig. 235). However, the vast majority of all known transparent Oregon sunstone gems fall into the labradorite category.

The productivity of Oregon sunstone is fairly low, estimated at about 1 pound per ton of dirt; and most of that is not gem-quality material. The most common varieties are still very reasonably priced and under $100 per carat. The rarest varieties, such as large eye-clean ruby reds, can command thousands of dollars per carat. Many Oregon sunstone gems also have a Schiller effect, which is caused by microscopic flecks of Copper aligned in layers inside the stone. This is particularly interesting in transparent stones where the Schiller flashes from only certain orientations but is otherwise invisible. Oregon sunstone is a gemstone that appears to be in transition. With an increase in mining operations, supply, and promotion, some varieties could eventually match tanzanite in popularity. Once the more available varieties become more popular in the gem industry, the value of the rarer varieties may benefit too.

There are some fine examples of Oregon sunstone gems in the Grainger Hall of Gems, including some fine faceted gems (figs. 240, 241) and a pendant designed by Karla Proud of Exotic Gems, LLC (fig. 242).

Benitoite

SYSTEM Inorganic

CLASS Silicate

GROUP Bentoite

SPECIES *Benitoite*

GEM VARIETIES blue Benitoite; colorless Benitoite

COMPOSITION Barium Titanium Silicate ($BaTiSi_3O_9$)

TRACE ELEMENT FOR COLOR unknown, possibly Iron (Fe)

HARDNESS ON MOHS SCALE 6½

Benitoite is an extremely rare and beautiful gem that is another relatively recent discovery, and it is the state gem of California. It is a transparent gem with color hues ranging from sapphire blue to colorless. Benitoite has exceptional optical properties, with particularly strong FIRE rivaling that of Diamond. This fire is somewhat masked by the color of the stone in blue gems and is more apparent in colorless examples. Benitoite forms fine natural flattened, triangular-shaped crystals, usually on a matrix of Natrolite or blue schist.

In 1907 Jim Couch and L. B. Hawkins, two prospectors working in San Benito County, California, each claimed to have discovered this mineral. The blue crystals were at first thought to be sapphire, but this identification was later brought into question because of the distinct crystal form, the strong DICHROISM, and softness relative to Corundum (Mohs 6½ vs. Mohs 9 for Corundum). The stones eventually went to George Louderback (1874–1957), a mineralogist at the University of California at Berkeley, who recognized it as something clearly new. He described it and named it Benitoite, after the place it was first discovered—in the area of the headwaters of the San Benito River. This small locality remains the only source of Benitoite gemstone to this day, and the mines there are becoming exhausted.

Traces of Benitoite have also been reported from Arkansas and Japan, but these deposits are so far only trace deposits that do not contain crystals large enough for gem production. Even in California, most Benitoite gems are very small due to the small size of the natural crystal. Fine faceted stones of over 2 carats are rare and over 3 carats extremely rare, and the largest known faceted Benitoite in the world is a 15-carat gem in a private collection that was exhibited at the Royal Ontario Museum in 2009.

Benitoite—like bixbite, hiddenite, and a few other varieties—is much rarer than its price suggests. It was estimated in 2007 that less than 10,000 carats of Benitoite had been cut and polished in the 100 years since its discovery in 1907. In comparison, there were over 90 million carats of gem-grade Diamonds excavated in the year 2006 alone. The finest Benitoites can sell for thousands of dollars per carat in sizes over 2 carats, but for stones less than a carat, the price is generally under $1,000 per carat. There is a fine bracelet in the Grainger Hall of Gems with a series of faceted Benitoite gems showing color hues ranging from sapphire blue to colorless (fig. 245).

For further reading on Benitoite, we recommend Sinkankas (1959) and O'Donoghue (2006).

FIGURE 243.
Top right: Natural crystals of blue Benitoite together with black neptunite on matrix from San Benito County, California. Specimen is 152 × 102 × 76 mm (FMNH M17541).

FIGURE 244.
Below: Large natural crystal of blue Benitoite 30 mm wide by 15 mm deep, with a few smaller crystals. From San Benito County, California (FMNH M11613).

FIGURE 245.
Right: Faceted Benitoite gems from San Benito County, California. *Right, top:* Bracelet measuring 185 mm in length with a series of 40 round brilliant-cut gems set in 14-karat white Gold, showing the range of Benitoite color from deep blue to colorless (FMNH H2361). *Right, bottom:* A single round brilliant-cut blue Benitoite gem weighing 1.1 carats and measuring 7 × 7 × 4 mm (FMNH H2187).

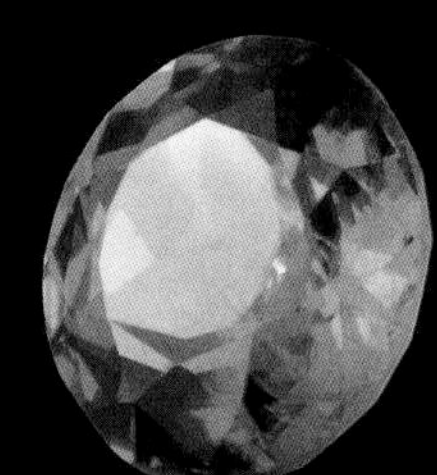

Turquoise

SYSTEM Inorganic

CLASS Phosphate

GROUP Turquoise

SPECIES *Turquoise*

GEM VARIETIES sky-blue Turquoise (Persian); spiderweb Turquoise; greenish Turquoise (Tibetan)

COMPOSITION Hydrated Copper Aluminum Phosphate [$CuAl_6(PO_4)_4(OH)_8{\cdot}4H_2O$]

AGENT FOR COLOR Blue color due to idiochromatic Copper (Cu) with occasional traces of Iron (Fe), for greenish tint

HARDNESS ON MOHS SCALE 5–6

Turquoise is a gemstone of great antiquity, worn by Pharaohs and Aztec kings. The name, derived from sixteenth-century French, means "Turkish stone." This does not refer to Turquoise being mined in Turkey, but rather that Turquoise first came from Persia (now Iran) to Europe by way of trade routes through Turkey. The opaque sky-blue beauty of Turquoise has been revered for millennia. It was mined as far back as 8,000 years ago on Egypt's Sinai Peninsula. Ancient Egyptians buried Turquoise sculptures and jewelry with mummies and also used it for inlay in Gold burial masks, including the iconic burial mask of Tutankhamen. Turquoise was also used by the Aztecs and by Native Americans, including the Zuni, Navajo, and Apache. The Navajo spoke of Turquoise stones as pieces of sky that had fallen to the earth. Aztec and other Native American mining of Turquoise began many centuries ago in what is now the southwestern United States. Turquoise is among the oldest known gemstones that today are still used as gemstones.

Turquoise is almost never found as distinct crystals, but is instead found as a massive opaque secondary mineral occurring in veins through volcanic

FIGURE 246. Natural vein of rough Persian Turquoise running through a rock from Razavi Khorasan Province, Iran, weighing 648 grams and measuring 85 × 90 × 70 mm (FMNH H184).

FIGURE 247. *Facing page:* Necklace of fine sky-blue Persian Turquoise cabochons set in 22-karat yellow Gold with six round brilliant champagne Diamonds and matching earrings (FMNH H2341).

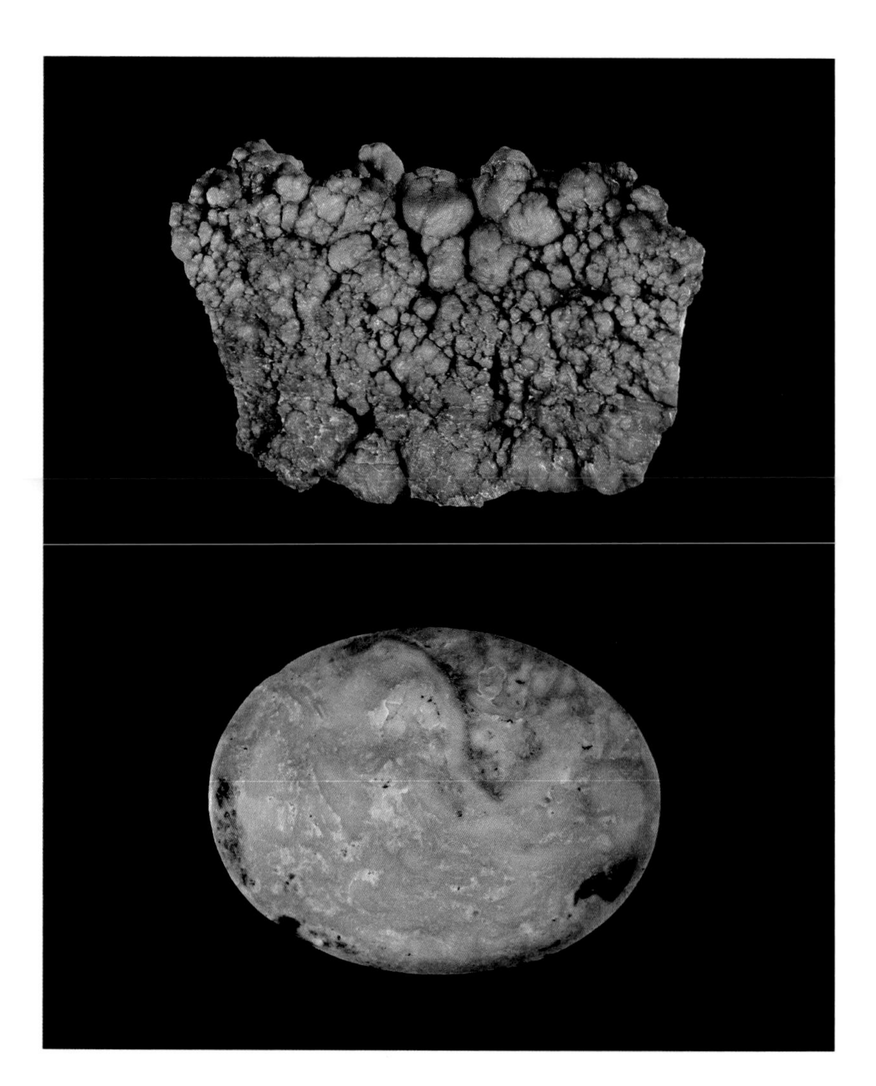

FIGURE 248.
Turquoise from New Mexico, including a piece of botryoidal natural rough weighing 20.1 grams and measuring 50 × 30 × x 8 mm (*top*); and a polished cabochon weighing 9.4 grams and measuring 40 × 30 × 4 mm (*bottom*) (FMNH H1903, H1763).

or sedimentary rocks, or as weathered nodules. It is formed as the product of a long-term chemical reaction by acidic groundwater seeping into phosphatic rocks containing Copper and Aluminum, and is usually found only in arid regions. Because of the way it forms, it is sometimes recovered as a by-product of mining Copper in the western United States.

Because of its opacity, it is not faceted for jewelry, but is instead finished into cabochons or sometimes used for small carvings. Some Turquoise, such as fine Persian Turquoise, is a solid, pure color, while other Turquoise is mottled with small veinlets of black Manganese or brown limonite and is called spiderweb Turquoise. Turquoise can range in color from sky blue to greenish-gray. The finest blue Turquoise is generally acknowledged as Persian Turquoise from Iran. Tibet has a variety that is greenish in color. Turquoise is also known from Afghanistan, Argentina, Brazil, China, Mexico, Tanzania, England, Victoria and

Queensland in Australia, Russia, France, Germany, Chile, Egypt, and China. In the United States, Turquoise is best known from New Mexico, Arizona, Nevada, and Colorado. There is an extremely fine necklace of fine Persian Turquoise set in 22-karat Gold in the Grainger Hall of Gems (fig. 247).

Finished Turquoise gems are often waxed, which is a finishing process that highlights the color and gloss of the stone and increases its durability and resistance to staining from oils and liquids. The waxing process does not diminish the value of the original stone. Occasionally, very soft and porous Turquoise stones are soaked with artificial resin to improve hardness and color. Such stones are worth less than natural high-quality stones. Turquoise is soft and heat-sensitive. Sky-blue Turquoise can be turned an unattractive dull-green color by heating it to 480°F (250°C). For this reason, grinding and polishing must be done carefully so that the friction-generated heat does not destroy a stone's good color. Good color can also be harmed by immersing stones in household detergent, oil, or by overexposing them to intense light or overly dry conditions.

Turquoise is frequently imitated with dyed chalcedony, dyed Howlite, dyed clay, glass, porcelain, and plastic. Synthetic Turquoise has also been widely used in the inexpensive gem market since the early 1970s. Occasionally small pieces of Turquoise are pulverized and baked with a glue mixture to create larger stones. These stones are obviously worth less than natural stones with good color. As one of the earliest minerals used as a gem, it is not surprising that Turquoise was also one of the earliest gems to be imitated. SIMULANT turquoise was created at least as far back as 6,000 years ago in Egypt and also by later Roman civilizations.

For further reading on Turquoise, we recommend O'Donoghue (2006) and Sinkankas (1959).

INORGANIC GEMS NOT DESCRIBED HERE

For the sake of completeness, we wish to make note of a few mineral species and varieties sometimes thought of as gemstones by various authors that are not covered in this book. As we indicated in our introduction, we do not discuss in detail inorganic minerals with a Mohs hardness of less than 6, because they are generally too soft to be considered practical as gemstones in our opinion (although they may be valuable as collector items to mineral collectors and are sometimes faceted for exhibits and collectors' showcases). In alphabetical order, some of these minerals (together with their Mohs hardness values) are Anglesite (3), Apatite (5), Aragonite (3½), Azurite (3½), Barite (also spelled Baryte) (3), Beryllonite (5½), Brazilianite (5½), Calcite (3), Celestine (3½), Cerussite (3½), Chrysocolla (2), Crocoite (2½–3), Cuprite (4), Datolite (5), Diopside (5½), Dioptase (5), Dolomite (3½), Enstatite (5½), Fluorite (4), Gypsum (2), Howlite (3½), hypersthene (5½), jet (4), Kyanite (4–7), lapis lazuli (5½), Lazulite (5½), Malachite (4), meerschaum (2½), moldavite (5½), Natrolite (5½), Phosphophyllite (3½), Rhodochrosite (4), Scheelite (5), serpentine (5), Smithsonite (5), Sodalite (5½), Sphalerite (3½), Sphene (5), Titanite (5), Vivianite (2), and Zincite (5). Turquoise can range as low as Mohs 5, but the best and hardest Turquoise is generally 6, so we included it in our gem chapters.

We also excluded discussion of a few specific minerals that are Mohs 6 or harder because they belong to species that mostly do not contain exceptional gemstone varieties and are not generally used by the fine-gem trade (most of those listed below), or they are not currently represented in the Grainger Hall of Gems (e.g., in the case of Andalusite). These include Amblygonite (6), Andalusite (7), Axinite (7), Danburite (7), Dumortierite (7), Epidote (6½), Euclase (7½), Hambergite (7½), Hauyne (6), Hematite (6½), Idocrase (6½), Kornerupine (6½), Microcline (6), Petalite (6), Prehnite (6), Rhodonite (6), Rutile (6), scapolite (6), Sillimanite (7½), Sinhalite (6½), Staurolite (7), Taaffeite (8), Tugtupite (6), and Vesuvianite (6½).

We included some varieties that are seldom used for gems today because they were used in the past or because they belonged to mineral species that include other varieties important to the gem trade. Examples include black Andradite Garnet and Schorl Tourmaline, both of which are members of important gem groups (Garnet and Tourmaline) and both of which were used in the past for mourning jewelry.

For further discussion of varieties left out of this book, see O'Donoghue (2006), Hall (2002), Lyman (1986), and Oldershaw (2004).

ORGANICALLY DERIVED GEMSTONES

Organically derived gemstones are those gemstones formed by living organisms. These include Pearl, Coral, Amber, and Ivory, discussed here, and a few varieties not discussed in this book such as mollusk shells (clams, oysters, snails), tortoiseshell (from the shells of marine turtles), and jet (a variety of coal). Additionally, in ancient times, people also included teeth and bone as items of adornment, and we have largely excluded these from our discussion. Organic gems were first used by humans because they were soft and easy to carve with the simplest of tools. Beads made from snail shells more than 80,000 years ago have been reported from an archaeological site in eastern Morocco, providing valuable documentation about the early development of abstract thought in *Homo sapiens* (humans).

ORGANIC GEMS are a product of life itself. There are only a few varieties of true gems that are of organic origin, and these include primarily Pearls, Coral, and Amber. Animals produce Pearls and Coral, and plants are the source of Amber. The sharp angles and geometric shapes that define most inorganic gemstone crystals are absent from organic gems, because their organic growth is not dictated by internal atomic structure. Instead, organic gems are the solidified product of biological activity, generally resulting in a less orderly development of external shape.

Organic gems from animals are among the most rapidly formed of all natural gems. They form within the lifetime of the animals that produce them. Pearls are iridescent layers of NACRE, a form of the mineral aragonite. Nacre is secreted by MOLLUSKS such as Pearl oysters, Pearl mussels, and some marine snails, around irritants or invaders (like small crustaceans or food particles) that the animals cannot expel. Layer upon layer of nacre surround the irritating nucleus until a finished Pearl is formed. This process can take only a few years in Marine Pearl oysters to several decades in certain Freshwater Pearl

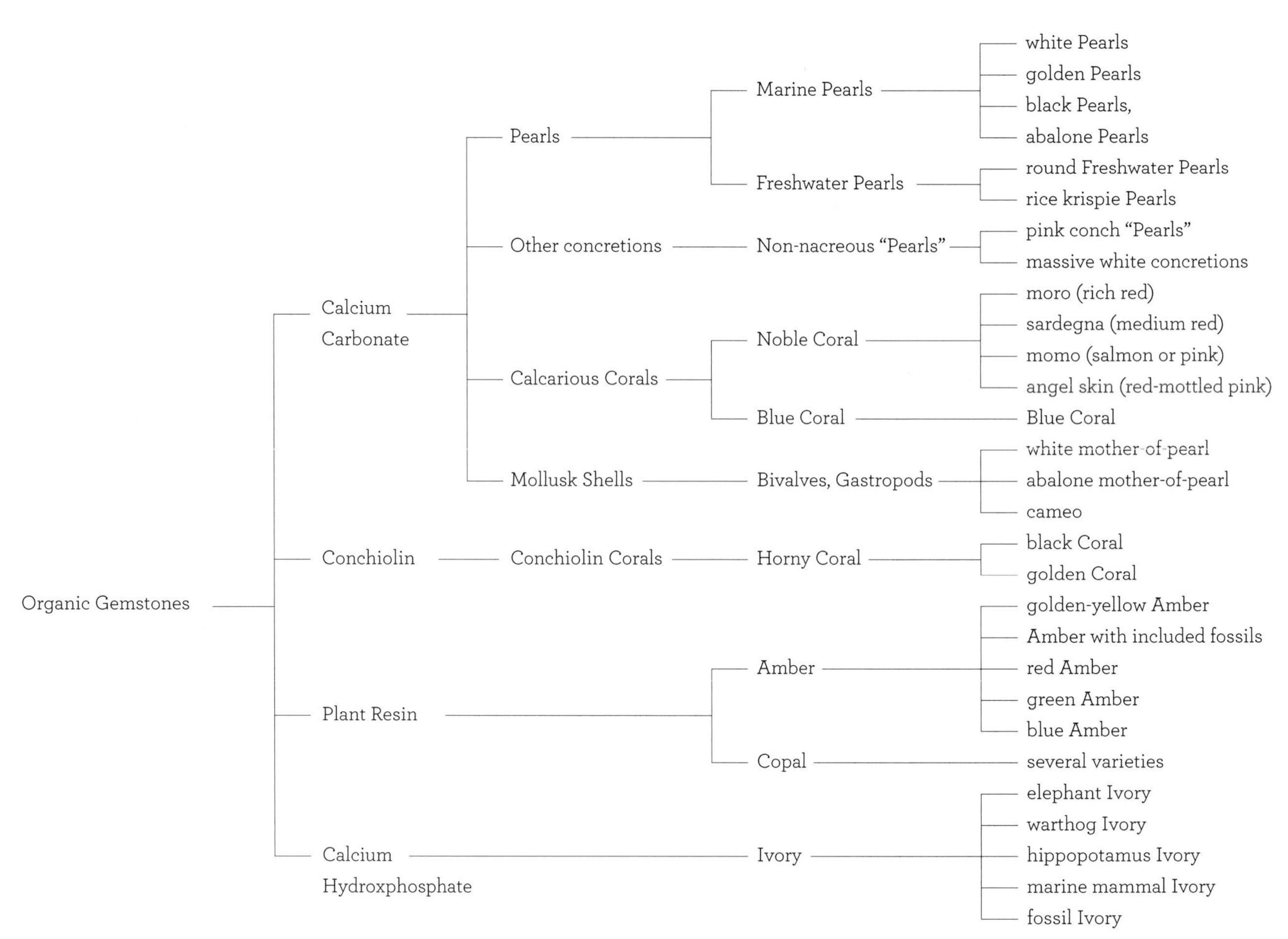

FIGURE 249. Classification of organically derived gemstone varieties discussed in this book.

mussels. Pearl farmers accelerate the process in Pearl oysters and Pearl mussels down to a couple of years by inserting large polished shell beads into mollusks that subsequently will need fewer layers of nacre to become harvestable-sized Pearls. Pearls that are the result of such beads or other seeds artificially inserted, or "nucleated," into mollusks are called CULTURED PEARLS.

Precious Coral is simply the skeletal remains of soft-bodied marine animals called Coral polyps. Colonies of polyps extract Calcium out of seawater, which they then use to collectively excrete a hard, branching framework that structurally supports the entire colony. Think of it as a type of solid internal shell. The soft-bodied polyps live on the surface of the supporting structure, adding to its thickness and length with time. This Calcium Carbonate structure can eventually become thick enough to be carved, made into beads, or cut and polished into cabochons. Although hard Coral structures come in many colors and each color is produced by a different species of polyp, it is mainly the red Coral from the Mediterranean species *Corallium rubrum* that is considered to be true gemstone material. Red Coral has been used as a gemstone for thousands of years. *Corallium rubrum* is often commercially harvested as young as 10 years of age for its precious skeletal structures, but these Corals can live up to 100 years or more. Very large pieces of red Coral from very old colonies of *Corallium rubrum* are very uncommon today.

Plants produce organic gems as well. Amber is fossilized tree resin secreted from the bark of trees that once lived in subtropical or tropical forests. The resin first solidifies into a hard brittle substance called Copal, which can take as little as a few hundred years. Amber is formed when the oils within the Copal slowly evaporate or leach out of the stone in a process estimated to take from 1 to 10 million years to form. Amber often contains fossil insects, moss, pine needles, lichens, and rarely small lizards or frogs. These special inclusions are the remains of ancient organisms that were trapped in sticky tree resin millions of years ago, and they can greatly enhance the value of the Amber pieces that contain them.

We know precisely how some organic gemstones form in nature because it is possible to observe the entire process over the course of their producer's lifespan. We can seed a Pearl oyster and watch the progress of a growing Pearl, or watch a colony of Coral polyps build their branching Calcium Carbonate support structure and then later "mine" the Pearls and Coral skeletons as gem material.

Overexploitation of the organisms that produce organic gems can be a significant ecological problem if not managed properly. Ivory is a prime example. Although Ivory is harvested from hippopotami, warthogs, walruses, sperm whales, narwhals, and even fossil mammoth and mastodon tusks, the primary source by far is elephant tusks. The often illegal overexploitation of elephant Ivory in the 1970s and 1980s led to major declines in the world's elephant populations, which would have led to the rapid extinction of the species had strict protective measures not been undertaken. Certain Corals have also been in danger of possible extinction due to overexploitation. For example, black and golden Corals are extremely vulnerable due to their very limited geographic and ecological distributions. Other Corals are at risk because of poorly managed harvesting techniques that use deep-bottom drag nets and other methods that are highly destructive methods to Coral ecosystems. Fortunately, management of Coral harvesting is now receiving significant attention. In 1994 the European Union banned the use of dredging equipment for the harvesting of Noble Coral in the Mediterranean, and numerous reserves have now been set up around the world to protect various Coral species. Still, much more careful management is needed to create a truly sustainable and responsible program for the harvesting of precious Coral.

Ivory and Coral have historically been very important gem materials, going back millennia in human culture. Some of the most beautiful examples of human artistry have been crafted from these materials. Hopefully, new resource management and farming techniques will allow responsible, humane, and sustainable collection of Ivory and Coral in the future. This has been done in the case of Pearls. Today there is an immense CULTURED PEARL industry that provides plentiful supply from farm-raised bivalves. There are also some organic gem varieties whose harvest have no appreciable impact on living animal or plant populations. These include Amber and also Ivory harvested from fossils such as mammoth and mastodon.

Our classification of organic gemstones is more informal than the one we use for inorganic gemstones, but no less important for the sake of organization. INORGANIC GEM varieties discussed in the previous sections are grouped primarily by specific chemical compositions, and they fall into mineralogical SPECIES categories that follow formal nomenclatural rules in order to be valid named species. In contrast, ORGANIC GEM varieties discussed in the following sections are grouped primarily according to their biological origin (e.g., Corals from living Coral, Pearls from living mollusks, Ivory from mammalian tusks, and Amber from tree resin). We call these groups of organic gem varieties "supervarieties." As with inorganic gem species, we treat organic supervariety-level, group-level, and class-level names as proper nouns and capitalize them.

ORGANIC GEM TYPES

PEARLS

SYSTEM Organic

CLASS Calcium Carbonate

GROUP Pearls

GEM SUPERVARIETIES

Marine Pearls; Freshwater Pearls

Irritation rarely takes the form of beauty, but Pearls are the exception. This, the most valuable of all organic gems, is simply the result of a MOLLUSK's defense against an intruding irritant. In most cases, a natural Pearl is the result of an oyster's or a clam's attempt to protect itself from the invasion of a foreign body. The foreign body (nucleus) can be organic material or even something as simple as a grain of sand. The intruding material is covered with concentric layer after layer of pearlescent NACRE secreted by the living part of the mollusk. Cultured Pearls involve artificial insertion of foreign bodies into Pearl-producing mollusks to stimulate the organism's defense mechanism to build a Pearl around the planted irritant. The inserted fragment is often a small piece of bead or shell, but sometimes the Pearl seeds consist of specific shapes, such as tiny Buddhas, to produce pearlized images. The process of Pearl formation can take several years.

Cultured Pearls have been produced for more than 700 years, particularly by Chinese and Japanese industries. The method of culturing saltwater Pearls, called periculture, was perfected in the early 1900s largely by Kokichi Mikimoto, who cultured the Akoya Pearl oyster. By 1938 there were 350 active Pearl farms in Japan producing millions of high-grade Akoya Pearls. Cultured Pearls are no less real than "natural" Pearls, because both the natural and the human-induced irritants stimulate the same Pearl-making process of the mollusk. Natural gem-quality Pearls have always been an extreme rarity, but when the Japanese developed Pearl culturing into a huge industry in the early twentieth century, the prices of Pearls came within reach of a much broader range of people. In fact, there was a "crash" of the Pearl market in 1930, when the price of Pearls dropped 85 percent in a single day. Today virtually all Pearls on the gem market are cultured Pearls.

Pearls have been produced by mollusks since their early evolution. Fossil traces of Pearls as old as 225 million years are known, and actual Pearls are known in rocks from the age of the dinosaurs. Pearls are made by a variety of different mollusk species, including marine bivalves and gastropods, and freshwater clams. The majority of natural Pearls come from marine bivalves of the genus *Pinctada*. The majority of cultured Pearls today come from Freshwater Pearl mussels in China.

True Pearls have a characteristic mother-of-pearl-like iridescence known as ORIENT OF PEARL or NACREOUS luster. Pearls can be white, golden, bluish-gray, black, or even multicolored. South Sea Pearls are generally the most valuable, especially those that are free of blemishes and are of even color, good spherical shape, and a diameter of 9 millimeters or more. Pearls are soft and are normally used for necklaces or earrings rather than rings and bracelets because of their susceptibility to damage. Besides being easily scratched, they are vulnerable to damage from acids, liquid chemicals, chlorinated water, and heat. Pearls should be wiped with a soft cloth after wearing, and they should be washed in mild soapy water to remove acidic oils and perspiration. Pearls should also be restrung periodically if worn frequently, with knots tied between the Pearls to prevent abrasion between Pearls and the loss of all Pearls if the string should break. For further reading on Pearls, we recommend Landman et al. (2001).

Marine Pearls

SYSTEM Organic

CLASS Calcium Carbonate

GROUP Pearls

SUPERVARIETY Marine Pearls

GEM VARIETIES white Pearls (Akoya and South Sea); golden Pearls (South Sea); black Pearls (Tahitian); abalone Pearls (awabi)

COMPOSITION Calcium Carbonate, conchiolin, and water ($CaCo_3$,$C_3H_{18}N_9O_{11}\cdot nH_2O$)

TRACE ELEMENTS FOR COLOR Iron (Fe), Magnesium (Mg), or Aluminum (Al)

HARDNESS ON MOHS SCALE 2½–4

FIGURE 250.
Facing page, top: The Japanese Pearl oyster *Pinctada fucata* (family Pteriidae), which produces cultured Akoya Pearls such as the ones in figure 251 (FMNH Z312405).

FIGURE 251.
Facing page, bottom: White Pearl necklace containing 97 Japanese Akoya Pearls. Each Pearl is about 7.5 mm in diameter. Total weight of necklace is 54 grams (FMNH H2265).

Marine Pearls rank with the finest of gems and have been highly valued for centuries. Even the name Pearl has become a metaphor for something of great value (e.g., "pearls of wisdom"). Nacreous Pearls are famous for their characteristic luster, from which the term "mother-of-pearl" is derived. They form from concentric layers of crystalline Calcium Carbonate ($CaCO_3$, largely in the form of aragonite) held together by thin layers of an organic, horn-like material called conchiolin. The combination of conchiolin and aragonite is what forms nacre, the peculiar iridescent compound responsible for pearly luster. The thinner and more numerous the concentric layers of nacre, the finer the luster of the Pearl. Nacreous Pearls are formed primarily by bivalves (Marine Pearl oysters and Freshwater Pearl mussels), although there are exceptions, such as abalone Pearls. The abalone is actually a large marine snail.

Marine Pearls come primarily from tropical and subtropical coastal marine waters. Gem-quality Pearls are extremely rare in nature, and the development of the cultured Pearl industry has probably saved several species of Pearl mollusks from becoming extinct long ago. Many thousands of Pearl oysters must be opened, and therefore killed, to find a single "wild" Pearl of gem quality, and for centuries this was the only way to harvest Pearls. This is one reason why Pearls were so much more expensive in the past. Modern cultured Pearls are genuine Pearls, every bit as beautiful as wild Pearls, and much more available. Consequently, natural Pearls, although still very valuable, are mostly just a curiosity today. Natural or "wild" Pearls can often be distinguished from cultured Pearls with the use of an X-ray machine. A radiograph of a natural Pearl will show concentric growth rings through the interior of the Pearl, while a radiograph of most cultured Marine Pearls will show a solid center that lacks growth rings because the bead used for nucleation of the Pearl lacks the concentric rings. Exceptions include most cultured Freshwater Pearls that are today nucleated with soft tissue rather than with beads, making it undetectable using X-rays.

The value of a Marine Pearl is determined by its shape, size, luster, and surface quality. The most valuable Pearls are completely spherical with luster evenly covering the entire surface. Pearls of 8 millimeters in diameter or larger are desirable.

Pearls come in many shapes, including round, semi-round, drop, oval, pear, button, circle, and baroque (irregular). Drop and pear shapes are often used for pendants or earrings, and semi-round Pearls are often used in necklaces where their lack of full roundness can be hidden by orienting them properly when stringing them. Round Pearls are the rarest and most valuable of shapes. In nature, only one or two per thousand Pearls are round. Round Pearls grow loosely within the soft tissues of the mollusk. Blister Pearls are those that grow between the soft tissues and the shell, usually attached to the shell. Blister Pearls are almost always of highly irregular (nonspherical) shape. Although

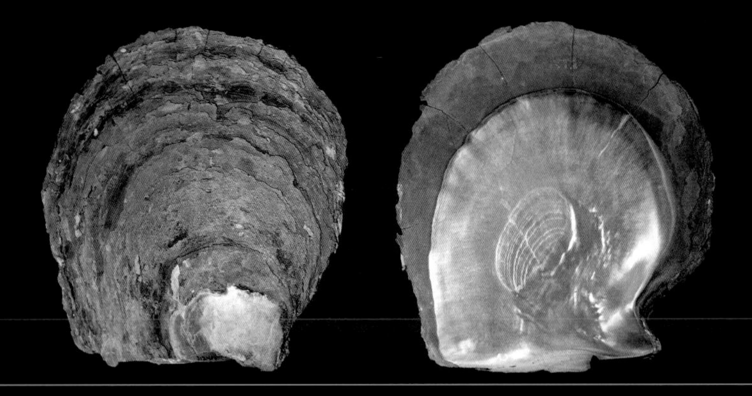

FIGURE 252.
Top: The gold-lipped Pearl oyster *Pinctada maxima* (family Pteriidae), which produces South Sea Pearls for the gem trade such as the white Pearls in figs. 253 and 254 and the golden Pearls in fig. 255 (FMNH Z312400).

FIGURE 253.
Middle: Brooch with 22 white South Sea Pearls set in 14-karat Gold. Average Pearl size is 11 mm in diameter (FMNH H2295).

FIGURE 254.
Bottom: South Sea baroque Pearl of 15 × 12 mm, set in 18-karat Gold with round brilliant-cut Diamonds. Designed by American Modernist goldsmith and jewelry designer Arthur King in the late 1960s. Top and side views (FMNH H2550).

FIGURE 255.
Golden South Sea Pearl necklace of 37 AA-quality Pearls with a clasp of 18-karat yellow Gold and .75 carats of Diamonds. Diameters of Pearls range from 10.5 to 12.0 mm (FMNH H2386).

round Pearls are the most valuable, irregular-shaped large natural Pearls have also been used in jewelry, often as inset parts of small sculptured ornaments. Sometimes irregular-shaped Pearls can reach great size. The largest nacreous Pearl we know of is the Arco-Valley Pearl, a 575-carat (4-ounce) irregular-shaped white blister Pearl dating back to the eleventh century. This Pearl, measuring 79 × 41 × 34 mm, was once owned by the Mongol emperor Kublai Khan, grandson of Genghis Khan, and was given as a gift to the famous Venetian traveler Marco Polo. After that it changed hands many times before ending up in the possession of the Arco-Valleys, a wealthy twentieth-century Austrian family after which the Pearl is named. Most recently, the Pearl was acquired by an Abu Dhabi jeweler who plans to sell it.

Pearl-diving for wild Pearls in centuries past was an amazing and extraordinarily dangerous feat. In the Persian Gulf, deep-diving for Pearl mollusks probably began more than 2,000 years ago. A diver had to go as deep as 30 to 100 feet while holding his or her breath for up to five minutes. They often had to be weighted down with heavy stones to descend quickly enough to have time to search the seabed and then return to the surface before drowning. The introduction of scuba gear in the twentieth century made the process much safer.

There are many varieties of valuable Marine Pearls. The most commonly marketed colors of Pearl in the gem trade today are white, golden, and black. The color of a Pearl usually matches the color of the inside shell surface of the mollusk that produced it. The color in Pearls comes from two different sources: organic pigment in the conchiolin layers, and trace elements such as Iron in the nacre.

Most Pearls are the product of the Pearl oyster family Pteriidae. Edible oysters belong to a different family, the Osteridae, and the once-popular idea that the next plate of oysters on the half-shell might contain a gem is a fallacy. The most valuable white Marine Pearls come mainly from Asian oysters of the species *Pinctada fucata* (Akoya Pearls) and *Pinctada maxima* (South Sea Pearls). Golden Pearls are also produced by *Pinctada maxima*, otherwise known as the gold-lipped or silver-lipped Pearl oyster. Black Pearls are produced by *Pinctada margaritifera*, the black-lipped Pearl oyster from Tahiti and the Cook Islands. There are fine examples of white, golden, and black Pearls in the Grainger Hall of Gems (figs. 251, 253, 254, 255, and 257).

Not all Pearls come from Pearl oysters. Abalone Pearls, sometimes called awabi Pearls, come from marine snail species in the genus *Haliotus*. Abalone Pearls are multicolored nacreous Pearls that are almost never round, difficult to use in perliculture, and consequently are not often used in the fine-gem trade. They are nevertheless quite beautiful, often with a multicolored iridescence strong in blue and pink hues. Even the shell has these iridescent colors (fig. 258).

There are also types of calcareous concretions that lack nacre, such as those

FIGURE 256.
Facing page, top: Tahitian Pearl oyster *Pinctada margaritifera* (family Pteriidae), which produces black Tahitian Pearls for the gem trade, such as the thirteen pearls in figure 257. From French Polynesia, Tuamotu Islands (FMNH Z278466).

FIGURE 257.
Facing page, bottom: Black Tahitian Pearl necklace with multicolored Tourmalines and 18-karat Gold. Pearl size ranges from 12 to 15 mm in diameter (FMNH H2348).

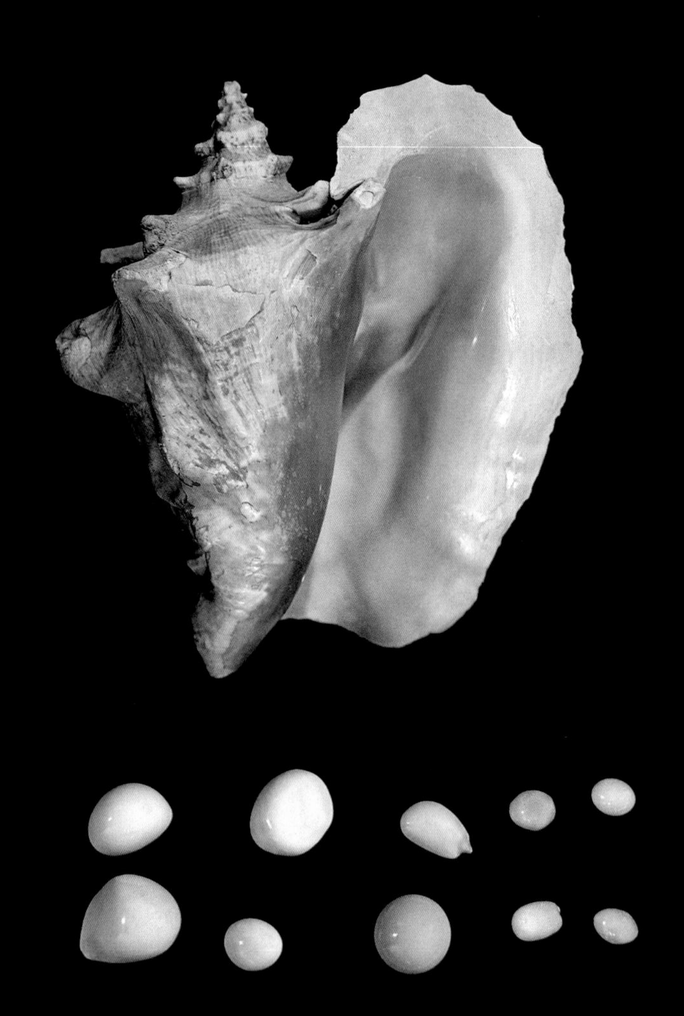

FIGURE 258.
Above: Shell from the green abalone, *Haliotis fulgens* (family Haliotidae) from Baja California, alongside a group of abalone Pearls produced by this species. Shell length is 178 mm (FMNH 312451, H668).

FIGURE 259.
Right: Shell from the queen conch, *Strombus gigas* (family Strombidae), from the Bahamas, above a group of pink conch concretions produced by this species ("conch Pearls"), ranging in weight from .47 to 6.79 grams each (FMNH Z312403, H661–H668, H2007, H2009).

that form in the gastropod *Strombus gigas*, otherwise known as the queen conch (fig. 259). They are often beautiful pink stones that exhibit a particular CHATOYANCY known as "flame structure" and are sometimes called "conch Pearls." These can be quite valuable as gems, but in the gem industry these are not generally accepted as "true Pearls" because they lack nacre. They are usually of irregular shape and only rarely found as perfectly round. Special care must be taken with pink conch Pearls because they can fade after prolonged exposure to sunlight. Non-nacreous concretions also form in other mollusks and can sometimes reach much larger sizes than nacreous Pearls. The Pearl of Lao-Tzu ("Pearl of Allah") is a white non-nacreous, irregularly shaped concretion weighing just over 14 pounds (6.4 kg). It was discovered in a specimen of the giant clam *Tridacna gigas* from the Philippines in 1934. Although it is commonly referred to as the world's largest "Pearl," it is considered by many or most gemologists not to be a true Pearl because it is non-nacreous.

Freshwater Pearls

SYSTEM Organic

CLASS Calcium Carbonate

GROUP Pearls

SUPERVARIETY

Freshwater Pearls

GEM VARIETIES

round Freshwater Pearls; rice krispie Pearls

COMPOSITION Calcium Carbonate, conchiolin, and water ($CaCo_3,C_3H_{18}N_9O_{11}{\cdot}nH_{20}$)

TRACE ELEMENT FOR COLOR

none

HARDNESS ON MOHS SCALE

2½–4

FIGURE 260. Shell from the freshwater mussel *Amblema plicata* (family Unionidae) above a group of Freshwater Pearls produced by this species from Wisconsin, showing a variety of shapes. Total weight for lot of Pearls is 9 grams. (Shell is FMNH Z150765; Pearls are FMNH H1996–H1999, H2001, H2005.)

The Freshwater Pearl industry is a relatively underappreciated one compared to the Marine Pearl industry. The quality of cultured Freshwater Pearls today, particularly from China, rivals some of the finest Marine Pearls. Freshwater Pearls form the same way that Marine Pearls do, with concentric layer upon layer of thin nacre around an irritant in the mollusk. White is the most common color, but a variety of pastel colors can also be found, including pink, salmon, red, bronze, lavender, green, blue, orange, and yellow. In spite of major qualitative advances in farming cultured Freshwater Pearls, Marine Pearls remain more highly valued than Freshwater Pearls in general.

Freshwater Pearl mussels come from the families Unionidae and Margaritiferidae, and include dozens of species worldwide, mostly in temperate climates. Sadly, many of these have become extinct in recent decades (including 35 species in North America) due to pollution and habitat destruction. Conservation measures are being implemented today to protect the species that remain. Unionid mussels are some of the longest-lived animal species, able

FIGURE 261.
Necklace of 530 Freshwater "rice krispie" Pearls from China, on six 36-inch strands with a 14-karat Gold clasp containing a single large half-round Pearl (FMNH H2286).

to reach 150 years of age or more. Historically, their shells were an important source of material for the fine button industry, which also led to their decline in some areas. Freshwater Pearl mussels occur in freshwater rivers and streams around the world, but the greatest diversity of species today is in the Mississippi River drainage of the eastern United States.

The cultured Freshwater Pearl industry became successful in Japan in the 1930s, using Pearl mussels native to Lake Biwa near Kyoto. Early efforts were successful, and production rose to a peak of 6 tons annually in 1971; but shortly thereafter environmental pollution brought the industry of Lake Biwa to extinction. Biwa Pearl production was later revived, mainly with many new farms cultivating Biwa mussels in Shanghai, China. Production in China rose to amazing levels, eventually exceeding 1,500 metric tons of Pearls per year. Early production consisted largely of raisin-shaped rice Pearls or rice krispie Pearls cultivated with the cockscomb Pearl mussel *Cristaria plicata*. There is a fine necklace of these Pearls in the Grainger Hall of Gems (fig. 261). Then in the mid-1990s, the Chinese industry switched to a different species of freshwater mussel, the triangle shell Pearl mussel *Hyriopsis cumingii*, and were eventually able to produce large quantities of perfectly round Pearls. China had an advantage over Japan in having a huge land mass with many available lakes, rivers, and streams; a near-limitless low-cost workforce; and a species of mussel that reliably produced a beautiful product. Chinese Pearls now dominate the Freshwater Pearl market. The quantity of Chinese Pearls keeps improving and will probably continue to change the market in the future. Some are now being made that are virtually indistinguishable from near-perfect "wild" Pearls, and they match the appearance of fine Akoya Pearls at a fraction of the cost.

There was also an attempt to start up a cultured Freshwater Pearl industry in the eastern United States in the early 1960s and '70s. Production of Freshwater Pearls in the United States remains at fairly low levels today, particularly when compared to the production in China.

Noble Coral

SYSTEM Organic

CLASS Calcium Carbonate

GROUP Calcarious Corals

SUPERVARIETY Noble Coral

GEM VARIETIES rich red Noble Coral (moro); medium red Noble Coral (sardegna); salmon-colored Noble Coral (momo); red mottled pink Noble Coral (angel skin)

COMPOSITION

Magnesium and Calcium Carbonate [$(Ca,Mg)CO_3$]

TRACE ELEMENTS FOR COLOR

organic

HARDNESS ON MOHS SCALE

3–4

The most popular and valuable Coral to the gem trade is Noble Coral, also called precious red Coral. The different varieties of Noble Coral are made by living Coral polyps belonging to the genus *Corallium*. These Corals have been harvested for gems for more than 5,000 years, and today this practice must be strictly managed in order not to push species of *Corallium* to extinction. Very large Noble Coral specimens are 100 years old or more, so the rate at which large, mature Corals can replace themselves is a slow one.

Living Coral consists of a colony of small soft-bodied organisms called polyps. The polyps live on the surface of a hard, branching, shrub-like support structure, which is actually produced by CARBONATE secretions from the colony. It is this support structure that is used as gem material. The most valuable species for gems are the red *Corallium rubrum* found mainly in the Mediterranean Sea, the red *Corallium japonicum* around the Pacific coasts of Japan and Taiwan, and the pink *Corallium secundum* off the coast of Oahu. *Corallium* lives on rocky-bottom deep-ocean habitats mostly at depths rang-

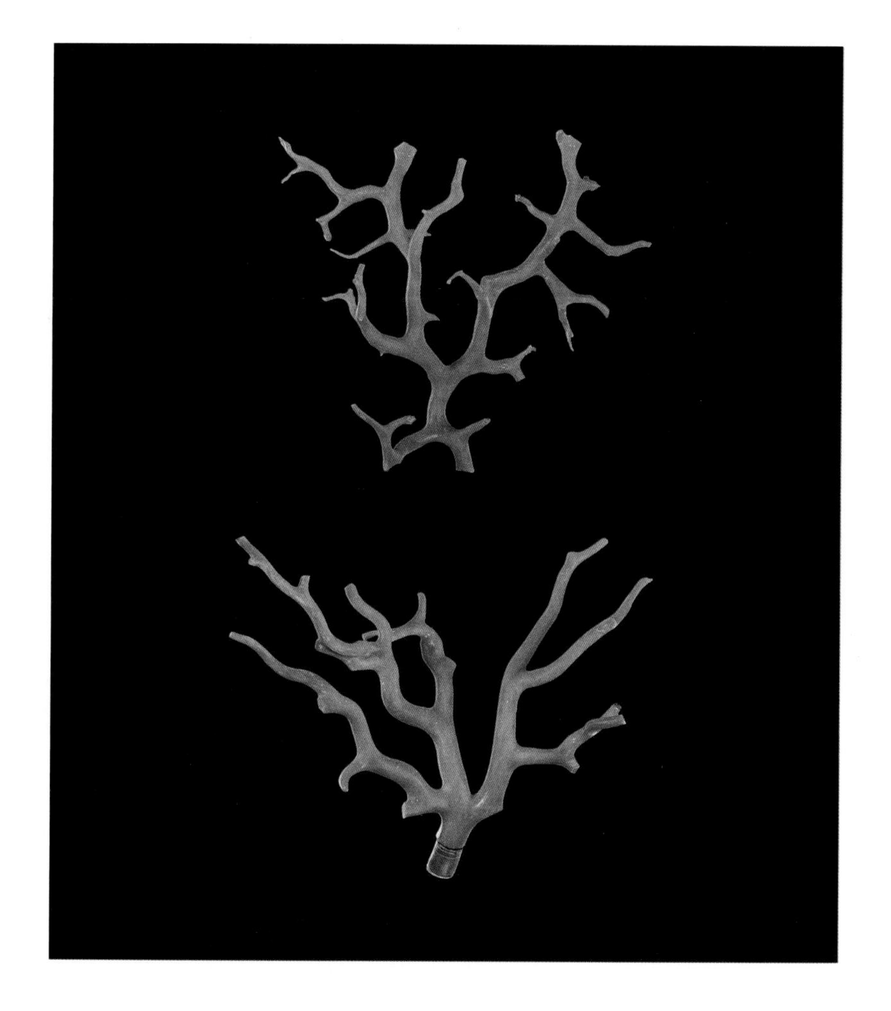

FIGURE 262.
Large pieces of precious red Coral (Noble Coral) from the Mediterranean region. Natural skeletal structures that make up the rough material cut and polished for the gem trade. Rough piece on top is 120 mm in height; polished piece on bottom is 123 mm in height (FMNH NM308).

FIGURE 263.
Above: Rich red (moro) Coral necklace made of 74 small polished Coral branches. Total weight 104 grams (FMNH H2362).

FIGURE 264.
Right: Intricately carved bracelet from the mid-nineteenth century made from salmon-colored Noble Coral (momo) (FMNH H1994, H1468).

ing from 150 to 5,000 feet below sea level. It is capable of living in shallower water, but most shallower-water *Corallium* colonies were harvested and depleted decades ago.

When the Coral skeletal branches are dried, they have a matte finish, but they are hard enough to take a high polish. Because of Coral's opacity and softness, it is not faceted for jewelry. Instead it is used to make cabochons, beads, and carvings for jewelry and ornamentation. The highest-quality Coral is of evenly saturated, rich red color and is free of notable porosity, holes, cracks, blotches, or striations. In general, the highest-quality Coral also comes from older, mature colonies with thick Magnesium-reinforced Calcium Carbonate, approaching a Mohs hardness of 4. Care should be taken with Coral jewelry. Many acids will etch the surface of Coral, and it is very easily scratched because of its extreme softness. Coral is sometimes imitated by plastic or dyed bone.

The use of Coral as a gemstone is an extremely ancient practice, going back at least 25,000 years. Coral jewelry has been found in ancient Egyptian tombs and prehistoric European burial sites. PLINY THE ELDER documented the use of precious Coral for trade between India and the Mediterranean in the first century AD. Coral was also used as an item of currency by the ancient Phoenicians and Romans. The ancient Greeks thought Coral was a gift from the gods that had special powers against poison, witchcraft, and robbery.

There are other species of Corals not described here nor featured in the Grainger Hall of Gems. Blue Coral, for example, is produced by the Coral species *Allopora subviolacea*. Although once used in Africa for ornamented carvings, it has never been widely used as a gemstone and is not used in the fine-gem trade today. It is of little commercial value.

There is another group of Corals with a different composition than the Carbonate Corals described above. These are the Conchiolin Corals, which are formed of a horn-like substance rather than a calcareous one. Conchiolin Corals include black Coral from Hawaii produced by Coral species in the genus *Antipathes*, and golden Coral, also from Hawaii, produced by Coral species in the genus *Gerardia*. Conchiolin Corals are much more restricted in distribution than Noble Coral, highly protected today from overexploitation, and less popular than fine Noble Coral. Trade in *Antipathes* Coral is today prohibited under international CITES (Convention on International Trade in Endangered Species) regulations.

Amber

SYSTEM Organic

CLASS Plant Resin

SUPERVARIETY Amber

GEM VARIETIES

golden Amber; Amber with fossil inclusions; red Amber; green Amber; blue Amber

COMPOSITION fossilized plant resins (mainly $C_{10}H_{16}O$)

TRACE ELEMENTS FOR COLOR

due to various organic impurities

HARDNESS ON MOHS SCALE 2½

Most polished Amber has a uniquely warm, golden glow, with soft smooth surfaces. It is a gem of free-form beauty that has long been valued by human cultures. Stone Age Amber artifacts have been found, the oldest of which are human-crafted beads from a cave in Cheddar, England. These beads are estimated to be 11,000 to 13,000 years old and were crafted from Baltic Amber most likely washed up on the eastern shores of England and Scotland. Paleolithic man is known to have stored Amber in caves 15,000 years ago, based on an archaeological site in Hautes-Pyrénées, France.

Amber, like Pearl, is an organic gemstone. Its origin is ancient coniferous tree resin that has fossilized into a hard natural "plastic" that can be easily polished. The age of the resin that formed today's Amber is between 320 million and 2.5 million years old, although the oldest Ambers are not generally suitable as gemstones and most gem-quality Amber is between 25 million and 40 million years old.

The process of Amber formation starts with a tree trying to defend itself against insects or disease, or reacting to injury. The tree produces a resin that

FIGURE 265.
Piece of natural rough Amber from Mexico, with a small window polished into part of the surface to reveal the quality of the material. Weight is 133 grams and dimensions are 128 × 68 × 37 mm (FMNH H2462).

FIGURE 266.
Amber necklace made with rough pieces of Baltic Amber from Lithuania (FMNH H1676).

can fill internal injuries or cavities within the tree or run down the external surface of the bark. Resins have compounds called terpenes, which link up and form a semi-fossilized solid substance. The resin material will continue to transform over time if it is buried or submerged in an Oxygen-free environment. Eventually, after decades or centuries, the resin becomes a solid called Copal; Copal resembles Amber but is not yet as permanently hardened as Amber. Copal will melt when placed near a flame or soften and become tacky when immersed in alcohol. After hundreds of thousands to millions of years, Copal becomes Amber, which is more resistant to melting from heat or dissolving in volatile liquids, and stable enough to qualify as a gemstone. In the gem industry, Copal is sometimes referred to as "young Amber."

Because the Amber-forming trees often exuded resin in defense against insect damage, the resin often trapped insects. For this reason, many pieces of Amber are found with fossil insects or other organisms inside them. These represent some of the world's best preserved insect fossils, because they are both detailed and completely three-dimensional. They are also prized by scientists, collectors, and gem dealers. More than 1,000 different species of insects have been identified from remains found in Amber. Very rarely, leaves, flowers, feathers, hair, and even small frogs and lizards are also found in Amber. Fine transparent Amber with aesthetically placed fossil insect fossils, such as moths or termites, makes both beautiful and very interesting pieces of jewelry (fig. 267).

The majority of Amber used in jewelry does not contain fossils, and the most common variety is yellow to reddish-orange. The best grade of Amber is transparent, but it also is found as cloudy, milky, or opaque. Amber can also be red (cherry Amber), green, or very rarely blue. Red and green Amber carry little or no value premium over fine golden-yellow Amber. Blue Amber is very rare and comes primarily from either the Dominican Republic or Mexico. It can look like ordinary golden Amber under incandescent light, but under sunlight it has a fluorescent blue glow. Fine blue Amber is worth a significant premium over golden Amber.

The origin of the name Amber is from the Arabic *ambergris*, referring to an oily perfumed secretion of the sperm whale. The word "electricity" is derived from the Greek *elektron*, which was the name used by the ancient Greeks for Amber. This reflects the ability of Amber to generate an electric charge when rubbed with a cloth. Such a charge on the stone will attract dust.

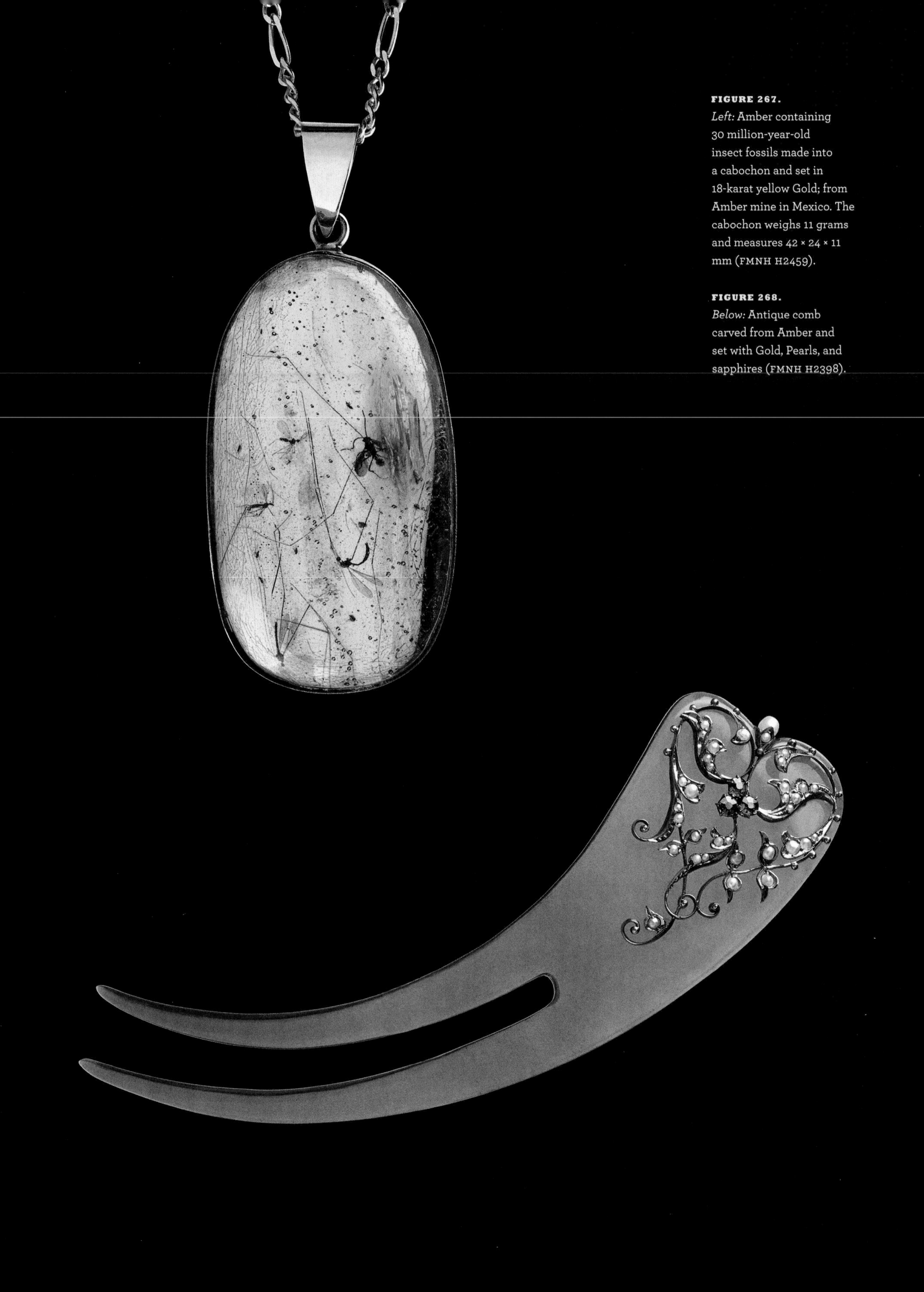

FIGURE 267.
Left: Amber containing 30 million-year-old insect fossils made into a cabochon and set in 18-karat yellow Gold; from Amber mine in Mexico. The cabochon weighs 11 grams and measures 42 × 24 × 11 mm (FMNH H2459).

FIGURE 268.
Below: Antique comb carved from Amber and set with Gold, Pearls, and sapphires (FMNH H2398).

FIGURE 269.
Goat figure carved from red Amber, Han dynasty (206 BC–AD 220), China (FMNH 126613).

The most famous principal localities for Amber used by the gem trade include the Baltic region (particularly along the coasts of Poland and the former Soviet Union), Sicily, the Dominican Republic, Mexico, and Myanmar. Although often mined from sedimentary deposits associated with the ancient remains of long-extinct forests, Amber is also found washed up along the seashore on and near the Baltic coast as far away as the coasts of England, Norway, and Denmark. Amber that is mined is called pit Amber, and Amber that is collected from the sea is called sea Amber. For further reading on Amber, we recommend Grimaldi (1996).

Ivory

SYSTEM Organic

CLASS Carbonate

SUPERVARIETY *Ivory*

GEM VARIETIES elephant Ivory; marine mammal Ivory; hippopotamus Ivory; warthog Ivory; fossil Ivory

COMPOSITION

Calcium Hydroxyphosphate $[Ca_5(PO_4)_3(OH)]$

TRACE ELEMENTS FOR COLOR

organic material

HARDNESS ON MOHS SCALE

2–3

Ivory is an ornamental material that has sometimes been used for making beads or carved jewelery, but it is more often seen as a carving medium for statues. The use of Ivory causes conflicting emotion in this day and age. Its creamy white color conveys a sense of purity; but we now know that much Ivory comes at a terrible cost to one of the largest animals living on the planet today, the elephant. Therefore, we do not endorse the purchase of new Ivory. Trade in new African and Asian Ivory is today prohibited under international CITES (Convention on International Trade in Endangered Species) regulations. That being said, there is much antique Ivory that has been fashioned into objects of great beauty. Antique Ivory (Ivory more than 100 years of age) can be legally imported with the proper CITES certification from the country where the Ivory originates. Although there is no Ivory in the Grainger Hall of Gems, there is antique Ivory in some of the anthropology halls and collections of The Field Museum (fig. 270).

Ivory is an organic mineral called dentine that grows as tusks and teeth in many species of mammals from both land and sea. Land mammals that produce workable Ivory include hippopotamus, wild boar, and warthog. Productive marine mammals include walrus, sea lion, sperm whale, and narwhal. Even fossil Ivory from mammoth and mastodon tusks is used. Tens of thousands of years have done little to alter the mineralogy of tusks from fossil mammoths, and they are still composed of true Ivory. But the majority of all harvested Ivory by far has come from elephant tusks. A single elephant tusk can weigh up to 200 pounds. Before the 1989 ban on the Ivory trade, the African elephant population had fallen from 1.3 million to 609,000 in just 10 years. Ivory exports, including much illegally harvested elephant Ivory, had risen to over 700 tons, or 75,000 elephants, per year.

Ivory has been used in carvings and jewelry for tens of thousands of years. Carvings in Ivory dated 32,000 years old have been found in French caves. In 2009 in the scientific journal *Nature*, a 35,000-year-old carving of a woman that may have once hung from a string was reported as being discovered from a cave in Germany. These Pleistocene carvings were done using the tusks from woolly mammoths, which were not yet extinct at that time. Ivory was a very important medium in Asian culture, used in fine carvings dating back at least to the Han dynasty (220 BC). In the early and mid-twentieth century, Ivory was probably best known as a veneer material for piano keys, chess pieces, combs, and small figurines. Today these items are simulated using plastics or synthetics.

FIGURE 270.
Facing page: Nineteenth-century Ivory carving from China, depicting Lu Dongbin, one of the eight Daoist (formerly spelled Taoist) Immortals. He commands charms to tame evil spirits and cure illness. Statue is about 220 mm high (FMNH A233357).

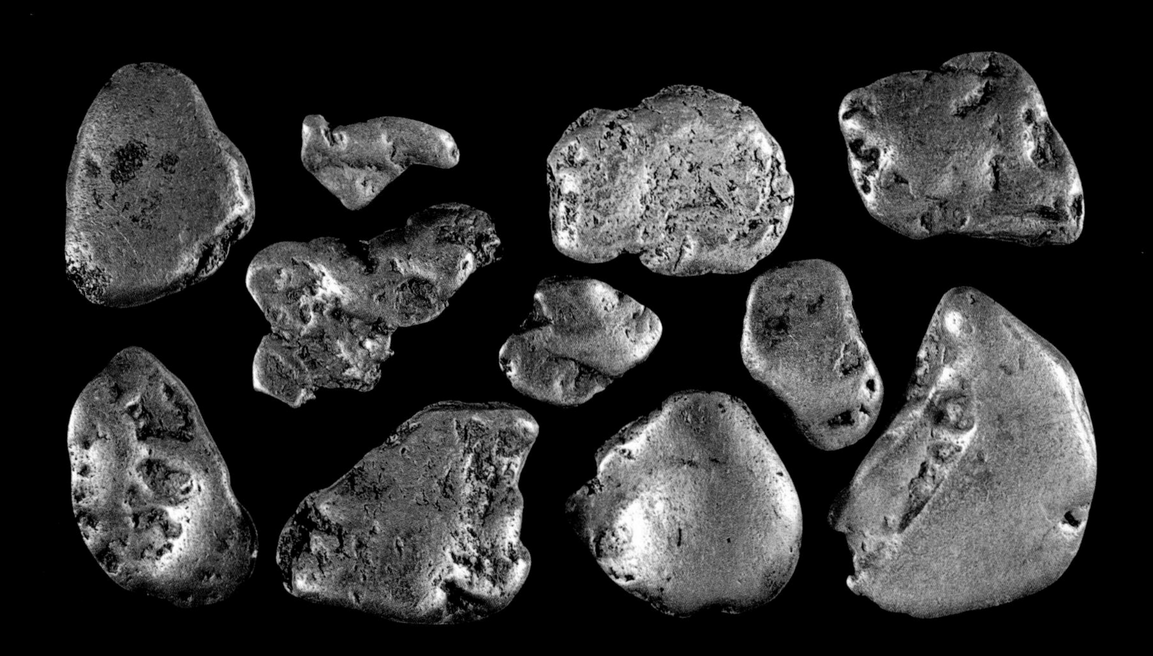

PRECIOUS METALS (PRIMARILY GOLD)

Precious metals commonly used to set gems and create fine jewelry include Gold, Platinum, and Silver. Of these three, the precious metal most commonly used for ornamentation is Gold, an ideal metal for jewelry and other items of adornment because its natural beauty does not tarnish or corrode. We therefore include a discussion of Gold in this book, and Gold is also highlighted in the Grainger Hall of Gems.

Pure Gold is an element (symbol Au) rather than a compound, but it can be mixed with other metals to form **ALLOYS**, giving it added strength and different shades of color. Pure Gold in its natural state is bright yellow in color. White Gold is a mixture of Gold and Palladium, Silver, or Nickel, although Nickel is no longer widely used for jewelry because of possible toxicity. Rose-colored Gold is made by mixing Gold with 25 percent Copper; greenish Gold is made by mixing Gold with Silver; blue Gold can be made by alloying Gold with Iron, and purple Gold can be made by alloying Gold with Aluminum. Gold is extremely **MALLEABLE** and can be drawn into ultra-thin wire or pounded into very thin sheets known as Gold leaf. A single ounce of Gold can be beaten into more than 250 square feet of Gold leaf or drawn into thin wire more than a mile long. Because Gold can be used in such thin foil and retain its aesthetic appearance, it has been used since ancient times for **GILDING** ornamental wood and metal surfaces. Gilding is a process that mechanically applys a thin Gold foil to the surface of another material. Although a number of gilding techniques can be used, the simplest is to glue a paper-thin Gold foil onto the other surface. Ancient Greeks, Egyptions, and Romans used this process to gild wooden statues and the ceilings of temples and palaces. Gold can also be electrochemically applied to the surfaces of other metals in extremely thin layers to give them the appearance of Gold through the process of **GOLD PLATING**. Fine jewelry generally relies on solid Gold rather than gilded or plated metal.

FIGURE 271.
Facing page, top: Large natural Gold nugget of nearly 4 pounds in weight. From central Victoria, Australia. Dimensions are 100 × 65 × 45 mm (FMNH H2316). *Facing page, bottom:* A group of waterworn nuggets from Kittitas County, Washington, with a total weight of 456 grams, or 14.7 troy ounces (FMNH H850.1–H850.8, H850.11, H852.1, H852.2).

FIGURE 272.
Peculiar leaf-shaped Gold nugget, apparently a distorted natural crystal piece, discovered in the Yukon Territory of Canada in 1906. Illustrated in top, side, and bottom view. Specimen weighs 3 troy ounces and measures 55 × 42 × 4 mm (FMNH H858).

The purity of Gold is measured in KARATS (not to be confused with CARATS, a measurement of weight). Twenty-four-karat Gold is pure Gold; twelve-karat Gold is 50 percent Gold; and so on. Pure Gold is too soft for ordinary use in jewelry (Mohs 2½) and is hardened by alloying it with Copper, Silver, or other base metals. The purity levels of Gold most frequently used in fine jewelry are 9-karat, 14-karat, 18-karat, and 22-karat, with 14 and 18 being the most frequently used. Gold jewelry is usually stamped or "hallmarked" to indicate its level of purity. Sometimes the stamp is in karats, and sometimes the stamp is a proportion of fineness or purity (e.g., ".900" instead of "18-karat").

Gold, in nature, is commonly found as veins in IGNEOUS ROCKS, Quartz, and other minerals. It also commonly occurs as shiny flakes and nuggets. Because it can be found in stream gravels and other PLACER DEPOSITS, it was undoubtedly among the first metals to be mined. Gold NUGGETS are often found in streambeds and other watercourses, but they are also found in areas where Gold-bearing host rock has weathered away or decayed. As a result of tumbling in water or weathering, nuggets often have a worn appearance. Nuggets

FIGURE 273.
Natural Gold crystal from Summit County, Colorado, showing a combination of octahedral crystals and natural wire. Specimen weighs 2.7 troy ounces (FMNH H847).

are never pure 24-karat Gold. Instead, they generally range between 20 and 22 karats (83–92% pure). The natural impurities alloyed with Gold in nuggets are most commonly Silver and/or Copper. Gold nuggets that contain significant amounts of Silver are also called ELECTRUM. Electrum was occasionally referred to as "white gold" in ancient times, but it is actually more of a pale shade of Gold. Some of the earliest ancient coins used for commerce were made of electrum.

The largest Gold nugget found to date was the "Welcome Stranger" nugget discovered in Victoria, Australia, in 1869. It weighed over 2,500 troy ounces and was refined to 2,284 troy ounces (about 157 pounds) of pure Gold. Victoria, Australia, is a well-known source of large nuggets, and another notable example from that region is a large Gold nugget in the Grainger Hall of Gems weighing nearly 4 pounds (fig. 271). The Grainger Hall display also has a number of smaller nuggets from North America. Native Gold is also found in crystallized form, but only rarely. Fine Gold crystal specimens are much rarer than nuggets, and consequently they are usually worth much more than the value

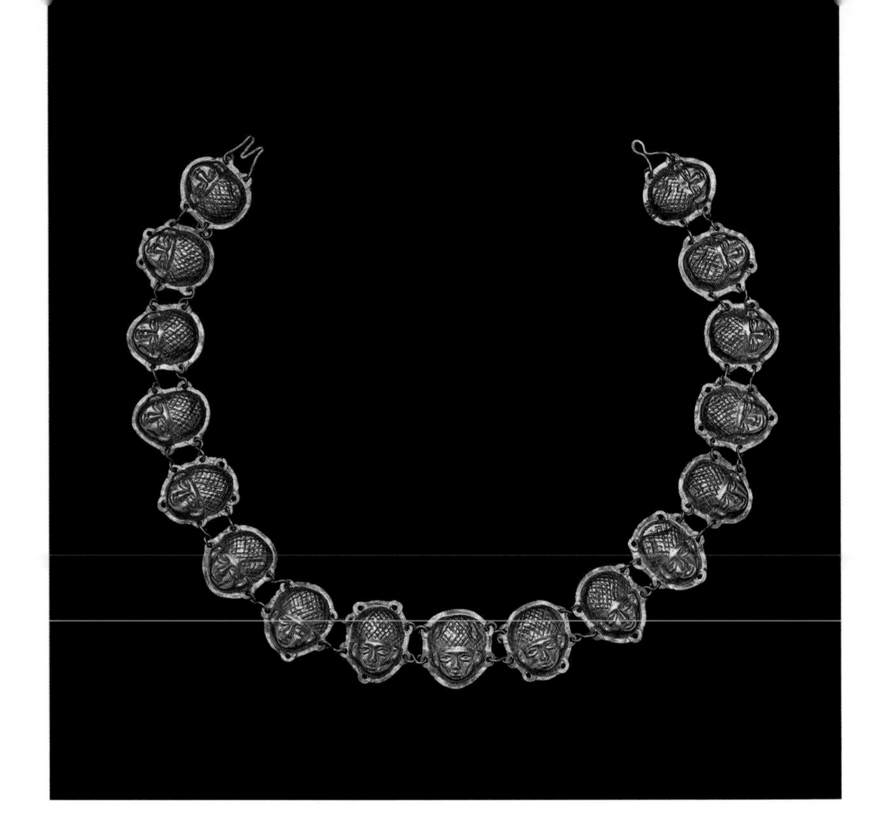

FIGURE 274.
Etruscan Gold necklace from Italy (probably between 800 and 500 BC). Length of necklace is 418 mm, and it weighs 35.5 grams (FMNH 239187).

of the Gold by weight. The most common type of Gold crystals are wire or leaf-like shapes, but crystals can also have the form of cubes, octahedrons, dodecahedrons, tetrahexahedrons, and trapezohedrons. These shapes are often found distorted in various ways. One beautiful distorted crystal specimen is a leaf-shaped piece in the Grainger Hall of Gems from the Yukon Territory of Canada (fig. 272). Another type of crystal specimen in the exhibition is a mixture of octahedral crystals and natural wire (fig. 273).

Gold has been a symbol of beauty and wealth for thousands of years. It has been used for ornamentation since the beginning of recorded history, worn by pharaohs and Aztec kings, and used to create artifacts of great religious significance for most every major religion. Several significant Gold cultural objects are on display in the Grainger Hall of Gems. One of the most impressive is the "Agusan Gold Image," a priceless thirteenth-century statue of a Hindu-Malayan goddess found in the Philippines in 1917. It was discovered on a silty riverbank of the Wawa River near Esperanza, Agusan del Sur, Mindanao, after a flood. It is one of the most spectacular discoveries in Philippine archaeological history. The statue is made of about 4 pounds of solid 21-karat Gold and stands about 7 inches tall (fig. 277).

Gold has long been the precious metal standard for fine jewelry. The oldest known Gold jewelry in the world is from an archaeological site in Varna Necropolis, Bulgaria, and is over 6,000 years old (radiocarbon dated between 4600 and 4200 BC). This prehistoric jewelry belonged to the NEOLITHIC Varna

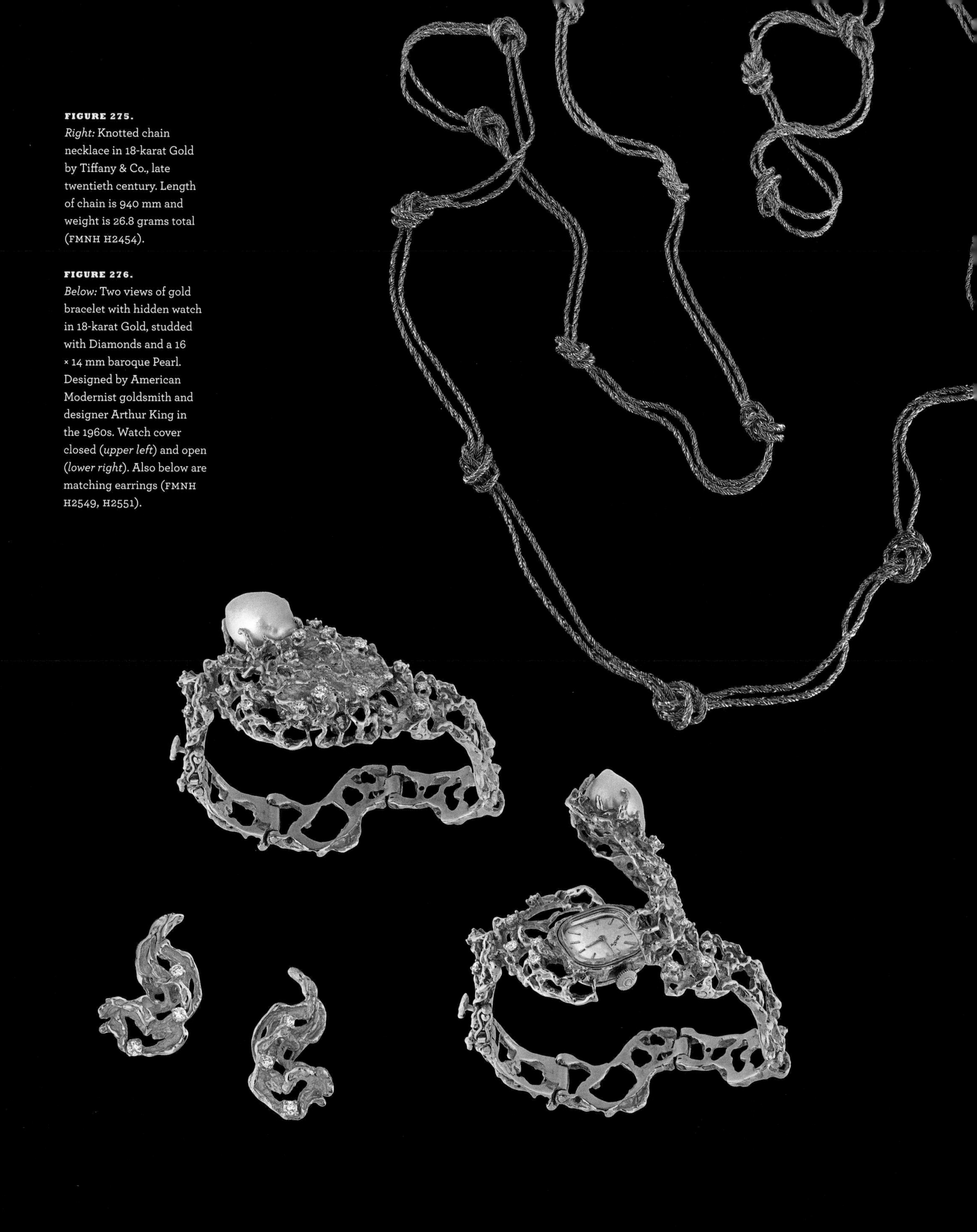

FIGURE 275.
Right: Knotted chain necklace in 18-karat Gold by Tiffany & Co., late twentieth century. Length of chain is 940 mm and weight is 26.8 grams total (FMNH H2454).

FIGURE 276.
Below: Two views of gold bracelet with hidden watch in 18-karat Gold, studded with Diamonds and a 16 × 14 mm baroque Pearl. Designed by American Modernist goldsmith and designer Arthur King in the 1960s. Watch cover closed (*upper left*) and open (*lower right*). Also below are matching earrings (FMNH H2549, H2551).

culture and shows stunningly advanced Gold-working artistry. The oldest known Gold jewelry in the Americas is from an archaeological site in Peru, near Lake Titicaca, dated at about 2000 BC. This material came from a hunter-gatherer society that predated the later Chavin, Moche, and Inca societies of the region. The beauty, workability, and tarnish resistance of Gold has made it the precious metal of choice for jewelry making to this day. Many of the jewelry pieces in the Grainger Hall of Gems are set in Gold, including necklaces thousands of years old as well as modern pieces (e.g., figs. 274–276, 278).

Gold has also been a medium of exchange and wealth accumulation for millennia. There were medals and non-circulating coins made in ancient Egypt, but these were used mainly as gifts and as items of adornment rather than as money. The first known metal coins for commerce were made out of electrum with high Gold content, and they date back to the seventh century BC in the ancient Greek area of Ionia, which is in present-day Turkey. These coins, called **STATERS**, were at first strictly utilitarian in design: irregularly shaped lumps of electrum or Gold with crude markings consisting of simple striations and punch marks. Although they were irregular in shape, these coins began to establish standards of weight and Gold purity, which facilitated trade and commerce. Using coins as currency served to streamline trade beyond a primitive barter system because it represented a compact form of commodity for transactions. Throughout history, Gold has established the monetary value of currencies and served as money longer than any other material.

The art of Gold coinage evolved rapidly after the issuance of the first staters. The Gold purity and weight of coins increased, and the later coins of ancient Greece and Rome became miniature artistic sculptures of deities, royalty, and political expression as well as the metal of choice for rendering idealized beauty (fig. 279). Gold coins have also been incorporated into jewelry since ancient times (fig. 279, *bottom*).

Over the centuries, Gold coinage continued to be a key part of trade and commerce. Until the early twentieth century, the Gold standard formed the basis for nearly all of the world's currencies. In the United States, Gold coinage was used as money from 1795 up until the early 1930s. U.S. Gold coins contained 90 percent Gold and 10 percent Copper, making it 21.6-karat Gold. (U.S. Gold coinage from 1795 to 1804 was slightly higher in Gold purity at 91.67 percent, or 22 karats.) Some of the most artistic coins of the United States were $10 and $20 Gold coins from the early twentieth century designed by the great American sculptor Augustus Saint-Gaudens. Examples of these works of **NUMISMATIC** art are on display in the Grainger Hall of Gems (fig. 280). Since 1986 modern Gold bullion pieces made for sale by the U.S. government have adopted the obverse design of the Saint-Gaudens $20 Gold piece, but these modern pieces are not really functional "coins." They are intentionally assigned a "cur-

FIGURE 277.
Facing page: The "Agusan Gold Image," a priceless thirteenth-century statue of a Hindu-Malayan goddess discovered in the Philippines in 1917, made of 4 pounds of solid 21-karat Gold. Statue is about 178 mm in height (FMNH A109928).

FIGURE 278.
Facing page: Ancient pieces of Gold jewelry. *Top:* A pair of Gold Etruscan earrings from Italy (seventh to sixth century BC), each measuring 36 mm in diameter (FMNH A239153). *Bottom:* Frog effigy made to wear as a pendant, Veraguas style, from Panama (twelfth to fourteenth century AD), weighing 1.08 Troy ounces and measuring 60 × 46 mm (FMNH A6388).

FIGURE 279.
Right: Ancient gold coins as a reflection of artistic expression. *Top:* Greek stater from around 336–323 BC, in obverse view depicting the Greek goddess Athena (17 mm in diameter, weighing 7.9 grams). *Middle:* Greek stater from around 359–336 BC depicting the god Apollo on the obverse, and a charioteer on the reverse (17 mm in diameter, weighing 8.5 grams). *Bottom:* Roman Gold ring with a Gold solidus set into it from somewhere between AD 364 and 375 (FMNH A259779, A259780, A239056).

rency" value far below the market value of Gold and not intended for general circulation as money. They are common bullion pieces intended primarily for precious-metal investment, and their value follows the fluctuating price of Gold on the commodities market. These pieces copy the Saint-Gaudens design on their obverse side for aesthetic reasons to enhance their sale.

Gold coins ceased being used as money in most of the world by 1933, as countries abandoned the Gold standard during the worldwide economic crisis of the Great Depression. In the United States, Franklin D. Roosevelt signed Executive Order 6102 in 1933, outlawing the hoarding of Gold coins, Gold bullion, and Gold certificates by U.S. citizens. Gold for jewelry and limited collector coins were exempt from this order. In the decades following Executive Order 6102, huge numbers of the Gold coins in the United States were melted down by the federal government and made into Gold bars. In 1974 the limitation on ownership of Gold bullion and coins was repealed by President

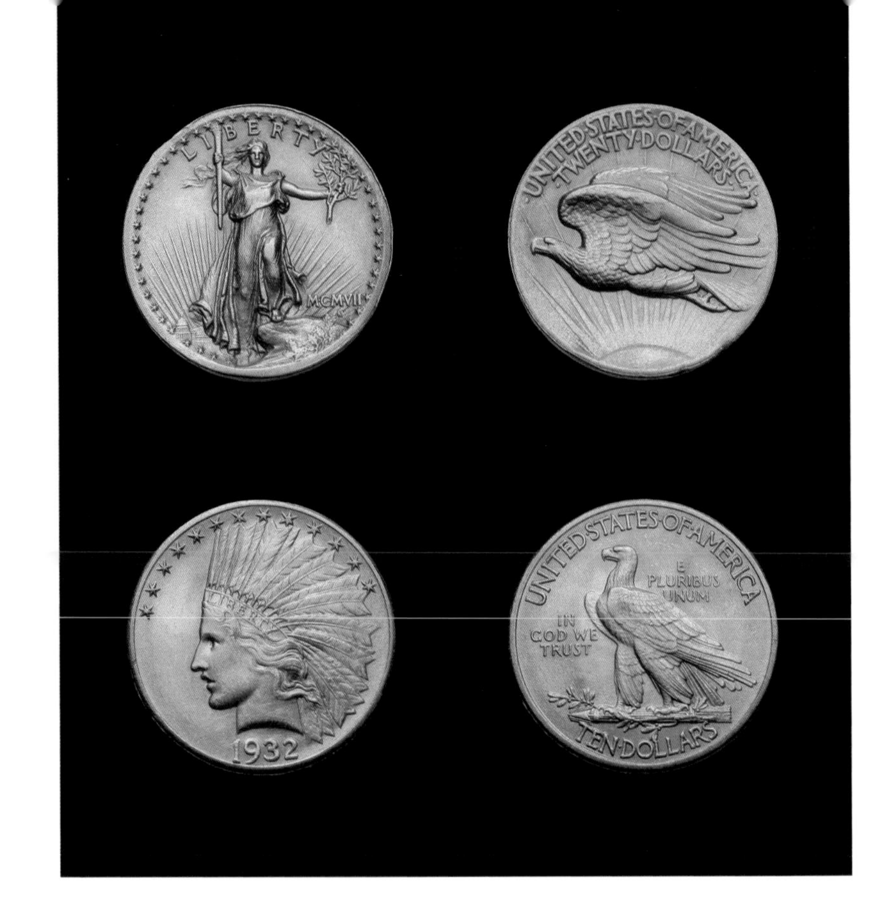

FIGURE 280. Early twentieth-century United States Gold coins designed by the famous American sculptor Augustus Saint-Gaudens (1848–1907). Top coin is obverse and reverse of a high-relief $20 Gold piece from 1907 with date in Roman numerals as MCMVII (34 mm in diameter, weighing 33.44 grams), and bottom coin is obverse and reverse of a $10 Gold piece from 1932 (27 mm in diameter, weighing 16.72 grams). These two coin types are considered by many to be the most beautiful circulating coins ever minted by the United States (FMNH H2580, H2578).

Gerald Ford, but the minting of Gold coins for use as normal currency has never been resumed. Although Gold coins are no longer used as currency, the specimens that escaped melting by the government are today highly valued by numismatists. These coins are all stamped with dates ranging from 1795 to 1933 (although nearly all of the 1933 pieces were recalled and melted down into gold bars). Some dates were protected in numbers in foreign banks and later sold back into the U.S. coin market. Other dates are extremely rare and command extremely high prices. The highest price paid for any coin was for a Saint-Gaudens $20 Gold piece dated 1933. In 2002 this coin sold for $7.59 million at auction in New York.

Gold has a long and interesting history in human culture and is still the focus of major mining operations around the world. Although Gold deposits are geographically widespread, most Gold production today is located in South Africa, Russia, Brazil, Colombia, the United States, Canada, and Australia. Today Gold is still unsurpassed as the metal of choice for mounting fine gems.

SYNTHETIC GEMS, SIMULANT GEMS, *and* AUGMENTATION

Gems may be created or imitated in a variety of ways, but here we focus on two major categories of man-made gems. The first category is the **SYNTHETIC GEM**. A synthetic gem is a real gem, created out of the same chemical elements you would find in nature. It is called a synthetic because scientists create the gem in a laboratory, which is obviously not a natural process. The second type of man-made gem is the **SIMULANT GEM**. A simulant gem is an imitation or fake, and can be created through a variety of methods. Occasionally, simulants are also natural stones of one mineral species used to mimic another. Both synthetic and man-made simulant gems are relatively inexpensive to create in a laboratory (in most cases), and both take much less time to create in the lab than they would take to form in nature.

While the art of *imitating* gems goes back thousands of years, the science of successfully *creating* gems began in the late nineteenth century. Imitation and created gems are used for a variety of purposes, ranging from being mounted in inexpensive jewelry to being used by industry to create items such as Diamond drill bits or watch bearings.

SYNTHETICS

Of the two types of man-made gems, synthetics are the most similar to gems grown in nature. In fact, it sometimes takes an expert to tell them apart. Although not "naturally" produced, synthetic gems are not "fake." They are real gemstones and precise copies that display the qualities of beauty and durability that define natural gems. Scarcity does not apply to synthetics, however, because synthetics can be manufactured at will

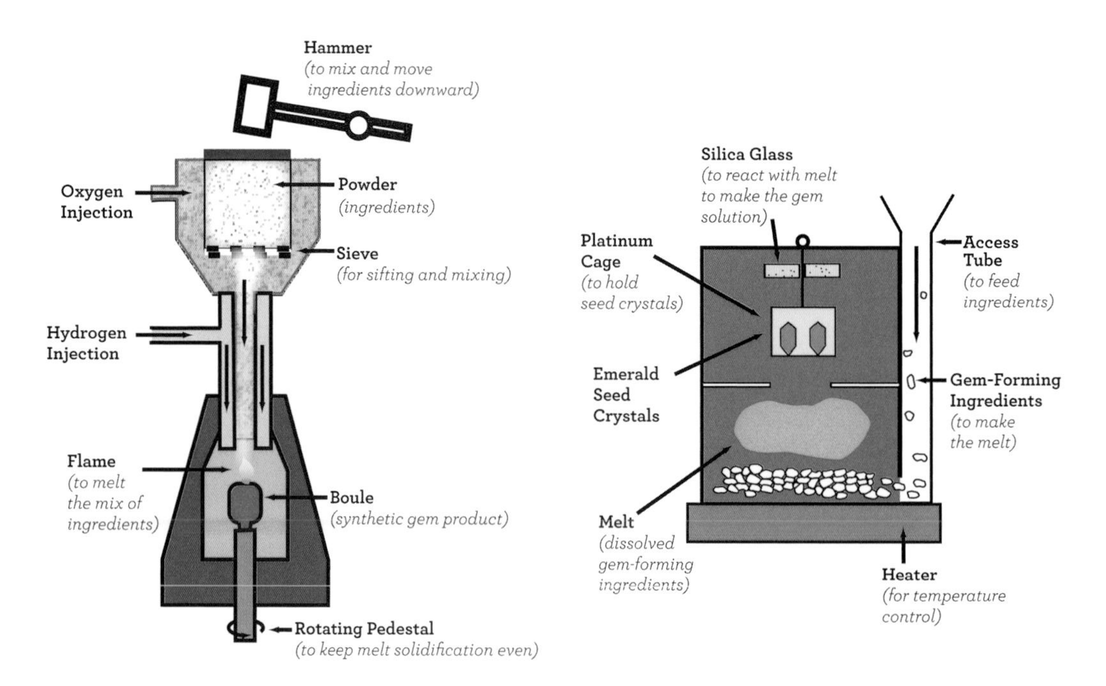

FIGURE 281. Processes of synthetic gemstone production explained diagrammatically. Flame-fusion method for production of rubies and sapphires *(left)*. Flux-melt method for production of emerald crystals *(right)*.

in a lab using common materials. This is the main reason that synthetic gems are usually much less expensive than their natural counterparts.

Although the process used to create a synthetic gem differs from nature's process, synthetic gems contain virtually the same chemical composition and structure as natural gems. Synthetics and natural gems have optical and physical properties that are extremely similar, and they are practically identical to the naked eye. They share common crystal structure, common colors, and common chemical elements. Despite the similarities, synthetics and natural gems are different in some ways. The most direct way to understand the differences is to understand a bit about how they are made.

Synthetic gems are created in laboratories and factories using a combination of mineral ingredients, high pressure, and high temperatures—as well as man's desire to copy the beauty of nature. The art and science of creating synthetic gems dates back to early attempts in the medieval ages. Those attempts were unsuccessful but laid a foundation for later improvements. In 1877 French chemist Edmond Frémy published a paper about his successful attempts to create clear, red rubies. These rubies were quite small, however. For 16 years afterward, Frémy and his assistant Auguste Verneuil worked together to create slightly larger rubies, as well as other colors of Corundum. His book *Synthèse du rubis*, published in 1891, explained how he and Verneuil were successful. Frémy and Verneuil recognized that Aluminum Oxide, a primary ingredient in making rubies, would start to crystallize when mixed with

FIGURE 282.
Synthetic emerald made with the flux-melt process. Crystal rough is 135.7 carats and faceted stone is 5.3 carats (FMNH H2162, H1718).

a variety of other chemicals and heated at high temperatures. The Aluminum Oxide compound turned into a vapor when reacting with water in moist air, and that vapor would recrystallize as ruby as it returned to room temperature. This is an early form of FLUX GROWTH, predecessor to a process used today know as FLUX-MELT (fig. 281, *right*).

Today's flux-melt process is different from what Frémy and Verneuil considered flux growth. The modern flux-melt process heats a saturated chemical SOLVENT (FLUX) with a small gem SEED CRYSTAL to start the crystal growth. First, gem-forming powdered chemical ingredients (such as Beryllium and Aluminum Oxides, used to make emeralds) are dissolved in a superheated solvent (flux). The dissolved chemicals and flux are together referred to as a MELT. Silica glass is floated on top of the melt, and the combination of melt and silica creates a gem-ready SOLUTION. Real gem seed crystals (such as very tiny emerald crystals) are placed into a Platinum cage and lowered into the gem-ready solution, and the temperature of the entire crucible is lowered. As the temperature drops, the solution becomes SUPERSATURATED. A solution is supersaturated when there is more dissolved material in the liquid than would occur in nature. In this case, very high laboratory temperatures create an environment for more material to dissolve in the liquid. As the solution continues to supersaturate, synthetic crystals form and grow on the seed crystals, creating sizable synthetic gems. Throughout the process, additional chemical ingredients are added to the melt through an access tube. This keeps the melt saturated with gem-forming ingredients; as the initial ingredients dissolve and turn into crystals, more ingredients are added to make more crystals. The process may take months but produces gems of market size and quality, and is particularly effective with emeralds (fig. 282).

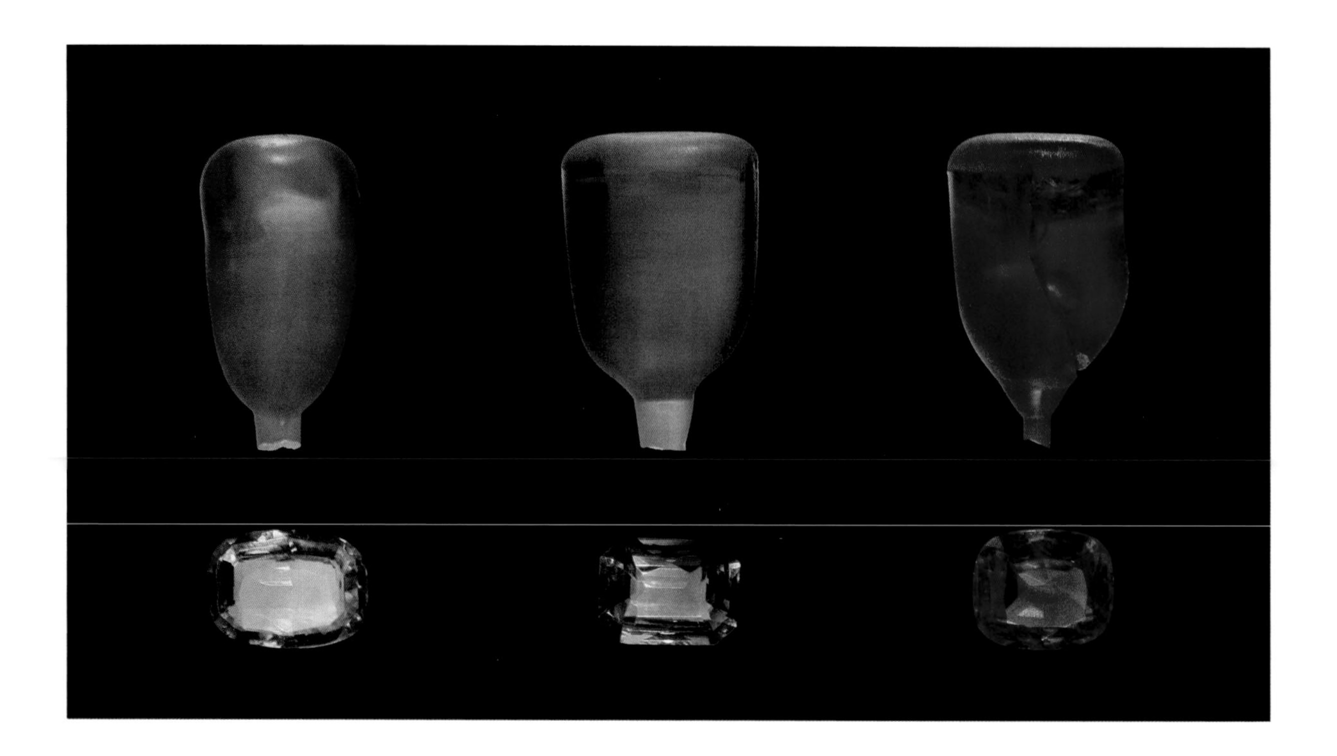

FIGURE 283.
Synthetic ruby and sapphire made using the flame-fusion process. Faceted gems along bottom with boules above. Boules weigh between 64 and 120 carats each, and the faceted stones are between 9.2 and 12.7 carats each (FMNH H752, H753, H766, H767, H770, H771).

Flux growth was one of the earliest effective synthetic methods, as Frémy and Verneuil recognized. But Verneuil was able to take gem synthesis even further than his teacher. His work with Frémy provided the foundation he needed to formulate his own synthesis method called FLAME-FUSION. Flame-fusion is so effective that it is still used today, and the process remains relatively unchanged. It is often referred to as VERNEUIL FLAME-FUSION (see fig. 281, *left*).

Flame-fusion fuses ingredients together in a superheated flame furnace. Elements of the process to produce Corundum (such as ruby and sapphire) include powdered Aluminum Oxide for the main body, traces of additional chemical elements to create different colors, an Oxygen-Hydrogen flame, and a hammer. The powdered gem ingredients are placed in a holding chamber above the flame and mixed with Oxygen. A hammer repeatedly taps the top of the chamber to ensure that the ingredients are constantly shifting and sifting in the proper amounts before dropping into the flame below. As the sifted chemical ingredients drop, they pass into an Oxygen-Hydrogen furnace, melting as they fall through a flame of more than 3,630°F (2,000°C). The ingredients melt into chemically saturated liquid drops, drip onto a rotating pedestal below, and crystallize. The pedestal is withdrawn downward out of the flame chamber, and a long, cylindrical crystal forms. This crystal is known as a BOULE. A gem can be cut and polished from a boule (fig. 283). Verneuil developed this technique primarily for rubies, but with different ingredients the technique creates other gems, including other colored Corundums like sapphires, and colored Spinel gems.

There are other synthetic methods, and new and improved synthetic methods continue to emerge as scientists learn more about the processes of gem formation. Additionally, some companies, including the French company Gilson and the American company Chatham, use methods that are closely guarded, akin to a "secret recipe." Both of these companies are known for creating high-end jewelry gems (emeralds, rubies, sapphires, lapis lazuli, Turquoise, and Coral, among others). Diamond giant De Beers also experiments with creating gem-quality Diamonds from inexpensive sources of Carbon, such as Graphite or very small Diamond crystals. They have succeeded in producing gems that can be used in jewelry. They are also active in developing instruments that differentiate between natural and synthetic Diamonds. Currently, some of the processes used to create gem-quality Diamonds are very costly, and the resulting synthetic stones may cost nearly the same as natural stones of equivalent size and quality.

Sometimes synthetics are used in unconventional ways. For example, in recent years, the symbolic timelessness of Diamonds and advanced synthetic Diamond technology has cultivated a rather unusual industry: the creation of synthetic Diamonds from the Carbon ash of deceased loved ones. This synthetic is chemically identical to a real Diamond and does not possess any particular human imprint or DNA. It relies on the Carbon ash from cremated remains. The cost of the process ranges from $4,000 to over $30,000, depending on the size of the Diamond and is currently more expensive than the cost of a natural Diamond of similar size and quality. Obviously, the price paid for a gem made from a loved one includes a great deal of personal sentiment. This idea has even been used for the ashes of cremated pets (the ultimate "pet rock"). Such an idea changes the traditional definition of "gem." But this strange example also underscores the degree to which people will use technology to replicate what they see as enduring, and exemplifies man's emotional investment in creating gemstones.

Yet for all man's concerted efforts to replicate natural gems, a skilled gemologist can still distinguish a synthetic stone from a natural stone by telltale indicators, including growth lines, color zoning, and particular types of inclusions specific to individual stone varieties. Flaws or inclusions, or sometimes the very lack of these imperfections, are primary synthetic indicators. If a stone is completely unflawed, it is often a good indication that it may be a synthetic, as flawless stones for most gem varieties are very rare in nature. If a stone is flawed, the flaws or inclusions sometimes differ between natural and synthetic stones. Synthetic inclusions may be typical of a process or of a synthetic gem species. Under magnification, a flame-fusion ruby reveals curved growth lines rather than the straight lines that occur in nature. This is because the ingredients in the synthetic stone have not mixed together as fully as in the natural stone.

The gradual mixing that occurs in nature requires a process much longer than what is possible in a lab. And in flux-melt emeralds, we find characteristic patterns that resemble veils or feathers deep within the internal structure of the stone. These are just a few examples of the ways to differentiate a synthetic from a natural gem.

SIMULANTS

Simulants are another type of man-made gem, but these are imitation or fake "gems." Simulants differ from synthetics because they are intended to fool the eye rather than replicate nature. For this reason, simulants are usually much easier to identify because the physical properties of simulants are quite different from those of natural stones. For example, man-made glass is a popular simulant, often used to imitate a Diamond. The internal chemical structure of each is very different, even though they sometimes look similar to the naked eye.

Because it is easier to fake than make a gem, people created simulants long before they created gems in a lab. In fact, simulants predate the manufacture of synthetic gems by over 5,000 years. Ancient Egyptians imitated gemstones with glass because most genuine gems were too rare or expensive for common people. And imitation beads simulating lapis lazuli have been found dating back to 4000 BC. Today some natural minerals such as Zircon, as well as man-made synthetics, are used to imitate Diamonds. The most common man-made Zircon compound is Cubic Zirconium Oxide, or cubic zirconia. Cubic zirconia (CZ for short) has a Mohs scale hardness of 8½ and is marketed in several countries under different names such as phainite, djevalite, and diamonesque. Because it has excellent **FIRE** and **BRILLIANCE**, it is the most popular Diamond simulant. Glass is also often used to replicate gems, but because it chips easily and is only singly refractive (not doubly refractive), the fire and light of the "stone" is greatly reduced and simplified, and not as attractive. Another example of a simulant is a **FABRICATION**, which often combines natural stones with man-made glass or other natural stones. For example, one type of fabrication is the **DOUBLET**, in which a natural stone is cemented to a colored glass base, which gives the upper stone additional color. Some white Opal doublets are made to look like expensive black Opals by fusing a thin white translucent Opal to a base of black stone. If doublets are set into jewelry, it may be difficult to distinguish between the two layers cemented together if examined from above. It is sometimes easier to see the two pieces glued together when viewed from the side. Creative techniques such as this can create well-disguised imitations, so it is wise to examine gems closely when purchasing so that you know whether or not you are buying a simulant.

It is also important to know about several AUGMENTATION practices that are increasingly accepted in the gem market. These practices fall into a gray middle ground when defining gems as natural, synthetic, or simulant. Augmentation practices include HEAT TREATMENT, IRRADIATION, STAINING, and OILING.

HEAT TREATMENT is becoming more common in modern practice, used to change the color or clarity of gems. For example, aquamarine is known for being pure blue but may be actually more of a green color in its natural state. When it is heated to about 450°C, the natural color often changes to the more familiar blue. Similarly, brownish-green Zoisite (tanzanite), when heated to around 370°C, becomes the more market-popular blue color. And poorly colored purple amethyst, when heated to about 450°C, becomes yellow citrine. Heat treatment is also widely used to enhance the color of ruby and sapphire, although much higher temperatures are required (1,000° to 1,900°C, depending on original color). Heat treatment can also improve the clarity of gems, as external heat can change the internal structure, improving existing flaws within.

IRRADIATION is a radiation process that changes the color of a gem. This process may occur naturally, but is often used in laboratories to improve the color quality of gems. Currently, blue Topaz is the principal gem variety that benefits from irradiation color-enhancement. Irradiation takes place either with gamma rays generated by a radioactive source, neutrons from a nuclear reactor, or electrons from a linear accelerator. Irradiation is often used on colorless or brown Topazes. The stones are heated to around 250°C, which produces a desirable blue color. The blue Topazes are then put on the market as pure blue Topaz. In this case, the color of the Topaz has been changed in a laboratory. But this same radiation process may have occurred naturally if the gem had more time to grow before it was removed from the ground. The likelihood of this happening is difficult to determine, which is why irradiation is a form of augmentation that falls into our gray middle ground and is readily accepted by the gem trade. Laboratory irradiation does not differ substantially from natural irradiation except for the duration of the process. Natural irradiation within the earth may occur gradually over millions of years, while irradiation in a lab takes only a few hours. There is therefore no easy, immediate way to distinguish a man-made irradiated blue Topaz from a naturally irradiated blue Topaz.

STAINING is the process of adding stains, dyes, or chemicals to alter the surface layer of a gem. This process works best with porous stones, such as Turquoise or jade. Poorly colored jade is sometimes dyed to simulate more valuable green colors. The gem is still real, but the staining process has improved the color. Staining is usually detected by microscopic examination of surface cracks to locate concentrations of dye. Dyes are used to simulate Turquoise, precious Coral, lapis, and other lesser gemstones. A SPECTROSCOPE, an instrument used to measure light, can be used to distinguish a dye from a natural color based on light reflection of the colors. A stone that is colored with a stain or dye is generally much less valuable than a naturally colored stone.

OILING is a process in which colored oil is applied to a gem to fill cracks in the surface, thereby hiding natural flaws and covering up blemishes. Oiling is most commonly used on emeralds. The stone is placed in an oil-based dye that has been colored to match the stone. Sometimes the stone is left to sit in the oil for a long time for the oil to achieve sufficient penetration into the deeper cracks, while other times a partial vacuum is used to draw the filler even deeper into the stone. It is important to remember that one should never put natural emerald jewelry in detergent or hot soapy water, as this can leach oil or other fillers out of the stone. Oiling can usually be detected under low-power microscopic examination of surface fractures, and oil-impregnated stones will sometimes develop a somewhat greasy feel over time. Other types of filler used in addition to or in place of oiling include epoxy resin or even colorless glass.

In some cases, augmentation techniques are permanent and irreversible, such as irradiation with certain varieties of Beryl, Corundum, Tourmaline, Topaz, and Quartz. But some irradiated stones will revert to their old color after some time. An oiled gem may lose its filler, or a stained colored stone may fade. Most of these telltale changes take time. So there are systems in place to help the buyer immediately recognize augmentation techniques. The World Jewelry Confederation (WJC) requires that non-permanent color-enhanced gems be designated as TREATED GEMS. The "treated" term is intended to prevent deception, making the buyer aware that the gem has been enhanced in some artificial fashion that may fade with time.

Given all the methods of creating or enhancing gemstones, one would think that man would now be capable of sidestepping nature completely, creating any gem that he needs or desires. But there is an inherent attraction to the purity of nature, to beauty that has not been altered. All other elements being equal, a manufactured or enhanced gem can never possess the wonder (or value) of a gem created by nature.

For further reading on synthetics and simulants, we recommend Read (2005).

MINING

The earth is in constant motion beneath our feet. A seemingly stable surface covers extreme geologic forces: movement of continental plates, earthquakes, volcanic activity, and chemical reactions that are possible only under unimaginable pressures and superheated temperatures. Over thousands to millions of years, these forces help to create gemstones. These same forces, combined with erosion, eventually move the gemstones closer to the earth's surface. Once within reachable distance, gemstones can be mined using a number of methods, including artisanal, placer, hard rock, and open pit mining.

ARTISANAL MINING comes from the word ARTISAN, meaning someone who creates a handmade object without the aid of machine manufacturing. Artisanal mining is simply mining done by hand using simple tools or sieves. Workers scratch the earth's surface, digging slightly into the crust, in riverbeds, anywhere they can reach with their bodies. Artisanal mining is the general method of SUBSISTENCE MINING, prevalent in poorer countries that lack readily accessible machinery or electricity, such as Sierra Leone or Myanmar.

PLACER MINING is a slightly more focused gem hunt, searching eroded geological pockets on the earth's surface, including alluvial, eluvial, and colluvial geological deposits. ALLUVIAL DEPOSITS are soil or sediments deposited by a river or other running water. ELUVIAL DEPOSITS are geological deposits left behind by the weathering of earth over time. And COLLUVIAL DEPOSITS are geological deposits created by the accumulation of eroded material at the base of an incline, such as a mountain. The eroded soil of these deposits makes it easier for miners to find gems. Placer miners often use artisanal methods to hand-dig the wet or dry deposits to find gems, or they can use water-pressurized machines to explore the surface of a dry dig. Artisanal methods of placer mining include stream-washing gravel in baskets, screens, or sieves, or "panning" for

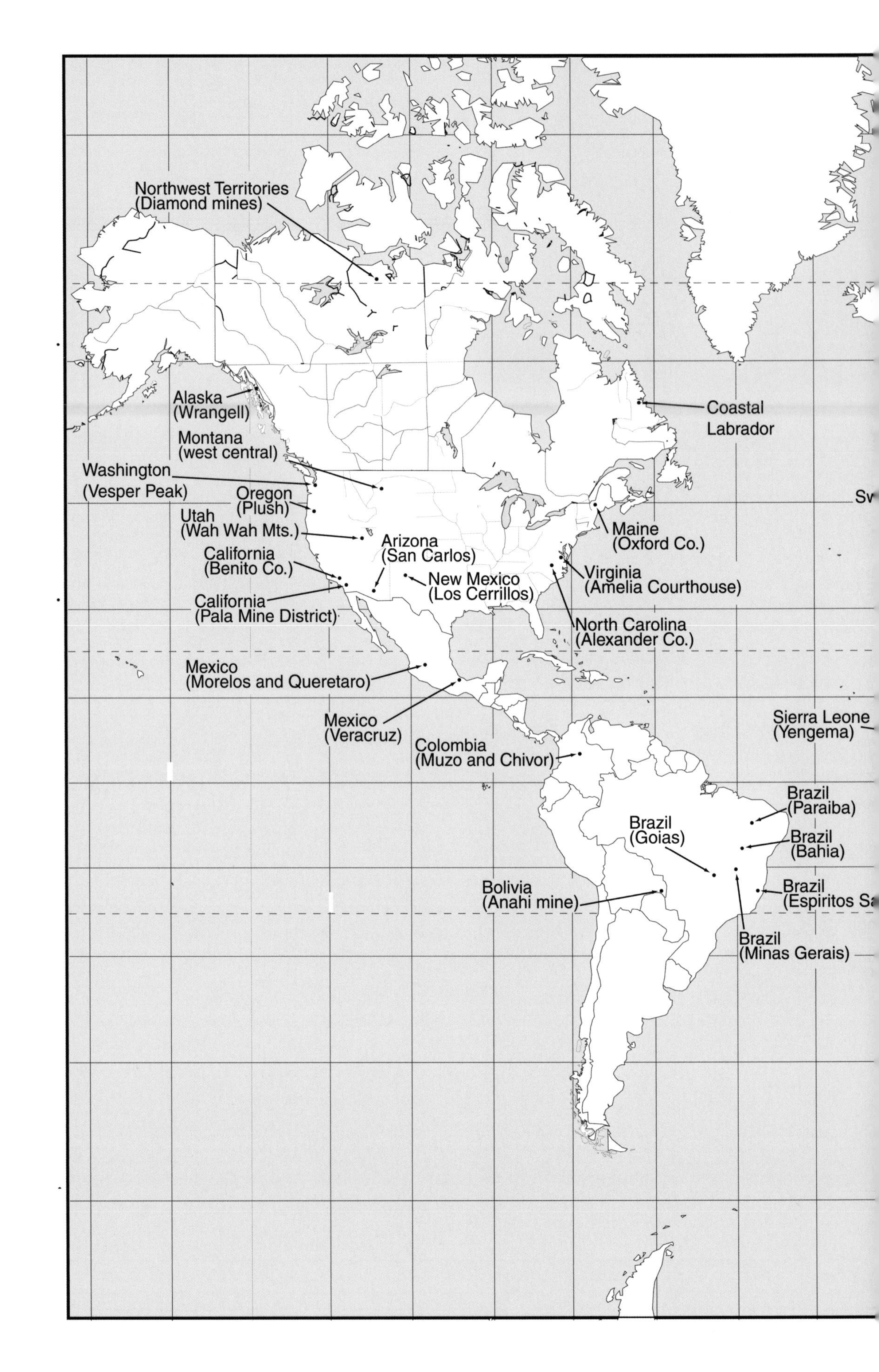

FIGURE 284.
Map showing the general locations of some gemstone mining deposits and regions mentioned in this book.

Russia (Siberian Region)
Russia (Yakutia)
Russia (Udachnaya)
Russia (Ural Mountains)
Russia (Alabashka)
Russia (Poldnevaya)
Russia (Yekaterinburg)
Czech Republic (Meronitz)
Italy (Lanzada)
Tajikistan
Afghanistan (Jagdalek)
Pakistan
Iran (Razari Khorasan)
Afghanistan (Nuristan)
Myanmar (Mogok)
Egypt (Zebirget)
India (Kashmir)
Vietnam (Yen Bai)
China (Fujian Province)
India (Golconda)
Myanmar (Tawmaw)
India (Madras)
Myanmar (Bago)
Cambodia (Palin)
Sri Lanka (Ratnapura)
Congo (Tshikapa)
Kenya (Voi)
Tanzania (Umba Valley)
Angola
Madagascar (Tsihombe)
Mozambique (Alto Lingonha)
South Africa (Kimberley)
Western Australia (Kimberley)
Queensland (Sapphire)
Queensland (Quilpie)
South Australia (Andamooka Opal Fields)
New South Wales (Inverell)
New South Wales (Lightning Ridge)
New South Wales (White Cliffs)

Gold and gems by swirling water over the gravel. The water separates the dirt from the gems, and the heavier Gold and gem material, such as sapphire and Diamond, sink to the bottom of the basket and can then be separated more easily. Miners may also rake the area of a dry dig, turning soil to expose gemstones.

OPEN PIT MINING is largely machine-based and most effective for uncovering large surface areas. It is not necessary in some mines to dig deep to find gemstones because they are located in veins or pockets near the surface. But in such mines, it is often necessary to excavate a large volume of material to find quantities of gemstones. Machines create large open pits on the surface of the earth, turning over large surface areas with shallow digs. These wide, shallow digs can unearth more gems than a single, specific gem tunnel would provide. Some open pits measure more than 2 miles across.

HARD ROCK MINING is used when gems are mined in hard rock. To get to the gems, miners dig a series of tunnels deep within the earth. Explosives, tractors, and other pieces of equipment are used to tunnel into the prospective gemstones deposits. If the rock is strong enough, the tunnels and room stand on their own. Otherwise, pillars are constructed to hold up the walls to prevent the room from collapsing.

We traveled to the Ocean View Mine, in the famous Pala mining district of Southern California, to explore an underground gemstone mine, prospecting mainly for Tourmaline, aquamarine, and Quartz. The Ocean View Mine uses both hard rock mining to dig underground, as well as a variety of artisanal methods to prospect within the tunnels. The mine has a single main tunnel with hard rock walls that have been blasted and dug into the side of a large hill (fig. 286A, B). There are a number of chambers and VUGS leading off of the main tunnel that were once hydrothermal pockets where crystals formed millions of years ago. The vugs over time have filled with dense, sticky clay that has to be removed by hand (figs. 286C–E). Heavy machinery is not practical because there is a chance that a machine can break the gem crystals. Human hands can feel through the clay, and hand-held tools have a better chance at extracting crystals without breaking them. On our trip into the mine, some of the vugs were not easily accessible from the ground level of the tunnel. This required the workers (and the authors) to climb a 30-foot incline with a rope to hold and guide us as we moved our way upward into the earth. Once the clay pockets were accessed, miners used hands, shovels, and other digging implements to slowly break apart the large, sticky clay deposits to expose the larger gemstone crystals. The clay was placed into buckets and carried outside of the mine cave to be screen washed and further examined for small crystals. Screen washing is simple: the crystal-containing clay is spread over a large mesh screen and rinsed with water from a standard garden hose to dissolve the clay and expose the crystals contained in the clay (fig. 286F). We managed to find a few crystals for our hard work (fig. 286G). The mine owners jokingly refer to the process as "primordial mining"—it is slow and can be physically painful, especially when encountering sharp gravel deposits within the clay.

The mining techniques mentioned above are used throughout different parts of the world. Each technique is effective in its own way and tailored to both the geology of the mine and the economic scale of the regional human culture. While there are many famous gem mines in the world, among the most famous are the Kimberley mines of South Africa, the Mogok Valley of Myanmar, Minas Gerais in Brazil, and many regions of Sri Lanka (see fig. 284).

Much of southern Africa produces Diamonds, but Diamond mining only

FIGURE 285.
Facing page: Mining and processing placer deposits from the Rock Creek sapphire mine in Philipsburg, Montana. *Top left:* Fresh washed alluvium on a screen-bottomed pan showing several raw sapphire pebbles in the center (enlarged in figure to the right). *Top right:* Close-up of raw sapphire pebbles discovered in the alluvium (arrow) (sapphire group is cataloged as FMNH H2599). *Bottom left:* A high-graded group of several hundred heat-treated raw sapphires and rubies from the Rock Creek mine, ranging from 1 to 2.5 carats each. This pile of prime gemstones represents about one week's yield from the Rock Creek mine at its peak production level in the late 1990s. *Bottom right:* A handful of faceted sapphires from the Rock Creek mine. Bottom two images courtesy of the Sapphire Gallery in Philipsburg, Montana.

FIGURE 286.
Ocean View Mine of Southern California, 2008. (A): The authors stand in the entrance to the Pala Mine. (B): We hike into the mine. Yellow tube along left ceiling corner moves fresh air deep into mine. (C and D): Once inside the mine, we find a pocket of soft reddish material in the hard rock wall indicating a clay-filled gem vug that was once a hydrothermal vent. We begin to remove clay to expose large crystals of Quartz and other minerals. (E): We remove the clay from the pocket and put it in buckets to search for smaller crystals. (F): We screen wash the clay in water outside the mine to dissolve the clay and expose loose crystals. (G): Two crystals discovered in the clay—a smoky Quartz crystal (*left*) and a Schorl Tourmaline (*right*) (FMNH H2583, H2584).

FIGURE 287.
The "Big Hole" open pit mine, with the business district of Kimberley, South Africa, in the background. This open pit mine, thought to be the largest hand-dug pit ever made, was dug by approximately 50,000 miners between 1866 and 1914 in search of Diamonds. At its maximum, it was about a mile wide and a half mile deep. Photo by Richard L. Jones, Kimberley, South Africa, 2009.

began in the late 1800s. The first Diamond found in Africa was located in 1867 in Cape Colony. Fifteen-year-old Erasmus Jacobs found a 21.25-carat transparent stone near Hopetown while playing on his father's farm. The discovery led to a Diamond rush across the entire region, enticing people like Cecil Rhodes to South Africa. Rhodes purchased Diamond shares and land tracts from a number of people, including two brothers named De Beers. Rhodes kept the name and founded De Beers Consolidated Mines Ltd. Rhodes found much of his vast fortune in the city of Kimberley, which would become known as one of the world's most famous Diamond mines for the sheer volume of Diamonds found there. By the 1870s and 1880s, Kimberley was home to mines that produced 95 percent of the world's Diamonds, including some of the world's most famous pieces like the Cullinan Diamond in 1905. The Cullinan weighed 3,106 carats uncut (p. 48). De Beers became the largest company to mine the area, making a fortune off of the land's rich deposits. The hill on the De Beers Brothers farm was slowly mined until it became a gigantic gouge in the earth. This gouge became known as the "Big Hole" and can still be seen in Kimberley today, along with a museum about the city's Diamond mining history (see fig. 287). The area continues to yield Diamonds and to date has produced 505 million carats.

Across the Atlantic Ocean in Brazil is another famous mining region known as Minas Gerais. When the Portuguese controlled the area in the 1700s, they named this region Minas dos Matos Gerais, or "mines of the general woods." The area has yielded some of the most impressive gems in the world, due to the fact that Minas Gerais has more gem-producing PEGMATITES (volcanic rocks) than any other country. These pegmatites produce the world's largest Topazes

FIGURE 288.
Garimpieros working the mines in Apui, in the state of Amazonas, Brazil. *Left:* Digging alluvium for processing for gems and gold. *Right:* Panning for gems and gold. Photos from Susan Schulman/Getty Images, 2007.

and a variety of impressive Beryls ranging from emeralds to aquamarines to heliodors, as well as sizable Diamonds, Quartzes, Tourmalines, and a vast number of other gems. Much of the area is mined in the artisanal style, hand sifting conducted by workers called GARIMPEIROS (fig. 288). Larger, open pit mining is also an option, but is usually conducted by large companies within the region. New and sometimes unexpected gem discoveries still occur here. For example, alexandrite was discovered in the region for the first time in 1987. The area continues to be profitable with its wide variety of known gems, as well as occasional new discoveries.

Myanmar (formerly Burma) is slightly smaller than the state of Texas, yet it produces the majority of the world's finest rubies, including the famous pigeon blood variety that are considered to have the most desirable red color of all rubies. Northern Myanmar's Mogok Valley is known as "the valley of rubies" and remains legendary in that regard. Here miners unearth unusually large specimens of the pigeon blood rubies, as well as particularly fine crystals of blue sapphire. In addition to ruby and sapphire, Myanmar also produces outstanding Chrysoberyl, Tourmaline, and Spinel. Middle Pleistocene–era tools of Jadeite have been unearthed in the mining areas of Burma, suggesting that people were mining the area much longer than the cities are known to have existed. Recent history is easier to track but fraught with conflict. In 1886 the British annexed Burma, taking control of the city of Mogok. By 1889 a famous British company called Burma Ruby Mines Ltd. mined the region until the company failed in 1925. By the 1930s, smaller local mines set up and remained mostly stable until 1962, when a military coup plunged the country into isolation. This led to the nationalization of gem mining, and the private trading of gems was outlawed. In 1988 anti-government riots racked the country. Ever since then, government restrictions have slowly loosened. Today's mining practices remain largely non-mechanized, relying primarily on human labor. Poor mining conditions and badly treated workers have created much controversy for the region. In 2008 the United States government banned the import of Burmese gems in protest of Myanmar's inhumane mining methods. Other governments have also restricted purchasing from Myanmar, and many international jewelers such as Tiffany & Co. and Cartier have refused to purchase gems from Myanmar until conditions improve.

Another impressive and historical gem-producing country is the tiny island of Sri Lanka, formally known as Ceylon. Gem mining and trading in Sri Lanka goes back some 2,000 years. The center of gem trades is the town Ratnapura (Ratnapoora), which is Singhalese for "gem town." The traditional mining areas of Ceylon were located in the same vicinity. The country is appropriately named "the island of gems," as Sri Lanka boasts an abundance of Corundums, Chrysoberyls and alexandrites, Garnets, moonstones, peridots, Spinels, Topazes, Tourmalines, and Zircons. Of all Sri Lanka's gems, perhaps the best known are the particularly fine sapphires. Sapphire from Sri Lanka occurs in a wide range of colors, including green, orange, pink, purple, yellow, and white. But it is the coveted orange-pink of the padparadscha and the rare cornflower blue known as Ceylon blue that drives the sapphire trade in Sri Lanka. Gem mining is conducted primarily in alluvial deposits found in ancient floodplains and streams that are now covered with productive farmland and terraced rice paddies. Mining pits are hand-dug by teams of workers who pump out groundwater as it enters the excavation. When they have dug down to the appropriate level, workers sluice the dirt and gravel, washing it in conical-shaped baskets to separate gems out of the mixture. As the basket is swirled around in water, the heavier stones such as Corundum settle to the bottom.

We only mention a few of the better known and historically significant mines here. In fact, there are hundreds of significant mines throughout place and time, ranging from the tiny artisan-mined bixbite (red Beryl crystal) mines in the Wah Wah Mountains of Utah, to the huge De Beers Diamond mega-mining operations of Canada, from the Middle Eastern emerald mines of Sikeit and Zabara going back some 3,500 years, to the paraiba Tourmaline mines in Brazil dating back only to the late 1980s. Economic, political, and ethical ramifications of gemstone mining are ever-changing, but mining itself is a very old and common form of natural resource exploitation and commerce throughout human cultures.

For further reading on mining, see O'Donoghue (2006).

ETHICS

When you purchase a gem, you are most likely curious about the "four Cs." They are color, cut, clarity, and carat. But there is also an important "fifth C" that can help you document both the quality and ethical provenance of a gem: CERTIFICATION. Certification is useful in authenticating size, quality, and gem species. Certification can also verify that your jewel is not a CONFLICT GEM, also referred to as a BLOOD GEM. Conflict gems (most often Diamonds) are used by rebel movements or their allies to finance armed conflicts aimed at undermining legitimate governments. The definition is sometimes broadened to encompass larger social issues, such as gems obtained illegally or through dangerous and inhumane mining conditions.

Gems are a convenient way to transfer massive amounts of wealth quickly and conveniently because they are small and can be easily transported. In some extreme cases, gems are smuggled across borders by people swallowing them or sewing gems just beneath the skin of farm animals to hide them. Once a gem finds its way to a major city, it can be sold quickly on the gem markets. If a gem passes through many hands, it can be difficult to trace its origin.

So how does one identify a conflict gem, or better yet a conflict-free gem? A warranty or certificate attesting to its conflict-free status is one possibility. In response to growing public concern, global markets have begun a series of protective measures to incorporate new ethical standards in the worldwide gem trade. The World Diamond Council (WDC) was formed in the year 2000 and includes representatives from Diamond manufacturing and trading companies all over the world. Its founding mission was to reduce the number of conflict Diamonds entering the market. It created a "System of Warranties," a measure pertaining to all Diamond transfers, requiring that all rough and polished Diamonds bought and sold by its members be conflict-free. Then, in 2003 the KIMBERLEY PROCESS CERTIFICATION SCHEME, otherwise known as KPCS and introduced by

United Nations resolution 55/56, was established. KPCS is an international certification scheme that requires member countries to pledge not to traffic in conflict Diamonds and that fully endorses the WDC System of Warranties. As of 2007, KPCS participants represented 74 governments that make up over 99 percent of the world's Diamond trade, including the United States, Canada, all of Europe, Brazil, China, India, the Russian Federation, and a substantial number of Diamond-producing countries throughout Africa, including South Africa.

While the spirit of these initiatives is admirable, there are still practical problems with such a voluntary regulatory system. First, there is no scientific process that can determine a Diamond's origin based only on examination of the gem. Therefore, verification possibilities are limited or nonexistent. The history and origin of the Diamond relies mainly on the truthfulness, accurate documentation, and inquisitiveness of the people who have mined it, carried it to market, and purchased it. Second, the KPCS in its own words "is not, strictly speaking, an international organization: it has no permanent offices or permanent staff." The KPCS is largely self-monitored, without impartial and regular reviews. The effectiveness of the organization relies solely on the countries involved to monitor their own practices and each other. Both the System of Warranties and the KPCS are essentially honor systems, which can be difficult to enforce on a global level.

In spite of these problems, the global attention to practices and markets is encouraging. The worldwide eye on the Diamond trade has persuaded some countries, companies, and organizations to find resolutions and enhance the ethical standards of the Diamond trade. Independent organizations and watchdog groups such as Global Witness and Partnership Africa Canada have become involved in monitoring the KPCS, as well as in reinforcing the ban on conflict gems. Meanwhile, KPCS claims immediate effectiveness, stating that conflict Diamonds now represent a fraction of 1 percent of the international trade in Diamonds, compared to estimates of up to 15 percent in the 1990s. KPCS also claims to improve economies by providing a legal monitoring system, which in turn provides legal funding to poor governments. For instance, some $125 million worth of Diamonds were legally exported from Sierra Leone in 2006, compared to almost none at the end of the 1990s.

The KPCS recommends that someone purchasing a gem ask the jeweler the following questions: How can I be sure your jewelry does not contain conflict Diamonds? Do your Diamond suppliers participate in the industry's System of Warranties? Can I see a copy of your company's policy on conflict Diamonds? By examining the provenance of gems and avoiding conflict stones, people can make a positive global impact. Collectively, individual consumer actions can impact companies in major ways, even international companies like the world's largest Diamond-mining operation, De Beers.

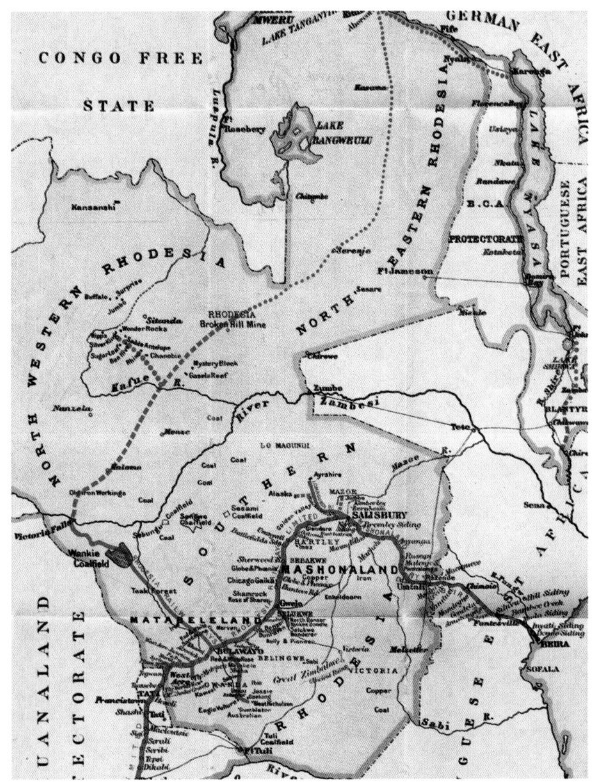

FIGURE 289.
Left: Cecil John Rhodes (1853–1902), a man with somewhat imperialistic political tendencies and the founder of De Beers. *Right:* Map of Rhodesia, the African country named for Rhodes after his British South African Company acquired the land for it in the late nineteenth century. Left photo by W. & D. Downey, ca. 1900, London. Map photo from Hall (1905).

De Beers has a somewhat checkered history and in recent decades has been implicated in the trade of conflict Diamonds. Some of its original bad press was due in part to its founder Cecil Rhodes, who established the De Beers Mining Company in 1880. Rhodes proved a shrewd and sometimes ruthless businessman. Only eight years after he founded De Beers, Rhodes had gained monopolistic control over the entire Diamond market. This is hardly surprising when you consider that his industrial strategy was no doubt influenced by his imperialist political tendencies. Rhodes said of the British, "I contend that we are the finest race in the world and that the more of the world we inhabit the better it is for the human race." Rhodes is also said to have treated his African workers as second-class citizens and been less concerned with the welfare of the people of the region than he was with expanding the British Empire and his own personal empire. Despite his uncompromising methods, it is well documented that Rhodes was successful in many of his ventures, gem and otherwise. In fact, a very large territory in Africa was officially named after Rhodes in 1895. "Rhodesia" consisted of land acquired by Rhodes's company, the British South Africa Company. Today we know Rhodesia as the countries Zambia and Zimbabwe. Perhaps the most famous summation of Rhodes was provided by American writer Mark Twain: "I admire him, I frankly confess it; and when his time comes I shall buy a piece of the rope for a keepsake."

Rhodes's early domination of the market and quest for total control allowed the De Beers company to manage much of the world's Diamond mining since its earliest years. To this day De Beers maintains a market stronghold, controlling when and where most of their Diamonds are released into the market. The combination of unrivaled success and majority control of the Diamond trade made De Beers a popular target for scrutiny in the 1990s, when the problem of conflict Diamonds became particularly acute in Sierra Leone. It was then discovered that De Beers was in fact purchasing conflict Diamonds. De Beers was also found guilty of price-fixing, a selling tactic intended to drive the cost of Diamonds as high as possible.

Reacting to public concern, the company has begun to change its public image and policy for the better. De Beers instituted a series of self-governing policies, beginning with a zero-tolerance policy. According to their policy, this means that all Diamonds sold by De Beers are certified as conflict-free Diamonds and are not mined using child labor. The company now also publishes an annual report, available for review on their website and providing insight into their operations and finances. De Beers is an example of how public scrutiny helps make everyone responsible, from the individual buyer to a major worldwide Diamond supplier.

For thousands of years, people have gone to extraordinary and sometimes deadly lengths to obtain gems. In today's market, the stakes are as high as ever. But we are equipped with better information. The sheer volume of online articles, newspapers, television programs, and books brings world conflict into our homes. With this increased awareness, we can take measures to adhere to ethical standards in our personal purchases of gems. The world is taking notice, and both dealers and consumers can make informed decisions that will help assure an ethical foundation for the future of the gem trade.

For further reading, see Zoellner (2006).

FOLKLORE, MYSTICISM, *and* MAGIC

Throughout history, gems have been considered by some people or cultures to have great **METAPHYSICAL** significance. Their mysterious beauty inspires the idea that humans might harness the forces of the universe and use those forces to mold the world to our liking. It is hardly surprising that gems have long been worn as protective **AMULETS** and **TALISMANS**, prescribed to cure aches and pains or used to harness powers of love and fortune. Even today, there are people who believe in the supernatural powers of gemstones, a faith that may seem superstitious to some. Many modern beliefs come from deeply rooted traditions and religions. Modern **VEDIC** astrology draws upon some of the oldest texts of Hinduism, linking gems with the sun, moon, and planets to interpret the influence of heavenly bodies on human affairs. Modern interpretation of the Chinese **ZODIAC** predicts how gems will align with certain star formations to bring prosperity to the gem owner. In the 1960s, **CRYSTAL POWER**, or **PYRAMID POWER**, was widely popular throughout North America and continues its popularity to this day. Crystal power builds on the belief that wearing certain gem crystals or certain geometric crystal shapes allows a person to channel, harness, and accentuate different abilities and healing powers.

Some scientific and aesthetic gem qualities mentioned in the prior chapters may begin to explain why people attribute magical and mystical properties to gems. The beauty of a gem is immediately recognized yet difficult to articulate, shrouding it in mystery. The durability of a gem gives it a quality of timelessness lacking in natural colorful objects of limited lifespan, such as flowers or birds. The scarcity of a gem makes its discovery and ownership something extraordinary. The general qualities of beauty, durability, and scarcity undoubtedly helped inspire the broad metaphysical attributes given to gems.

Like any folklore or cultural practice, documentation varies in detail from culture to culture. Some gem lore was written down, some was recorded in the

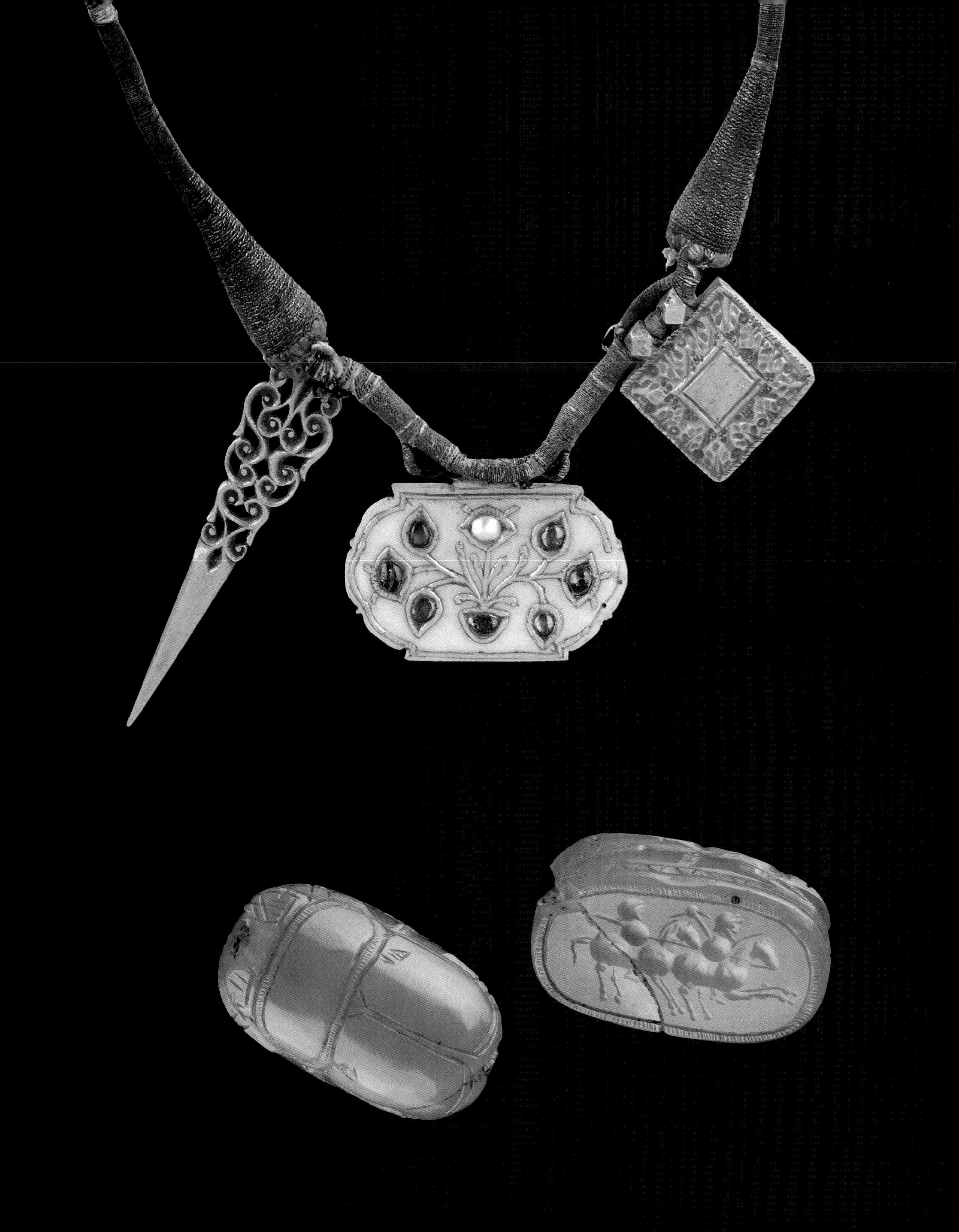

form of pictures, and some is simply the result of stories passed down from generation to generation. There are very few universal properties associated with specific gems: Babylonian beliefs about rubies are not the same as those of Greeks, Chinese, or Persians. One general belief common throughout many cultures, however, is the protective talismanic powers of gems.

Talismans and amulets are objects worn for protection against evil forces, supposedly channeling good, natural forces for the wearer's benefit. We find talismans and amulets dating back thousands of years, made out of everything from wood and shell beads, to carved Carnelian and crudely faceted gemstones. Gemologist George Frederick Kunz wrote: "Folklorists are wont to wonder whether the custom of wearing gems in jewelry did not originate in the talismanic idea instead of in the idea of mere additional adornment." If pretty ornamental objects attract supernatural powers, it is no surprise that gems especially would also be worn for their protective and controlling qualities, given their unique and beautiful attributes. Indeed, most gem species possess folklore unique to that species. We list only a few examples below.

FIGURE 290.
Facing page, bottom: Amulet from the Sassanian Empire, third century AD. Carved from carnelian with a scarab beetle engraving on one side and two horses with warriors on the other. Top and bottom view (FMNH A238208).

FIGURE 291.
Facing page, top: Talisman from nineteenth-century India, probably worn as an armband or necklace. The three charms (identified by Dr. Rahul Oka in 2009) include the following, from left to right: a dagger with ornate hilt (in Gold) used in South Asian iconography as a weapon against ignorance, evil, and malevolence; a white nephrite plaque, 36 mm wide, inlaid with Gold, rubies, emeralds, and Pearls, depicting an image of a plant with eyes and leaves on the branches, to foresee ill fortune; a box of Gold (a "Taweez"), to contain items associated with spiritual power (FMNH A82175).

DIAMOND

Diamonds were said to enhance the love of a husband for his wife; a variation on that theme is reflected today in the custom of the Diamond engagement ring. It has been speculated that the placement of an engagement ring upon the hand is based on an ancient Roman belief that the vein in the left-hand ring finger was connected directly to the heart. The Latin phrase VENA AMORIS, meaning "vein of love," refers to this practice, although that exact phrase appears in print much later in history, in the 1600s. Diamond is also often associated with lightning and control of that energy, possibly because of the brilliant "flash" of a high-quality Diamond. In the fourteenth century, Diamonds were said to drive away night spirits, presumably because of their ability to light up the darkness. It was also believed that Diamond could change color if the wearer was sinful, becoming blacker with every new sin. The Hindus classified the mystical properties of Diamonds according to the four VARNAS, following traditional Indian class categories. *Brahmin* Diamonds gave power, friends, riches, and good luck; *Kshuatriya* Diamonds prevented old age; *Vaisya* Diamonds brought success; and *Sudra* Diamonds brought all manners of good fortune. While it is implied that any Hindu could use Diamonds in this way, it is likely that these properties were intended primarily for those people who could afford the Diamond. Hindus also tasted Diamonds, as they tasted many gems, categorizing a Diamond into six flavors: sweet, sour, salty, pungent, bitter, and acrid. Because it was believed to possess this entire range of flavors, Diamond was thought to cure all ills. Because a Diamond is so hard, it was considered by many cultures to be among the most effective of talismans, withstanding great evil powers and driving them into submission. Also for that reason, Diamonds were, at one time, ground into powder and eaten to cure ailments. But as patients tended to die after such a valuable treatment, the medicinal use of Diamond powder was eventually discontinued, and it was instead used as an expensive way to dispatch enemies. Diamond was difficult to detect, caused severe internal damage that was often fatal, and was undoubtedly used for political reasons. Perhaps the most famous example of this unique assassination method comes from Catherine de Medici, the queen consort of King Henry II of France from 1547 to 1559, who reportedly used Diamond to eliminate a number of her enemies. Despite the lethal medieval overtones, many cultures have considered Diamonds to be of a good nature, including our own culture today. At the very least, Diamonds represent the good fortune of wealth.

RUBY

Many cultures believed that a ruby held a fire within its crystal, which explained its vibrant red color. According to Hindu culture, a ruby's internal fire could cause water to boil if the fire inside was strong enough. Many ancient cultures believed that gems grew in the ground, much like a vegetable. In Burma, rubies were believed to have a growing period. The reddest of rubies were "ripe," while the paler colors were "underripe." Flawed rubies were said to have been "overripe" and left in the ground too long. European tradition held that the owner of a ruby would live contentedly and that his rank in society would be secure. The Chinese archaeologist and Burmese specialist Taw Sein Ko reported that Burmese warriors would wear rubies by inserting them into their flesh; rubies were then part of the warrior's bodies, imbuing them with invincibility.

SAPPHIRE

People in Ceylon (modern-day Sri Lanka) believed that star sapphires could protect the wearer by "seeing" evil, a reaction to the **ASTERISM** in the stone that makes it look like an eye. Star sapphires were also considered to be guiding stars in both philosophical and traditional travel, guiding the way through a metaphysical journey of enlightenment or through more tangible winding roads and stormy seaways. In the 1500s, **NECROMANCERS** (sorcerers who summoned spirits of the dead) believed that they could communicate with oracles through a sapphire because it was as blue as the heavens where the oracles resided. Sapphires were also thought to be powerful healing stones. Blue sapphire was believed to be an antidote against poison, and pink sapphires were also worn for good health but needed to have direct contact with the skin in order to be beneficial. In fourteenth-century medieval Europe, sapphires were said to attract the favor of princes and royalty to the wearer. The Roman Catholic Church under Pope Innocent II created sapphire rings that Catholic bishops wore to demonstrate their connection to God and the heavens. Medieval belief also extended to the earthly realm: people believed that, if ingested, sapphire could cure scorpion bites, boils, and pustules. Another prominent belief was the spider test: if one were to place a spider in a vessel and suspend it over the sapphire, the venomous spider could not resist the power of the stone, making the spider lose its poison.

OPAL

In ancient Roman times, Opal was a popular gem known as the *cupid paederos*, or "cupid stone," a symbol of hope and purity. Native Americans and Australian aboriginal SHAMANS (magic healers and priests thought to be intermediary between the natural and supernatural worlds) believed that Opal allowed them to traverse the dream realms, aiding them in their dream quests toward enlightenment and understanding. Opal's glittering qualities were often associated with eyes or vision. In Norwegian mythology, it was said that the god Volomer (Vulcan) would steal the eyes of children and fashion them into Opal. In the 1800s, Opal's popularity took a turn for the worse. Opal suffered great unpopularity due to its fragile nature. It often shattered when worn in jewelry, and so the gem developed a reputation of attracting bad luck. Despite its negative reputation at the time, Queen Victoria was said to have been enchanted with the gem. She wore a brooch of it upon the back of her dress at her coronation. Unfortunately, the Opal proved to bring bad luck after all when the new queen's Opal clasp came undone during the ceremony, and she was "presented" in more ways than intended. This added further to the theory that Opal brought "bad luck." Today precious Opal has once again recovered its station as a highly valued gem as the superstitions surrounding it have subsided.

AMETHYST

Amethyst's deep purple color has long been associated with wine, as well as behavior brought about by drinking too much wine. The ancient Egyptians and Greeks believed that amethyst could cure drunkenness. If one were to wear an amethyst, it would curb poor drunken behavior. Placing an amethyst directly on or under the tongue was the most effective and direct way to bring about sobriety. Additionally, amethyst was believed to control passion (perhaps the passion inspired by wine). The Romans believed that women who wore amethyst would be better able to keep their husbands and lovers.

EMERALD

The earliest written reference to emerald comes from Aristotle, who wrote that emerald brought victory in business and legal affairs, and that emeralds would prevent epilepsy, therefore suggesting that children wear the gem to prevent the "falling sickness." Aristotle's claims persisted through the medieval ages with slight modifications; emeralds were believed to prevent epileptic fits, but an emerald that could not prevent fits was weak, and would break and lose its power. In the twelfth century, the poet and popular gem theorist Marbod wrote that emeralds made a speaker's speech more persuasive, echoing an earlier claim of Aristotle. The gem's protective properties were also extended: if a woman hung an emerald from her neck, it would drive off vanity and evil. If hung from her hip, it was said to speed childbirth. Hindus believed that emerald would protect the wearer from all poisons. Yet the romanticism of emerald was not only medicinal: ancient Egyptians believed that emerald would increase love. Hundreds of years later, Germans in the 1800s believed that Beryl crystal, such as emerald, could reawaken the love of married people who had fallen out of love.

AQUAMARINE

Aquamarine is from the Latin: *aqua* for "water" and *mare* for "sea," although ancient texts do not recognize aquamarine by that specific name until the 1600s. Before then, aquamarine was often called "sea green Beryl," an obvious reference to its water-like coloring. The specific coloring of aquamarine clearly inspired the association between supernatural properties and water; sea green Beryl was said to come from the treasure chests of mermaids, and sailors used them as protective talismans against shipwrecks and sea monsters.

GARNET

There are many different color varieties of Garnet, but most of the folklore surrounding the gem is connected to red varieties such as Almandine and Pyrope.

The ancient Romans said that Garnet symbolized fire, and the gem was associated with Mars, the god of war. In medieval Europe, Garnets were popular and often worn as rings. We know this from the German writer Johannes de Cuba, who published his *Hortus sanitatis* ("The Garden of Health") in the late 1400s. It was a popular medical book that described a test to prove whether or not a Garnet was genuine. The owner of the ring was to take off his clothes and smear his body with honey. He would then lie down with wasps and flies. If he was left unstung, the Garnet was genuine. One wonders how many owners were parted from "non-genuine" Garnets (and how many stings were imparted). Lucky for Johannes he didn't live in today's age of litigation. Later in the nineteenth century, warriors from India would shoot Garnets out of their guns at enemies, believing that a Garnet would inflict a more deadly wound than a bullet due to its fiery, war-like nature.

COLORLESS QUARTZ (ROCK CRYSTAL)

The Greeks believed that Quartz was petrified ice. In the Yucatán Peninsula in Mexico, Quartz was believed to channel the power of water to arid regions. Quartz crystals were also used for divining purposes. A Quartz crystal could be fashioned into a ball to help a seer tell the future (fig. 292). In the 1600s, a Danish chemist gave detailed instructions about how to ingest liquid rock crystal with wine, after heating cracked crystal with tartaric salts (organic acid). The potion was said to cure all sorts of various ailments. Today, Quartz crystals are

FIGURE 292.
Crystal sphere made from Quartz or "rock crystal" from Mount Antero, Colorado. Usually referred to as a "crystal ball," these were (and still are) used by mystics for divining purposes. Sphere measures 138 mm in diameter and weighs 3,380 grams, or about 7½ pounds (FMNH H1189).

still used to channel clearer vision and foresight into the future, but ingesting pulverized Quartz for medicinal purposes is fortunately less popular.

JADE

Jade, a vernacular name that refers to two minerals, Jadeite and nephrite, has often been considered a gem with HOLISTIC properties. These superstitions were especially popular in Asian, Polynesian, and Central and South American cultures. In China, if a parent were to attach jade to a child's neck, it connected him to life and kept him from disease. Jade was also considered to be an effective treatment for a variety of other body ailments, particularly of the feet and torso. The Chinese believed that jade was a musical stone, and that one could strike jade and it would ring true; if jade rang falsely, the stone was flawed. Jade was sometimes carved into the shape of butterflies as symbols of successful love. Butterflies were also often used in funeral rites, placed within the mouth of the dead to protect them from harm in the afterlife. The Maori people, indigenous to New Zealand, would conduct jade-hunting parties, which required the assistance of a TOHUNGA, or wizard. The tohunga would fall into a trance and contact a spirit who would guide the group to a location for jade. George Frederick Kunz pointed out, "Of course the wizard had previously assured himself of the presence of the stone in the place indicated."

MOONSTONE

Moonstone has a limited ancient folklore but has gained in popularity in more recent history. Moon worshippers use moonstone to connect with the heavens directly, and the relation of the moon and lunar cycles also create sexual overtones. In India, the moonstone is a sacred stone and believed to bring good luck and fortune. The stone is displayed against a yellow cloth, a color sacred to India. Some scattered stories across European cultures say that moonstone incites passion and that if lovers place a piece of moonstone into their mouths, they can read the future of their relationship.

TURQUOISE

Turquoise has been valued by many cultures throughout the centuries, including the Egyptians, Babylonians, and the Aztec, Mayan, and Incan peoples. Turquoise is also very important in Native American and Tibetan traditions.

Arab writers in the twelfth century believed that staring at Turquoise early in the morning would improve a person's vision. In fourteenth-century eastern Europe, it was believed that Turquoise protected horses from feeling sick if they drank too much cold water after riding too hard. Turkish peoples therefore attached Turquoise to the bridles of their horses. The association with horses persisted for centuries. In the 1700s Turquoise was thought to protect the wearer from falling off a horse. People across the Americas reference Turquoise in very fundamental ways. Turquoise appears in many myths, legends, and beliefs concerning the creation of the world and its creatures. Because of its rich blue hue, some Native American legends say that Turquoise is a piece of sky that fell to earth.

PEARLS

Pearls, like Diamonds, have traditionally symbolized wealth. There is a story that Cleopatra drank a toast to Mark Antony that contained a dissolved Pearl. This was meant to impress him with the vastness of Egypt's wealth. Similarly, this action was later repeated in the sixteenth century during a toast to honor Queen Elizabeth I at a banquet. In ancient Rome, the Pearl was so highly valued that Julius Caesar created a law that only aristocrats could wear Pearls. The Pearl also had divine overtones. The Roman author Pliny the Elder wrote that Pearls were the "dews of heaven" that fell into the sea, and Roman women wore Pearls in their hair to ensure love. In the European Renaissance, Pearls were so popular that they were sewn into dresses as accents. Throughout Asia and Europe, Pearls were believed to cure many medical maladies, including fever, insomnia, dysentery, whopping cough, and measles. Legitimate medical uses continue today: in modern medicine, **NACRE** (the shiny material known as mother-of-pearl) is used in bone implants because it is strong and resilient.

AMBER

Amber is an organic gem material found in trees that is often more easily attainable than inorganic gems residing deep in the earth. For this reason, Amber gems, carvings, and amulets appear very early in history. Some Amber amulets date as far back as the Mesolithic period, around 7000 BC. Later in history, Roman babies were decorated with Amber beads to protect them from evil spirits, and adult Romans carried pieces of Amber that would heat up and release balsamic scent, masking sweat. In Norway, Amber was carved into animal shapes to enhance its power; a bear-shaped piece of Amber was thought to harness the fierce qualities of the bear itself. The ancient Greeks believed that Amber was solidified sunshine that had fallen to earth. The Chinese believed that just as bees create golden-colored honey, so too did they create the honey-colored Amber. These honey-like qualities might explain why Amber pieces were often ground into powder for medicinal purposes, a belief that persisted through medieval times and even today in certain cultures.

CORAL

The ancient Persians believed that Coral was only powerful once it was taken from the ocean; only then was it "activated." Red and white Coral was believed to protect people who had to travel far or on broad rivers; the Coral could also calm bad storms. The Romans believed that Coral could stop the flow of blood from a wound, cure insanity, and cultivate wisdom and knowledge. Coral's powers were believed to be heightened and extended if it was wrapped in the skin of a cat. If emeralds were combined with Coral, they would cure fever. Hindu physicians ate Coral for its beneficial effect on the mucus and liquids associated with digestion. In the Middle Ages, it was believed that Coral would reflect the internal spirit of a person. Today Coral is believed to incite passion and strengthen imagination.

* * *

Although these examples represent only a small sample of gem folklore, the sheer variety and creativity represented here demonstrate the extraordinary power of gems and gemstones to inspire the human mind. For further reading about gem folklore, try Kunz (2003a).

BIRTHSTONES

The idea of a **BIRTHSTONE** (or **NATAL STONE**) has evolved into somewhat of a marketing opportunity much like anniversary tokens and engagement rings. The Western world's current promotion of birthstones is largely a tool of consumerism and the gem industry. But in other cultures and times, the gems that we call "birthstones" have held a deeper metaphysical significance. A gem associated with one's birth was, and still is to some, a powerful **TALISMAN** associated with folklore, astrology, and magic. The supposed "power" of a birthstone is that it brings good fortune or energy by forging a mystical connection with the heavens, drawing upon the constellations of the zodiac or the alignment of celestial bodies to balance the universe in the wearer's favor. Birthstone lore stems from a rich cultural and religious history that dates back thousands of years. In the next few pages, we can only touch on this history in the briefest of ways.

DEVELOPMENT OF BIRTHSTONE LORE

The origin and history of birthstone lore is long and complex, and much of it remains poorly documented or unknown. Yet this history clearly involves two major types of gem–birth date associations: the first connects a person's birth date with one of the twelve stations or "signs" of the **ZODIAC**, and the other connects the birth date to one of the twelve months of the solar (Gregorian) calendar. The two types of associations are different in many respects. The zodiac stations do not exactly correspond to the Gregorian months, and many of the gem-time associations differ between the two systems (figs. 293, 295). To further complicate this history, the biblical Hoshen, or "breastplate of

Aaron," may also be involved in the deep history of modern birthstones. The Hoshen, referenced in Exodus, contains twelve different stones, one for each of the founding sons of Israel (fig. 294). One common thread in all of these elements is a cycle of twelve: the twelve signs of the zodiac, the twelve months of the solar year, and the twelve stones of the Hoshen. We will briefly discuss these factors and their role in the history of birthstones.

Tying gems and their "powers" to the **ZODIAC** is a practice that may have dated back thousands of years. Modern zodiac charts were derived from ancient Babylonian astronomy, probably around the seventh century BC. A zodiac

FIGURE 293.
A simplified modern Western interpretation of the zodiac-gem connection based on the tropical zodiac (zodiac based on a twelvefold equal division of the earth's orbit in space or the year in time). The dates are for 2009 and may vary slightly from year to year. The specific gem connections are based on traditional Hermetic correspondences that have been in use since at least the early fifteenth century. See Thomas and Pavitt (1977) for a discussion of restrictions and timing for "Fortunate Gems."

ZODIAC SIGN AND SYMBOL	DATES*	FORTUNATE GEMS**
Aries the Ram ♈	 March 21–April 19	Diamond, bloodstone
Taurus the Bull ♉	 April 20–May 20	sapphire, Turquoise
Gemini the Twins ♊	 May 21–June 21	agate, chrysoprase
Cancer the Crab ♋	 June 22–July 22	moonstone, Pearl, emerald, cat's-eye, colorless Quartz
Leo the Lion ♌	 July 23–August 22	sardonyx, Amber, Tourmaline, chrysolite, peridot
Virgo the Virgin ♍	 August 23–September 22	carnelian, Jadeite, nephrite
Libra the Balance ♎	 September 23–October 23	Opal, Coral, lapis
Scorpio the Scorpion ♏	 October 24–November 22	aquamarine, pale green Beryl, Garnet en cabochon, lodestone
Sagittarius the Archer ♐	 November 23–December 21	Topaz, chrysolite
Capricorn the Goat ♑	 December 22–January 20	ruby, Spinel, black onyx, jet, Malachite
Aquarius the Water-Bearer ♒	 January 21–February 18	Zircon, Garnet
Pisces the Fishes ♓	 February 19–March 20	amethyst

**May vary slightly from year to year. ** Thomas and Pavitt (1977).*

charts the sun, moon, planets, and stars as they seemingly move across the sky when viewed from earth. These movements of stars and planets follow a regular annual pattern. Each month is represented by one of twelve star constellation symbols, a system the Greeks adapted into what we recognize today as the modern zodiac (fig. 293). Over time, astrologers assigned certain gems to each of the zodiac signs to help people influence the "heavenly bodies" in their favor. The signs of the zodiac and their corresponding gems represent complex powers and weaknesses depending on the position of celestial bodies.

While modern birthstones also associate particular gems with certain months, it is important to note that birthstones gems are not precisely the same as zodiac gems. Nevertheless, modern birthstone charts were probably strongly influenced by older zodiac-gem associations (note some similarities for some months between figs. 293 and 295). Today's birthstone gems simply correspond to the months and hold no great mysticism; and they follow our modern international standard of timekeeping, the Gregorian calendar. The Gregorian calendar, which follows the solar cycle, was first proposed by Aloysius Lilius, a sixteenth-century Italian doctor and astronomer. It was decreed by Pope Gregory XIII, after whom the calendar was named, in 1582. The solar month is slightly less than 30½ days, with a twelve-month year of about 365 days (a month is approximately 1/12 of the time it takes the earth to orbit the sun, although the months are of irregular length). Most modern birthstone charts are based on the Gregorian calendar. Because so many different cultures had their own different lists of birthstones, jewelers' lists are often inconsistent with regard to what constitutes a traditional birthstone. Birthstones are today most often thought merely to reflect the birthday month of their wearer. But older traditions going back hundreds of years suggested that a person should wear all of the birthstones, because the powers of each stone were heightened during its month. In order to get the full effect of birthstone power, individuals needed to own an entire set of twelve gemstones and rotate them accordingly.

GEMS, RELIGION, AND ORDER

There are two important works that further our understanding of how gems became associated with the twelve-month calendar, and ultimately with birth dates. The first is the biblical book Exodus, and the second is the work of Roman historian Flavius Josephus. These two works confirm that different ancient cultures were using gems symbolically in their daily lives in conjunction with the calendar, zodiac, and religion.

Exodus, generally accepted to have been written sometime around 1450 BC, tells the story of Aaron, High Priest of the Israelites. According to Exodus,

God designed a breastplate for Aaron that was called a **HOSHEN** in Hebrew (fig. 294). God asked Moses, brother of Aaron, to make the Hoshen he had designed. The Hoshen was to be set with twelve stones that represented the twelve founding sons and twelve tribes of Israel:

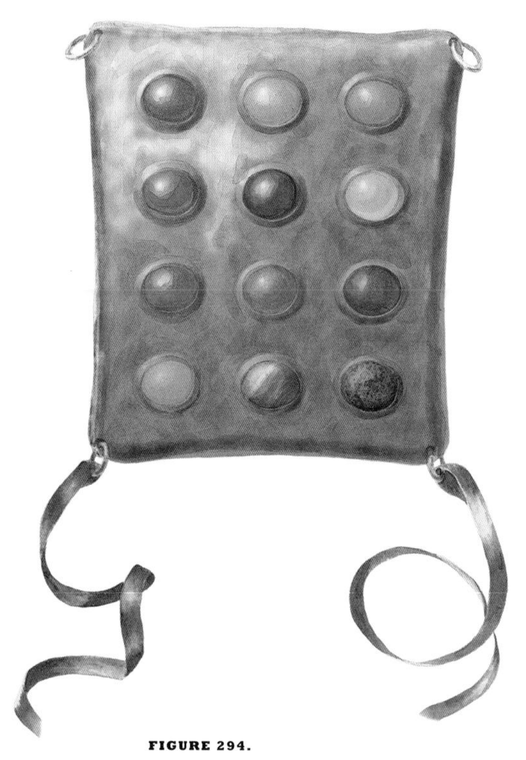

FIGURE 294.
Reconstruction of a Hoshen, also known as the breastplate of Aaron, referenced in Exodus (1450 BC). The Hoshen contains twelve different stones representing the twelve founding sons of Israel. The exact type of mineral species for some of these stones is controversial (see text).

And you shall make a breastplate of judgment, in skilled work; like the work of the ephod you shall make it; of gold, blue and purple and scarlet stuff, and fine twined linen shall you make it. It shall be square and double, a span its length and a span its breadth. And you shall set in it four rows of stones. A row of sardius, topaz, and carbuncle shall be the first row; and the second row an emerald, a sapphire, and a diamond; and the third row a ligure, an agate, and an amethyst; and the fourth row a beryl, an onyx, and a jasper; they shall be set in gold filigree. There shall be twelve stones with their names according to the names of the sons of Israel; they shall be like signets, each engraved with its name, for the twelve tribes. (Exodus 28:15–21)

It should be noted that this translation of the Bible is the King James Version. Other versions give slightly different translations of the stones used in the breastplate, but even biblical scholars do not agree on precise translations of the texts, and the exact types of mineral species for some of these stones is controversial. Regardless of the stones, we know that there were twelve sons of Israel and a different gem to represent each son; there were also twelve months in the calendar. Exodus confirms that ancient people were likely aligning gemstones with the twelve months of the calendar, a precursor to our modern attempts to align gems with the months of the year.

This is later confirmed in AD 93, when the Roman Jewish historian Flavius Josephus provided a detailed account of Judaic practices in his book *Antiquities of the Jews*. This is the first written account that decidedly connects the gems of the biblical breastplate with the calendar and also with the Greek zodiac. Josephus wrote:

And, for the twelve stones, whether we understand by them the months, or whether we understand the like number of the signs of that circle which the Greeks call the Zodiac, we shall not be mistaken in their meaning.

Josephus was saying that people who associated gems with months or the zodiac were using the stones incorrectly, and that the biblical importance of the twelve sons of Israel should not be forgotten. What this passage also tells us is that the practice of associating gems with the calendar or zodiac was common enough that Josephus wrote a pointed comment about it. Thanks to Exodus and Josephus, we now know that some cultures associated gems with religion based on the Bible, while others associated gems with the calendar months or zodiac. Gems clearly had unique significance and were used in a particular calendar order depending on religious or pagan beliefs.

During this time, another important Roman historian named Pliny, also known as **PLINY THE ELDER** (AD 23–79), was creating his own extensive written work on natural history. Pliny created an encyclopedia about the world, including everything that he could find on gemstones. He contradicted some common mystical uses and observed new uses in other gems, and his written work further proves that there were common cultural associations between gems and supernatural powers. His work was referenced and refuted for hundreds of years, even into the late 1800s. Many modern interpretations of gem lore and history are based on Pliny's comprehensive, although unconfirmed, writings.

GEMS AND RELIGION INTO THE MIDDLE AGES

The history of gems grew even more complicated in the first century AD, when European Christians added a new religious interpretation of their own in the New Testament. John, generally believed to be the author of the book of Revelation, wrote about a set of gems called "foundation stones." These stones were most likely similar to the breastplate stones from Exodus, but differed in that each stone was inscribed with the name of an apostle, rather than a son of Jerusalem. Note that Revelation does not tell us what gem belongs to what apostle:

The wall of the city had twelve foundations, and on them twelve names of the twelve Apostles of the Lamb. . . . The foundation of the city's wall were adorned with all kinds of precious stones. The first foundation was jasper; the second, sapphire; the third, chalcedony; the fourth, emerald; the fifth, sardonyx; the sixth, sardius; the seventh, chrysolite; the eighth, beryl; the ninth, topaz; the tenth, chrysoprasus; the eleventh, jacinth, and the twelfth, amethyst. (Revelation 21:14, 19)

Revelation does two things: first, it shows us that Christians believed that gems were significantly and symbolically important, suggesting that they were building a new tradition based on old practice. Second, Revelation opened a debate that never truly ended. Because John never indicated which gem belonged to which apostle, Christian scholars debated the topic purely on interpretation for hundreds of years, well into the thirteenth century. Some Christian authors created **BESTIARIES**, sweeping religious anthologies gathering vast information about the natural world (animals, plants, minerals), creating religious associations between certain gems and the twelve apostles. Meanwhile, pagan and Christian rites had been blending together over time, and some authors began to write **LAPIDARIES**. Lapidaries were books similar to bestiaries but were dedicated primarily to stones and gems, and to explaining their symbolism within the context of the natural and religious worlds. Some lapidaries were meant to be scientific, combining alchemy, mineralogy, and chemistry. Others were concerned with magic or astrology, while some were written primarily to explain gems in a Christian context. Marbod, Bishop of Rennes, wrote a much-cited medieval lapidary in the eleventh century called *De lapidibus*. He wrote three versions of *De lapidibus*, one focusing on the Christian tradition of foundation stones from Revelation, one providing more information about the foundation stones, and one focusing on the healing properties of gems. Although many historians point to Marbod as a definitive source, his work is hardly indicative of the era. With the rise of the Catholic Church, other lapidaries of the time introduced purely religious associations, including those between certain gems and angels, saints, and the Virgin Mary.

Authors continued to write lapidaries all the way through the Renaissance, as people continued to debate the identities, uses, religious significance, and magical powers of gems. Some authors continued not only the Christian debate, but turned their minds to science as they discussed the work of Pliny. The Renaissance encouraged scientific exploration and investigation. Throughout the medieval ages, the Renaissance, and beyond, the debate over the meaning of gems, their uses, and their role in the universe raged on, with no definitive conclusions.

BIRTHSTONES, MODERN MYTH, AND INDUSTRY

As time progressed, the debate over gems and their meaning died down substantially. People were still interested in gem lore and used gems in significant ways, but they did not theorize and debate as intensely as they had before. Scientific observation began to trump speculation, religious or otherwise. The cultural history of gems became spotty at best, as there were fewer and fewer written records regarding the common cultural uses of gems. It was not until the early 1900s that an author took an avid interest in how gems were used culturally and filled in some of the historical gaps.

American gemologist George Frederick Kunz provided the first written account of the modern term "birthstones" when he drew a connection between the ancient Judaic Exodus breastplate and modern Judaic practice in his 1913 book *The Curious Lore of Precious Stones*:

We have no instance of the usage of wearing such stones as natal stones until a comparatively late date; indeed, it appears that this custom originated in

Poland some time during the eighteenth century. It seems highly probable that the development of the belief in natal stones that took place in Poland was due to the influence of the Jews who settled in that country shortly before we have historic notice of the use of the twelve stones for those born in the respective months. The likely interest always felt by the Jews regarding the gems of the breastplate, the many and various commentaries their learned men have written upon this subject, and the fact that the well-to-do among the chosen people have always carried with them in their wanderings many precious stones, all this seems to make it likely that to the Jews should be attributed the fashion of wearing natal stones.

Kunz's quote is constantly used by modern gem historians to establish birthstones as originating in Poland in the 1700s. Although Kunz never fully explained why he thought this to be true, he may be right. We know that there was a large Jewish population migration to Poland throughout the 1700s. We also know that Jewish Poles continued religious traditions based on the Bible, including the passages from Exodus. High priests wore replicas of the breastplate set with gems in high ceremonies. If Kunz is correct, Jewish Poles were also wearing birthstones during this time.

This practice then had to make its way to America. The 1800s and early 1900s saw the largest wave of Polish immigration to America. Officially, more than 1.5 million Polish immigrants were processed at Ellis Island between 1899 and 1931. Polish Jews with their cultural and religious practices intact (and likely with their beliefs about birthstones) came to America and practiced their beliefs here.

Sometime during the Polish-Jewish migration to America, the gem industry presumably sensed an opportunity to standardize the industry and increase sales among popular audiences. In 1912 the National Association of Jewelers (NAJ) made the first attempt to establish a "definitive" birthstone chart. The list was not widely accepted across the industry. A more popular list emerged in 1937 across the pond in Britain, presented by the British Goldsmiths. The Jewelry Industry Council in America adopted this list in 1952, and the rest of the American industry soon followed (fig. 295).

The birthstone campaign was a success and has remained so ever since. Over the years, birthstone lists have been updated based on the popularity of gems in the market, as well as a few gems new to the trade. Today the Gemological Institute of America (GIA) presents a chart that is as good a representation as any of what is generally accepted in the Western Hemisphere.

As you may have noticed, different birthstone charts change depending on the industry. For example, tanzanite, discovered in 1967, was never considered a birthstone until 2002, when the American Gem Trade Association (AGTA) added it as an option for December. Today tanzanite is an expensive, beautiful gem

MONTH	1450 BC BOOK OF EXODUS	1912 NATIONAL ASSOCIATION OF JEWELERS	1952 JEWELRY INDUSTRY COUNCIL	2009 GEMOLOGICAL INSTITUTE OF AMERICA
January	onyx	Garnet	Garnet	Garnet
February	jasper	amethyst	amethyst	amethyst
March	ruby	bloodstone or aquamarine	bloodstone or aquamarine	aquamarine
April	Topaz	Diamond	Diamond or colorless Quartz ("rock crystal")	Diamond
May	"carbuncle" (generic red gemstone)	emerald	emerald	emerald
June	emerald	Pearl or moonstone	Pearl, moonstone, or alexandrite	Pearl
July	sapphire	ruby	ruby	ruby
August	Diamond	sardonyx or peridot	sardonyx or peridot	peridot
September	hyacinth (jacinth)	sapphire	sapphire or lapis lazuli	sapphire
October	agate	Opal or Tourmaline	Opal or pink Tourmaline	Opal
November	amethyst	Topaz	Topaz or citrine	Topaz
December	Beryl	Turquoise or lapis lazuli	Turquoise or Zircon	Turquoise

FIGURE 295. Simplified birthstone chart comparing the different formal lists of from Exodus (1450 BC), the National Association of Jewelers (1912), the Jewelry Industry Council (1952), and the Gemological Institute of America (as of 2009). The connection between the Hoshen stones mentioned in Exodus and the Gregorian calendar were not made until the nineteenth century, as far as we know.

that is largely promoted for jewelry, as well as December's birthstone. Clearly, these decisions were driven by sales, not mysticism. This no doubt takes away a bit of the magic of birthstones, as Kunz recognized when he wrote:

Once we allow the spirit of commercialism pure and simple to dictate the choice of such stones, according to the momentary interest of dealers, there is grave danger that the only true incentive to acquire birthstones will be weakened and people will lose interest in them. Sentiment, true sentiment, is one of the best things in human nature.

The long history of our fascination with gems—astrological, religious, or otherwise—is born of a place that is obviously close to our emotions, or else gem lore would not appear so often and broadly throughout cultures worldwide. Although birthstones in the Western world today may be a modern construct, human society stills seems to crave holding on to a bit of that magical history that carries some of the ancient gem traditions of the past into the future.

For further reading on birthstones, see Knuth (1999). For further reading on gems of the zodiac, see Agrippa (1651) and Thomas and Pavitt (1997).

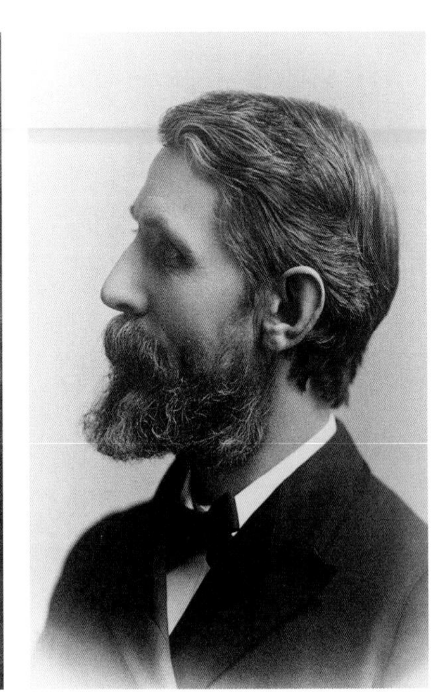

FIGURE 296.
Three important figures in the early history of The Field Museum's founding gem collection. *From left to right:* Charles Lewis Tiffany (1812–1902), George Fredrick Kunz (1856–1932), and Harlow Niles Higinbotham (1838–1919). Photo of Tiffany courtesy of Tiffany & Co.; photo of Kunz from Library of Congress, Prints & Photographs Division.

HISTORY *of* THE FIELD MUSEUM'S GEM HALLS

The Field Museum's gem collection is a treasure trove in many ways: a collection of priceless gems and a unique history. The story of the collection begins even before the founding of the museum in 1894, starting with a business not known for gems in its early years. That business was known as Tiffany & Co.

In 1837 Charles Tiffany (1812–1902) (see fig. 296, *left*) partnered with childhood friend John P. Young to build a stationery goods shop, specializing in paper. The paper business was fair, but their luck would soon change for the better due to an unlikely source. In 1848 the French people revolted again King Louis Philippe, sending the country into turmoil. French aristocrats fled the country, taking with them what they could carry, including gems. Tiffany and Young purchased a series of Diamonds from the fleeing aristocrats. They then made a profit by breaking the aristocrat's large gems into smaller gems, selling more pieces to more people. Their profits grew in ways that the sale of stationery could not match. Tiffany's business soon had a reputation for specializing in gems. But since Tiffany did not have any in-depth gemological knowledge, he realized that he needed someone with a mineralogical background to advise his company in this new venture.

The timing could not have been better for George Kunz (1856–1932) (see fig. 296, *middle*) to show up at Tiffany's front door. Kunz—a young, ambitious gemologist—had assembled a collection of North American gems and gemstones that he hoped to sell to Tiffany. Kunz got more than he expected. Tiffany recognized the beauty of Kunz's well-assembled collection and also Kunz's keen mineralogical eye. Tiffany purchased the collection and hired Kunz to be a company gemologist. Once hired, Kunz proved himself repeatedly, and Tiffany allowed him great autonomy as he traveled and purchased gems on behalf of the company. Tiffany's faith in Kunz paid off with a series of financial successes, and within three years of being hired, Kunz was vice president and chief gemologist of Tiffany & Co.

Tiffany and Kunz proved to be a good match. The Tiffany collection continued to grow over the years, guided by Kunz's expertise. Then Tiffany was invited to present at the 1893 World's Columbian Exposition, also known as the Chicago's World Fair. Kunz assembled a special Exposition collection that featured some of the best gems Tiffany had to offer, including a series of significant pieces from the famed Hope family. It was this Exposition gem collection that would prove to be pivotal in The Field Museum's history.

The Tiffany collection was breathtaking, even capturing the attention of royalty. Princess Eulalia and Prince Antonio of Spain were visiting the World Fair and were so impressed with Tiffany that they created a royal appointment based on the collection. Tiffany's Chicago Exposition manager reported this telegram back to headquarters in New York:

World's Fair Grounds, Chicago, June 10, 1893. To Tiffany & Co., Union Square, New-York. Princess Eulalia and Prince Antonio and suite have just visited our pavilion, expressed surprise, and were so pleased with our remarkable display, announced in person appointment of Tiffany & Co. Jewelers and Silversmiths to Her Highness.

Near the end of the six-month Exposition, many prominent Chicagoans considered converting the World's Fair collections into a permanent natural history museum for the people of Chicago. Harlow Higinbotham (1838–1919) (see fig. 296, *right*) was not only president of the World's Fair but also one of the men and women who sought to found a natural history museum. He purchased Tiffany's Exposition gem collection for $100,000 and donated it to help start the new Chicago natural history museum, thereby becoming a founding member of The Field Museum. The sum of $100,000 was a great deal of money in 1893 and noteworthy enough that the *New York Times* reported the event:

Among these gems is the famous Hope opal which was sold in London four years ago for $26,310; the Hope aquamarine, the famous Hope green tourmaline, and the diamond engraved by Devries of Amsterdam, Holland, which attracted much attention at the Paris Exposition in 1878.

Higinbotham also purchased a collection of books and articles assembled and written by Kunz, many of which are now part of The Field Museum's rare book collection, a special vaulted and carefully climate-controlled area within the museum's library. Higinbotham went on to serve as the museum's president from 1897 to 1909, a time during which he saw the rise in popularity of the Higinbotham Hall of Gems. The Hall of Gems became one of the museum's most popular attractions and remained so for 48 years. It even drew attention from international audiences; in 1933 Prince M.U.M. Salie of Galle, Ceylon, donated over 55 gem specimens representing gem mining in Ceylon. This collection is known as the Prince M.U.M. Salie Collection. Then in 1941 Mrs. Richard T. Crane, daughter of Harlow Higinbotham, purchased an entirely new set of gems from Tiffany & Co. and presented it to the museum. Like the founding gem collection, the 1941 Crane Collection was also assembled by the famous George Kunz. The Crane Collection was outstanding, featuring beautiful and unusual stones. In particular, the collection included the Crane Ruby Topaz, at 97.45 carats one of the world's largest, and the Crane Aquamarine, a 341-carat gem that is among the finest in the world. These subsequent donations allowed the Hall of Gems to grow in size, stature, and value.

The hall remained relatively unchanged for many years, consistently popular with visitors but in need of an update. Then in 1985 David and Juli Grainger decided to lend a hand. The Graingers have been strong advocates of the museum with a commitment of their time and resources; Juli Grainger has been a member of The Field Museum Women's Board since 1979 and is currently a Life Trustee. In 1985 they made a very generous donation to the museum through The Grainger Foundation to completely renovate and update the gem hall. It was the first time that the hall had been changed in over 40 years. The renovation was complete in 1986, and it then became known as the Grainger Hall of Gems. Additionally, the South Lounge, the area immediately outside the Grainger Hall of Gems, was also renovated and named the Grainger Gallery. The new Grainger Hall of Gems and Grainger Gallery featured everything from updated structure and new display cases, to modern lighting and security measures to properly display and protect The Field's beautiful gem and jewelry collection. Once the Grainger Hall of Gems was renovated, the Graingers remained patrons of the space, continually updating the hall and gallery.

Even with constant care and upkeep, styles and fashions change, and new material is acquired. After the turn of the new millennium, The Field Museum and The Grainger Foundation decided it was time once again for something new. In 2006 the museum assembled a team of scientists, exhibitions staff, educators, and researchers, including the two authors, to begin the planning for a massive overhaul of the Grainger Hall of Gems. This time, not only would the hall be renovated, but the exhibit collection would be updated as well. The hall would undergo massive structural changes, a complete redesign to display the gems to better advantage, including new casework, lighting, and interior design. The gems would now be displayed in groups, featuring like gemstones in matrix, as cut gemstones, and as jewelry pieces. With this new display in mind and after much research, the museum began to purchase new items for

HIGINBOTHAM HALL.

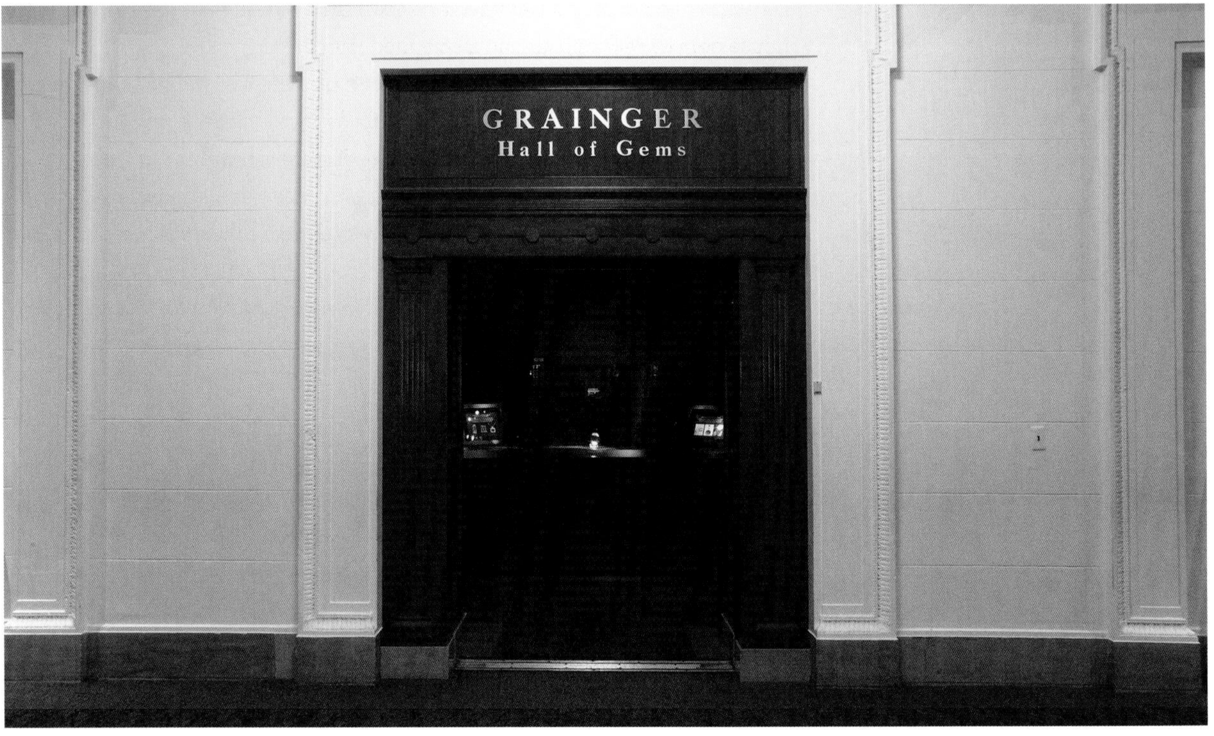
GRAINGER
Hall of Gems

FIGURE 297.
Facing page: The Higinbotham Hall of Gems in The Field Museum, 1893 (*top*), and the Grainger Hall of Gems in The Field Museum, 1985 (*bottom*). This hall is now being completely transformed and upgraded for a new Grainger Hall of Gems opening in October 2009.

FIGURE 298.
Right: The Grainger Foundation, longtime patrons of the Grainger Hall of Gems, made the 2009 exhibit renovation possible. *From left to right:* David Grainger, Juli Grainger, and Richard Winston (nephew of Harry Winston, the onetime owner of the Hope Diamond). Photograph taken in 1985 at The Field Museum Women's Board Gala, celebrating the opening of the original Grainger Hall of Gems.

the collection, visiting sites like the world-famous Tucson Gem, Mineral and Fossil Showcase, the largest gathering of gem dealers in the world. The team also explored working gem mines in California, going underground to see how gems are taken out of the earth, and visited famous jewelry houses across the country, including Tiffany & Co., Verdura, Bulgari, Lester Lampert, Inc., and others. The team spent time with Smithsonian gem specialists at the National Museum of Natural History, home of the Hope Diamond, and met with the Carnegie Museum of Natural History in Pittsburgh. By the end of 2008, The Field Museum had not only researched and examined what already existed of the in-house gem collection, but had also added many exciting specimens and modern external knowledge to bring the exhibit into the twenty-first century.

Additionally, the collection has continued to grow in the recent past through the generosity of other patrons, such as the Chalmers family, with their famous gigantic Chalmers Topaz, and Nancy Wald, with her valuable and unusual Arthur King Gold and Pearl jewelry set. Another very important donor to the collection is Mrs. Thuy Nguyen. The Nguyen Collection began in 1991 and includes such pieces as the historically important and unique 1904 Diamond necklace (fig. 16), a Diamond and emerald necklace (fig. 106), and a number of other very significant pieces of jewelry. Many new Nguyen pieces were added to the exhibit.

The museum also undertook an exciting project to have new jewelry created out of particularly fine gems from the collection. Designers like Lester Lampert of Chicago, Mish New York, and Tiffany & Co. have all been involved in the creation of pieces unique to The Field Museum and its dazzling collection.

The collaborative efforts of The Field Museum, The Grainger Foundation, and countless others ensure that the Grainger Hall of Gems is a truly spectacular display, befitting the best and brightest collection in the midwestern United States. And, as time goes on, the collection and its history will undoubtedly grow and change further, a living tribute to the generosity inspired by the beauty and love of gems.

EXHIBITION TEAM FOR THE GRAINGER HALL OF GEMS }

The planning, development, and construction of the 2009 Grainger Hall of Gems was an extensive project involving many talented museum staff and outside consultants. These included Laura Sadler, Senior Vice President for Museum Enterprises; Anne Underhill, Chair, Department of Anthropology; David Foster, Director of Exhibitions; Álvaro Amat, Exhibition Design Director; Jean Cattell, Graphic Design Director; Jaap Hoogstraten, Exhibition Operations and Media Director; Ray Leo, Exhibition Production and Maintenance Director; Matt Matcuk, Exhibition Development Director; Todd J. Tubutis, Exhibition Project Director; Beth Crownover, Public Programs and Operations Director; Neil Keliher, Exhibitions Maintenance Manager; Tony Stepovy, Media Services Manager; Clarita Nuñez, Collections Manager for Geology; James Holstein, Collections Manager for Geology; Jamie Kelly, Collections Manager for Anthropology; Christine Taylor, Collections Manager for Anthropology; Marianne Klaus, Assistant Conservator for Anthropology; Anna Huntley, Project Manager for Exhibitions; Eric Frazer, Exhibition Production Supervisor; Gloria Chantell, Graphic Designer; Teresa Murray, Exhibition Designer; Sarah Sargent and Allison Augustyn, Exhibition Developers; Thatcher Waller, Lighting Design; David Mendez, Mount Shop Supervisor; Pamela Gaible, Mount Shop Supervisor; Nel Fetherling, Graphics Shop Supervisor; John Weinstein, Head Photographer; Alison C. Neumann, Graduate Gemologist, Content Specialist; Janet Zapata, Decorative Arts Consultant, Content Specialist; and Lance Grande, Senior Vice President for Collections and Research, and Exhibit Curator and General Content Specialist for the 2009 Grainger Hall of Gems exhibit team.

FINAL WORDS AND ACKNOWLEDGMENTS

We would like to thank the many people who assisted us in the completion of this book. John Weinstein, lead photographer, did a superb job of image preparation for most of the images. For reading various drafts of the manuscript or parts of it and giving critical advice, we thank Mark Alvey, George Rossman, Riccardo Levi-Setti, Sarah Sargent, Donna Prestel, Harlan Berk, Janet Berres, and Jamie Dobie. Rudiger Bieler provided useful comments on the Pearls section, and Jim Phillips and Chap Kusimba provided useful discussions about Middle Eastern and South Asian archaeology. Terry Ottaway and Cathy Jonathan also provided valuable help from GIA. Karen Bean helped make color print proofs of most images. Marlene Donnelly and Peggy McNamara prepared some of the non-photographic figures. Elaine Zeiger helped with word processing. Clarita Nuñez and Jim Holstein were instrumental in cataloguing the collection. For helping us gain entry into a working gem mine, we thank Bill Larson, Josh Hall, Jeff Swanger, and Steve Koonce. For testing some of the gemstones for us to identify their composition, we thank Cap Beasley and the American Gemological Laboratories, and Laure Dussubieux of The Field Museum.

Again we thank David and Juli Grainger and The Grainger Foundation for their generous support of the beautiful new exhibit hall, which helped inspire us to write this book. We also thank Thuy Nguyen for her generous donations of jewelry to the Grainger Hall of Gems over the years. Others who generously donated gems, precious metals, and/or design and mounting services to mount Field Museum gems into innovative settings include Lester Lampert and David Lampert of Lester Lampert, Inc., Mish Tworkowski of Mish New York, Tiffany & Co., Ellie Thompson of Ellie Thompson & Co., Marc Scherer of Marc & Co., Richard Cheng of Merry Richards Jewelers, Harlan Berk of Harlan J. Berk, Ltd., Sam Ciccione and Oak Park Jewelers, Karla Proud of Exotic Gemstones, Juli Grainger, Cyril L. Ward, Chester Tripp, Mathilde Wilde, Anna Sophrenea Graham Gurley, Vida Woley, Mrs. E. Tolling, Tiffany & Co., Amit Jhalani of MySolitare.com, Edward Kramer, Nancy Wald, and Beth Crownover.

Finally, the first author thanks his wife, Terry, and daughters, Lauren and Elizabeth, for their patience during the long night and weekend hours he spent on this book. He would also like to thank his daughters for being stand-in models in figures 93 and 91. The second author would like to thank her family, including her parents, James and Elizabeth Augustyn, and her sister, Jessie, and brother, James Ryan, for their encouragement, patience, and unfailing humor during this project.

GLOSSARY }

Note: All terms in this glossary are defined in the context of gems and minerals.

adamantine luster. An extremely rich, submetallic luster that occurs on certain TRANSLUCENT or TRANSPARENT materials. A prime example of a mineral with adamantine luster is Diamond.

adularescence. See SCHILLER EFFECT.

alexandrite effect. Significant color change when going from daylight to incandescent light. Such gems are said to be ALEXANDRITIC.

alexandritic. See ALEXANDRITE EFFECT.

allochromatic. Gem coloration is derived solely from trace elements and impurities. *See* IDIOCHROMATIC.

alloy. A compound made from two or more combined metals.

alluvial deposits. Sediments (silt, soil, sand, and gravel) deposited by a river or other moving water.

alluvium. See ALLUVIAL DEPOSITS.

amorphous. A gemstone without crystal structure (such as Amber).

amulet. An object or protective charm worn to ward off evil.

anion. An atom or molecule that has more electrons than protons, giving it a negative charge. Anions such as O_3 bond to positively charged cations such as Al_2 to form stable compounds such as Corundum (Al_2O_3).

artisan. Person who creates a handmade object without the aid of machine manufacturing.

artisanal mining. Small-scale mining methods consisting of hand digging with simple tools and without machinery.

asterism. An effect that creates a star in a gem, created by intersecting bands of light crossing the surface of the stone (e.g., star sapphire). Stars can have 4, 6, 12, or 24 rays, but the most common number is 6.

astrological calendar. Calendar that marks time by charting the movement of the stars and planets. The Babylonian astrological calendar was divided into twelve astrological months and corresponded to a series of twelve symbols representing those months called the ZODIAC.

augmentation. Enhancement of a gemstone's qualities through HEAT TREATMENT, IRRADIATION, STAINING, or OILING.

avoirdupois ounce. From the English and American system of weights in which an ounce is 28.35 grams. This is the common weight used for produce and most other items in North America except for precious metals. *See* TROY OUNCE.

beauty. A subjective aesthetic quality influenced primarily by a gem's form, color, clarity, and reflection of light. The perception of beauty is also often influenced by current fashion and trend.

bestiaries. Medieval religious anthologies cataloging vast information about the natural world, including animals, plants, and minerals (gems). *See also* LAPIDARIES.

birefringence. A term for the splitting of a ray of light into two as it passes through a crystal or other transparent material, causing a double image of features viewed through it.

birthstone. Gem corresponding to an individual's birthday, birth month, or zodiac sign, said to bring the wearer good fortune.

blend. A mix of two mineral species that readily combine with each other to form a variety of INTERMEDIATE composition.

blood gem. See CONFLICT GEM.

bort. Non-gem-grade Diamond used in industry.

botryoidal. A texture in which a mineral has a globular external from resembling a "bunch of grapes" (etymology of the Greek word for this form).

boule. Long, cylindrical gem crystal created in a laboratory.

brilliance. A combination of light reflected from the surface of a gem (LUSTER) and light reflected from within a gem (FIRE).

brilliant cut. A faceting style that maximizes BRILLIANCE. The ROUND BRILLIANT CUT contains 57 or 58 facets.

c-axis. A term in CRYSTALLOGRAPHY that refers to the "vertical" axis (generally the long axis running the length of the natural crystal). This axis runs 90 degrees to the perpendicular A and B axes that run through the horizontal plane of the natural crystal.

cabochon. A gem cut with a polished domed upper surface (see fig. 2).

cabochoned. To be made into a CABOCHON.

carat. A unit of weight for a gem that today equals 1/5th of a gram.

Carbonate. A mineral in which CO_3 is the principal ANION, such as Coral ($CaCO_3$).

cat's-eye effect. An effect of CHATOYANCY that results in a single sharp line of reflected light across the surface of a stone, creating an "eye" effect.

cementation. The chemical process that binds particles together in the formation of SEDIMENTARY ROCKS.

certification. A process of official documentation certifying size, quality, type, origin, authenticity, or some other aspect of a gemstone. Certification can also verify that a jewel is not a CONFLICT GEM.

chatoyancy. A CAT'S-EYE EFFECT of a gem or gemstone, where light is reflected back in a single band across the surface of the stone (see fig. 55).

chatoyant. See CHATOYANCY.

class. A high-ranking category in our gem classification containing gem GROUPS, SPECIES, and VARIETIES (see fig. 12). Classes of minerals are based on their general chemical characteristics.

classification. A system of organization that groups gems according to shared feature.

clean. The highest grade of clarity in a gemstone, with no visible inclusions even under 10x magnification.

cleavage. The clean breakage of a mineral along a plane of weakness related to internal atomic structure.

cleaving. The breaking of a gemstone by carefully striking it parallel to the CLEAVAGE plane.

colluvial deposits. Geological deposits accumulated over time at the base of an incline such as a mountain or hill, due to a combination of erosion of the hillside and gravity. These deposits may be referred to as COLLUVIUM.

colluvium. See COLLUVIAL DEPOSITS.

colored stone. In GEMOLOGY, a trade name for all gems other than Diamonds, even those that are colorless.

common gem. An imprecise term, generally referring to relatively common minerals used for inexpensive jewelry, such as agate or rock crystal.

concoidal fracture. A rounded, uneven type of fracture surface that shows convex and concave surfaces.

condition rarities. Gem varieties that may not be extremely rare as a variety but are extremely rare in near-flawless condition (e.g., flawless emerald).

conflict gem. A gem used by rebel movements or their allies to finance armed conflicts aimed at undermining legitimate governments.

crown. The upper facets of a faceted gem above the GIRDLE (see fig. 2).

cryptocrystalline. Rock or mineral with crystals so small that even under microscopic examination they are only vaguely discernible.

crystal. A geometric solid bounded by smooth planar external surfaces.

crystal power. Modern belief that wearing certain gem crystals or certain geometric crystal shapes allows a person to channel, harness, and accentuate different abilities and healing powers. *See* PYRAMID POWER.

crystallography. The study of the complex way in which minerals form natural crystalline shapes in regular geometric patterns.

cubic zirconia (CZ). A man-made Diamond simulant with similar FIRE and BRILLIANCE but far less hardness (Mohs 8½).

culet. An optional small flat surface at the bottom of a ROUND BRILLIANT CUT (rbc) stone. Sometimes it is omitted in an rbc, and there is a sharp apex instead.

cultured Pearl. A Pearl grown in a mollusk by artificially seeding the animal with a bead of mother-of-pearl. A form of controlled stimulation of a natural biological process within the animal in which foreign particles are coated with NACRE resulting in a Pearl.

dewatering. Squeezing all water out of sediments as the result of compaction.

Diamond grit. Pulverized Diamond, used as a polishing agent for hard gems.

dichroic. A gem that appears as two different colors, each of which is dependent on viewing the stone from a particular direction.

dichroism. See DICHROIC.

dispersion. The splitting of white light into a rainbow of colors due to the refraction of white light bending each light frequency at a slightly different angle.

doublet. A composite gem made by gluing two pieces together.

druse. A type of crystalline coating found in the inside of rock cavities such as geodes. *See* DRUSY CAVITIES and GEODE.

drusy cavities. Spaces within rocks that are lined with crystals growing toward the center of the cavity.

durability. A combination of hardness and toughness in gems. *See* HARDNESS and TOUGHNESS.

electrum. A naturally occurring alloy of Gold and Silver, generally with 70–90% Gold. Trace amounts of other metals can also be present such as Copper, Platinum, and other metals.

Elemental. In gemstones and precious metals, Elemental species are those composed of a single type of atom (e.g., pure Diamond contains only Carbon atoms, pure Gold contains only Gold atoms).

eluvial deposits. Geological deposits accumulated over time IN SITU through weathering of host material.

eluvium. See ELUVIAL DEPOSITS.

emery. A mix of Corundum sand and other substances used to make sandpaper, grinding wheels, fingernail files, and other abrasive tools.

extrusive igneous rocks. Rocks solidified from LAVA or other material expelled from volcanoes.

eye-clean. A gem that looks free of inclusions to the naked eye (i.e., without magnification).

fabrication. The attachment of a gem to another stone or man-made product that makes the gem resemble a more expensive gem. *See* DOUBLET.

faces. The flat surfaces of a gemstone.

facet. The man-made cut-and-polished surface on a finished gem.

faceted. Cut and polished with FACETS.

faceting. The cutting and polishing of facets on a gem. *See* FACETS.

faceting grade. A qualitative category of natural gemstones, rough stones, or crystals, indicating suitable clarity and color for faceting into gems.

fire. The colored flashes of dispersed and refracted light in a transparent, faceted gem (not to be confused with PLAY OF COLOR). Gem fire is the result of white light being separated into its spectral colors of red, orange, yellow, green, blue, indigo, and violet within the stone.

flame-fusion. Method of creating gems in a laboratory by dropping chemical gem components through a superheated flame, producing a boule in the process. *See* BOULE.

flux. A solvent used in the creation of laboratory gems. *See* SOLVENT.

flux growth. Early method of creating gems in a laboratory, precursor to flux-melt technology. *See* FLUX-MELT.

flux-melt. Method of creating gems in a laboratory by melting gem-forming powdered chemical ingredients into a superheated solvent (FLUX), and allowing small gem crystals to grow in the solution into larger, marketable-sized gem crystals. *See* MELT.

fracture. Cracking, chipping, or breaking of a stone.

garimpeiros. Brazilian gem miners, often using artisanal methods. *See* ARTISAN.

gem. A gemstone that has been cut and polished. *See* GEMSTONE.

gemology. The science and study of GEMS and GEMSTONES.

gemstone. A mineral, stone, or organic matter that can be cut and polished or otherwise treated for use as jewelry or other ornament. *See* GEM.

geode. A rock containing a cavity whose inner surface is lined with crystals (DRUSES) growing toward the center of the cavity.

geologic time. The time of the physical formation and development of the earth, dating back about 4.6 billion years; deduced from the study of radioactive element decay, principles of uniformity, stratigraphy, and other scientific methods and philosophy.

GIA. Gemological Institute of America.

girdle. The ridge around the widest part of a faceted gem, where the PAVILION meets the CROWN (see fig. 3).

Gold nugget. A Gold nugget is a naturally occurring piece of native Gold, often concentrated in watercourses and recovered by PLACER MINING but also found in Gold-bearing rock formations in which the host rock has decayed or eroded. Nuggets with significant amounts of Silver are known as ELECTRUM.

greasy luster. A type of luster in which a mineral appears to have a greasy or oily surface.

group. A category in our gem classification that contains closely related gem species.

hard rock mining. A method of mining involving the digging of tunnels deep into the earth, often with explosives and heavy equipment, in search of pockets or veins of mineable material.

hardness. See MOHS SCALE OF HARDNESS.

heat treatment. The process of enhancing the color or clarity of gems through exposure to extremely high temperatures.

holistic. A system that cannot function as a whole without all of its parts, each part dependent on the other.

Hoshen. Hebrew word referring to a breastplate referenced in the biblical book of Exodus. According to Exodus, God designed a Hoshen for Aaron and asked Moses to make it. The breastplate was set with 12 stones corresponding to the 12 founding sons of Israel.

hue. The particular color shade or shades of a gem.

hydrothermal solution. A hot solution of mineral-rich water that provides the material for the formation of gems in underground rock cavities.

idiochromatic. Gem coloration derived from elements that are major parts of a gemstone's essential chemical composition. *See* ALLOCHROMATIC.

igneous rocks. Rocks whose origin is from solidified MAGMA or molten rock. Some igneous rocks solidify underground from magma and some solidify aboveground after a volcanic eruption. *See* INTRUSIVE IGNEOUS ROCKS and EXTRUSIVE IGNEOUS ROCKS.

inclusions. Internal marks or foreign bodies (flaws) in a gem.

inorganic gems. Gems derived from rocks, or from gravel derived from rocks.

interference. Reflection of light off of structures or inclusions within a gem.

intermediate. As formally used here, a classification rank between SPECIES and VARIETY, usually a BLEND of two species.

intrusive igneous rock. Igneous rock that solidifies from MAGMA below the earth's surface.

irradiation. An enhancement process for gems that uses high energy, sometimes followed by heat-treating, to enhance the color of the gem.

karat. A measure of purity for precious metals. Pure Gold, for example, is 24 karat, and half-Gold/half-alloy is 12 karat.

Kimberley Process Certification Scheme (KPCS). An international certification scheme that regulates and protects the legitimate trade of rough Diamonds, while attempting to prevent the trade of conflict Diamonds. Supported by the United Nations (UN).

kimberlite. A volcanic rock containing material from deep within the mantle of the earth, 150 to 450 kilometers deep. This material was erupted some time in the past, forming kimberlite "pipes," which are an important source of Diamonds today.

labradorescence. A particular "sheen" effect of light seen in rainbow moonstone that results from a combination of SCHILLER EFFECT and light INTERFERENCE effects from thin internal films within the stone.

lapidaries. Books popular in the Middle Ages that explained the symbolism of stones and gems within the context of the natural and religious worlds. *See also* BESTIARIES.

lava. MAGMA that has erupted onto the surface of the earth, usually through volcanoes.

life. A quality in a transparent gem that is a combination brilliant LUSTER and FIRE.

lithification. The process by which sediments are turned to rock through pressure, DEWATERING, and CEMENTATION.

lunar calendar. Calendar that marks time by charting the phases of the moon. Our current 12-month calendar is a lunar calendar.

luster. The visual property of light reflected from the surface of a gemstone. For example, luster can be VITREOUS (glassy as in most transparent gemstones and Jadeite), ADAMANTINE (hard highly polished surface of Diamond), METALLIC (Gold), WAXY (such as Turquoise), or GREASY (Nephrite).

mabe Pearl. A dome-shaped half-round CULTURED PEARL, with a flat back, grown on the inner shell of a MOLLUSK.

magma. Rock that is in a melted state beneath the earth's surface. Once erupted it becomes LAVA.

malleability. The ability to be hammered into thin sheets or be pressed or drawn into different shapes without breaking.

malleable. See MALLEABILITY.

market manipulation. Using the laws of supply and demand to promote value through promotion of product and control of distribution.

matrix. The base rock in which a gemstone crystal is found; the parent rock.

melt. In synthetic gem processes, a melt is a combination of chemicals dissolved in a FLUX (solvent).

metallic luster. A type of luster resembling the reflective surface of metals.

metamorphic rocks. IGNEOUS or SEDIMENTARY ROCKS that have been altered by heat and/or pressure to form new rocks and minerals.

metaphysical. Pertaining to metaphysics, the study of the reality of the world independent from any particular science. Metaphysics attempts to explain the ultimate nature of the world and of the individual, or being.

microcrystalline. Rock or mineral with tiny crystals visible only through microscopic examination.

mineral. A naturally occurring inorganic solid, usually crystalline in structure, whose composition can be expressed by a chemical formula.

mineralogy. The study of the chemistry, crystal structure, and other physical properties of minerals.

modifier. An overtone or secondary color hue of a stone that sometimes decreases the value of a stone (particularly gray or brown).

Mohs scale of hardness. A measure of a mineral's comparative hardness based on its ability to resist scratching using a graded set of standards from Talc (the softest at Mohs 1) to Diamond (the hardest at Mohs 10). Created by Frederick Mohs. See table 4.

mollusks. A large group (phylum) of invertebrate animals that include clams, oysters, and snails.

nacre. An iridescent form of the mineral aragonite. The natural substance of Pearls that gives it that mother-of-pearl finish.

nacreous. Having nacre. *See* NACRE.

natal stones. See BIRTHSTONES.

necromancers. In early history, necromancers were people skilled in attempts to contact the spirit world. The term took on a darker meaning after the medieval ages, when necromancers were believed to summon demons using black magic.

Neolithic. Sometimes referred to as the "New Stone Age," the Neolithic is generalized by the development of agriculture and spans a time of roughly 8000 BC to 1700 BC.

numismatist. A specialist in or collector of coins, paper money, or medals.

oiling. An enhancement process for fractured gems in which colorless oil is infused into the fracture to make it less visible, used mainly for emeralds.

old mine cut. Precursor to today's BRILLIANT CUT, containing 50 facets.

opaque. Not transparent or translucent (no light can pass through the stone).

open pit mining. A method of mining involving heavy machines to dig large open pits exposing large surface areas for prospecting and/or extraction of materials.

organic gems. Gems formed as a product or part of a living organism.

orient of Pearl. The characteristic, mother-of-pearl-like iridescence and luster of Pearls.

overall rarity. Gem varieties that are rare regardless of condition (e.g., bixbite).

Oxides. Minerals in which Oxygen is the principal ANION, often bonding with a metal element. In minerals of this class, the Oxygen symbol "O" or the hydroxyl symbol "OH" appears at the right end of the chemical formula, as in Corundum (Al_2O_3).

pavilion. The lower part of a faceted gem, below the GIRDLE (see fig. 3).

pegmatite. A coarse-grained INTRUSIVE IGNEOUS ROCK (and a type of granite) formed as MAGMA slowly cools under the earth's surface, often forming large crystals.

Phosphates. Minerals in which the ANIONS are principally Phosphates, with "PO_4" as part of their chemical makeup, as in Turquoise [$CuAl_6(PO_4)_4(OH)_8 \cdot 4H_2O$].

piezoelectric. Crystals that have the ability to produce an electrical charge when compressed.

placer deposits. Concentrated secondary deposits of minerals weathered out of their host rock, often found as gravel and sand in rivers or other moving water.

placer mining. A method of mining that focuses on searching PLACER DEPOSITS.

play of color. Display of blazing multicolors due to interference and diffraction of light in Opal and labradorite.

pleochroic. Exhibiting PLEOCHROISM.

pleochroism. A property in gems of showing different color when viewed at different angles. This is due to light traveling through a gem at different speeds along different directions within the crystal structure.

Pliny the Elder. An ancient Roman naturalist and author (AD 23–79) who was one of the earliest known writers on geology and mineralogy.

point. A unit weight for gems that is 1/100th of a carat.

polar bicolor (color zonation). A type of color zonation where one end of a crystal is a different color from the other (see fig. 143). POLAR TRICOLOR crystals can have a third color or a colorless section in between the two different colored ends. *See* RADIAL BICOLOR (COLOR ZONATION).

polar tricolor (color zonation). A type of color zonation in which a third color or a colorless section lies between the two different-colored ends. *See* POLAR BICOLOR (COLOR ZONATION) and RADIAL BICOLOR (COLOR ZONATION).

precious gem. A term traditionally fixed to gem varieties such as Diamond, ruby, emerald, and sapphire, but today having a much broader (and somewhat more ambiguous) connotation due to the discovery of many new rare and valuable gem varieties. *See* SEMIPRECIOUS GEMS and COMMON GEMS.

pyramid power. Belief that the internal geometric structure of a gem crystal allows the owner to harness and channel mystical properties. *See* CRYSTAL POWER.

pyroelectric. Crystals that develop an electrical charge in response to temperature change in the crystal are said to be pyroelectric.

radial bicolor (color zonation). A special type of color zonation in a crystal or gem that has one color in the center of the stone and another in the outer edge surrounding the center, such as watermelon Tourmaline (fig. 147). *Also see* POLAR BICOLOR (COLOR ZONATION).

rarity. Scarcity due to TRUE RARITY, CONDITION RARITY, and MARKET MANIPULATION.

refraction. The bending of light as it passes from air into a transparent gem.

refractive index. A mathematical relationship between the angle at which light strikes a gem and the angle that the light is bent within the stone.

replacement fossil. When a fossil is dissolved out of its host rock, leaving behind a three-dimensional mold inside the rock, sometimes minerals are later deposited in the mold from groundwater seeping through the rock over thousands or millions of years. The resulting natural cast is called a replacement fossil.

round brilliant cut. Also abbreviated as "rbc" and sometimes referred to as the "ideal cut," the round brilliant cut is the faceting pattern given a gem to maximize its BRILLIANCE by maximizing the amount of light reflected back through the top (TABLE and CROWN) of the gem. The modern round brilliant cut contains 58 facets (or 57 if the CULET is excluded).

saturation. The intensity of color in a gem.

scarcity. One of the factors that make a fine gem. If the demand exceeds the supply, a gem's market value rises. Partly a factor of rarity and partly a factor of popularity.

Schiller effect. A whitish or bluish-white sheen or shimmer of light reflected off internal features of a gem. A form of iridescence that is softer and more diffuse than CHATOYANCY. Also called ADULARESCENCE.

scintillation. The "sparkle" of a gem.

sedimentary rocks. Rocks formed by consolidation and hardening of fragments of rocks, organic remains, or chemical precipitate from water.

seed Crystal. Small gem crystal used to create a larger synthetic gem.

semiprecious gem. An imprecise term, traditionally referring to gem varieties such as Tourmaline, Garnet, aquamarine, and Topaz. This term is of little use today because some varieties within these groups would qualify as precious gems today, such as paraiba Tourmaline, demantoid Garnet, and others. *See also* PRECIOUS GEMS and COMMON GEMS.

shaman. Throughout different cultures, a shaman is either a magician, healer, or warrior, or a combination of these types.

Silicates. Minerals in which silica combines with other elements to form the ANIONS. All Silicates have a Silicon-Oxygen combination "SiO" as part of their chemical makeup, as in Beryl [$Be_3Al_2(SiO_3)_6$].

silk. Fine, parallel, hair-like inclusions in transparent gems. If the silk effect is extremely dense, ASTERISM or CHATOYANCY can result.

simulant gem. An imitation or fake gem made to look like a more expensive natural gem. A simulant does not have the same composition or physical properties as the natural stone it is imitating and should not to be confused with a SYNTHETIC GEM.

solidus. A Gold coin from ancient Rome that was first issued about AD 301.

solution. A mixture (often a solid dissolved in a liquid) composed of two or more components.

solvent. A solvent is a liquid (or gas) that contains a dissolved solid, liquid, or gas, thereby creating a SOLUTION.

species. In mineralogy, a mineral type with a specific chemical composition, arrangement of atoms, and other factors (e.g., Corundum). Officially recognized mineral species are listed in Back and Mandarino (2008). Many species have different varieties. *See* VARIETY.

spectroscope. Instrument used to measure light, often reflection and REFRACTION.

staining. Process of adding stain, dyes, or chemicals to alter the surface layer of a gem.

star. See ASTERISM.

stater. An ancient Greek or Lydian coin that circulated from about 700 BC to AD 50. First struck in ELECTRUM, some were later struck in purer Gold and in Silver.

stone. The most common term used to apply to all solid constituents of the earth's crust, as well as extraterrestrial matter such as meteorites.

subgroup. As formally used here, a classification rank between GROUP and SPECIES.

subsistence mining. Mining conducted over a specific region that sustains that region. Often uses ARTISANAL MINING methods and is prevalent in poorer countries that lack machinery or electricity.

supersaturated. When there is more dissolved material in a liquid than it can hold in solution, ultimately leading to crystal formation. *See* SOLUTION.

supervariety. A classification rank between GROUP and VARIETY used here for organic gems.

synthetic gem. A laboratory-made gem that contains the same chemical composition, structure, hardness, and physical properties as its natural counterpart. Although it is man-made, it is real and not to be confused with a SIMULANT GEM (a man-made imitation or fake).

system. The highest systematic rank in our classification of gemstones (see fig. 12), dividing the various classes of gemstones into either "Inorganic" or "Organic."

table. The central FACET on the top or CROWN of a faceted gem (see fig. 3).

talisman. A charm or object worn to give the wearer supernatural powers or influence over human feelings or actions.

tohunga. A wizard, native to the Maori people of New Zealand.

tone. The degree of lightness or darkness of color in a gem.

toughness. A mineral's resistance to FRACTURE, shattering, or chipping.

translucent. When light is transmitted through a stone but images cannot be distinguished through the stone. *See* TRANSPARENT and OPAQUE.

transparent. When an image seen through a stone is perfectly distinct. *See* TRANSLUCENT and OPAQUE.

trichroic. A gem that appears as three different color hues, each of which is dependent upon viewing the stone in a particular direction.

trichroism. The property of being TRICHROIC.

triplet. A composite gem made from gluing three pieces together.

troy ounce. From the ancient French system of weights still used for precious metals such as Gold, in which an ounce weighs 31.1 grams. *See* AVOIRDUPOIS OUNCE.

true rarity. Truly rare gems fall into two major categories: OVERALL RARITIES and CONDITION RARITIES (see definitions for those two terms).

variety. A subcategory or type of species (*see* SPECIES). For example, ruby and sapphire are varieties of the species Corundum.

varnas. Sanskrit word, referring to a type of classification grouping. Varnas may categorize gems, while other varnas categorize people.

Vedic. Pertaining to the Aryans who settled in India around 1500 BC.

vena amoris. Latin term meaning "vein of love." Refers to the vein in the ring finger of the left hand that was believed to extend directly to the heart.

Verneuil flame-fusion. See FLAME-FUSION.

vitreous luster. A glassy luster on TRANSPARENT or TRANSLUCENT material. Prime examples of minerals with vitreous luster include aquamarine, Quartz, Spinel, Chrysoberyl, Tourmaline, and sapphire.

vug. An open cavity in a rock, sometimes lined with crystals.

waxy. A type of luster resembling a waxed coating, characteristic of some minerals such as most Turquoise or jade.

zodiac. An imaginary belt of the heavens around the earth within which are the apparent paths of the sun, moon, stars, and principal planets. The zodiac contains 12 constellations that divide it into 12 equal zones of "celestial longitude," corresponding to the 12 signs of the zodiac. The zodiac concept goes back as far as 10,000 years in human culture.

zonation of color. The arrangement of color in zones or bands within a gemstone or gem, corresponding to varying composition or trace elements.

REFERENCES

Agrippa, H. C. 1651. *Three Books of Occult Philosophy*. London: printed by R.W. for Gregory Moule.

Back, M. E., and J. A. Mandarino. 2008. *Fleischer's Glossary of Mineral Species*. 10th ed. Tucson: Mineralogical Record, Inc. Pp. 1–146.

Ball, S. 1935. "A Historical Study of Precious Stone Valuations and Prices." *Economic Geology* 30:630–42.

Bauer, M. 1968. *Precious Stones*. Vols. 1 & 2. New York: Dover.

Berk, H. J. 2008. *100 Greatest Ancient Coins*. Atlanta: Whitman Publishing.

Birch, B. 1987. Gold in Australia. *Mineralogical Record* 18:5–32.

Boyer, B. 1993. *The Natural History of the Field Museum*. Chicago: The Field Museum.

Bruton, E. 1978. *Diamonds*. 2nd ed. London: NAG Press.

Cirlot, J. E. 2002. *A Dictionary of Symbols*. New York: Dover Publications.

Dietrich, R. 1985. *The Tourmaline Group*. New York: Van Nostrand Reinhold.

Drucker, R. B., ed. 2008. *The Gem Guide to Wholesale Gem Pricing: Color*. Glenview, IL Gemworld International. Pp. 1–140.

Evans, J. 1998. *The History and Practice of Ancient Astronomy*. New York: Oxford University Press.

Fales, Martha. 1995. *Jewelry in America: 1600–1900*. Suffolk, UK: Antique Collectors' Club.

Finlay, V. 2007. *Jewels: A Secret History*. New York: Random House.

Francis, C. A. 2004. Gold crystals: A primer. *Rocks and Minerals* 70(1):22–29.

Friedrichsen, G. W. S., R. W. Burchfield, and C. T. Onions. 1996. *The Oxford Dictionary of English Etymology*. New York: Oxford University Press.

Grimaldi, D. A. 1996. *Amber: Window into the Past*. New York: Harry Abrams.

Guastoni, A., and R. Appiani. 2005. *Minerals*. Buffalo, NY: Firefly Books.

Hall, C. 2002. *Gemstones*. 1st ed. New York: Dorling Kindersley.

Hall, R. N. 1905. *Great Zimbabwe: Mashonaland, Rhodesia*. London: Methuen.

Hoover, D. B. 1992. *Topaz*. London: Butterworth-Heinemann.

Hughes, R. W. 1990. *Corundum*. London: Butterworth-Heinemann.

———. 1997. *Rubies and Sapphire*. Boulder, CO: R. W. H. Publishing.

———. 1999. "Burma's Jade Mines: An Annotated Occidental History." *Journal of the Geo-Literary Society* 14, no. 1:15–35.

Hurlbut, C. S. 1971. *Dana's Manual of Mineralogy*. 18th ed. New York: John Wiley & Sons.

Hurlbut, C. S., Jr., and G. S. Switzer. 1979. *Gemology*. New York: John Wiley & Sons.
Jackson, B. 2006. "Garnets." in *Gems: Their Sources, Descriptions, and Identification*, ed. M. O'Donoghue, pp. 195–237. 6th rev. ed. of Webster (1962). Oxford: Elsevier.
Jones, B. 2004. *The Tucson Show: A Fifty-Year History*. Tucson: Mineralogical Record, Inc.
Kaner, S. 1995. *The Last Empire: De Beers, Diamonds, and the World*. New York: Farrar, Straus and Giroux.
Knuth, B. 1999. *Gems in Myth, Legend, and Lore*. Eastlake, CO: Jeweler's Press.
Kunz, G. F. 1917. *Rings for the Finger*. Philadelphia: J. B. Lippincott.
———. 1968. *Gems and Precious Stones of North America*. New York: Dover.
———. 1997. *The Magic of Jewels and Charms*. New York: Dover.
———. 2003a. *Curious Lore of Precious Stones*. Whitefish, MT: Kessinger.
———. 2003b. *Shakespeare and Precious Stones*. Whitefish, MT: Kessinger.
Landman, N. H., P. M. Mikkelsen, R. Bieler, and B. Bronson. 2001. *Pearls: A Natural History*. New York: American Museum of Natural History.
Leechman, F. 1978. *The Opal Book*. Sydney: Ure Smith. (Reproduction of 1961 edition.)
Liddicoat, R. 1993. *Handbook of Gem Identification*. 12th ed. Carlsbad, CA: Gemological Institute of America.
Loring, J. 1999. *Tiffany Jewels*. New York: Abrams.
———. 2007. *Tiffany Colored Gems*. New York: Abrams.
Lyman, K., ed. 1986. *Gems and Precious Stones*. New York: Simon & Schuster.
Marbode de Rennes. 1933. *De Lapidbus*, trans. C. W. King and J. Riddle. Stuttgart: Franz Steiner Verlag.
McLintock, W. F. P. 1983. *Gemstones in the Geological Museum: A Guide to the Collection*. 4th ed. London: HMSO.
Mounce, R. 1997. *The Book of Revelation (New International Commentary on the New Testament)*. Grand Rapids, MI: Eerdmans.
O'Donoghue, M. 1987. *Quartz*. London: Butterworth-Heinemann.
O'Donoghue, M., ed. and reviser. 2006. *Gems: Their Sources, Descriptions, and Identification*. 6th rev. ed. of Webster (1962). Oxford: Elsevier.
Oldershaw, C. 2004. *Firefly Guide to Gems*. Ontario: Firefly Books.
O'Neil, P. 1983. *Gemstones*. Arlington, VA: Time-Life Books.
Parise, F. 2002. *The Book of Calendars: Conversion Tables from Sixty Ancient and Modern Clandars to the Julian and Gregorian Calendars*. Piscataway, NJ: Gorgias Press.
Post, J. E., ed. 1997. *The National Gem Collection: Smithsonian Institution*. New York: Abrams.
Pough, F., and R. Peterson. 1998. *A Field Guide to Rocks and Minerals (Peterson Field Guides)*. 5th ed. Boston: Houghton Mifflin.
Read, P. G. 2005. *Gemology*. 3rd ed. Oxford: Elsevier.
Rees-Mogg, A. 2005. "Move over Diamonds, It's Tanzanite's Turn to Shine." *Money Week* (December 15). http://www.moneyweek.com/file/5384/tanzanite-0912.htm.
Roberts, J. 2004. *Glitter and Greed: The Secret World of the Diamond Cartel*. New York: Disinformation Company.

Rouse, J. 1986. *Garnet*. London: Butterworth-Heinemann.
Rudoe, J. 1997. *Cartier*. New York: Harry N. Abrams.
Schumann, W. 2006. *Gemstones of the World*. 3rd ed. New York: Sterling.
Sinkankas, J. 1959. *Gemstones of North America*. Princeton, NJ: D. Van Nostrand Co.
———. 1981. *Emerald and Other Beryls*. Radnor, PA: Chilton.
Solenhofen, A. 2003. "Rock properties and their importance to stoneworking, carving, and lapidary working of rocks and minerals by ancient Egyptians." http://www.geocities.com/unforbidden_geology/rock_properties.htm.
Stein, N. 2001. "The De Beers Story: A New Cut on an Old Monopoly. The Company that Has Ruled Diamonds for a Century Wants to Polish Its Image . . . and Dominate as Never Before." *Fortune* (February 19). http://money.cnn.com/magazines/fortune/fortune_archive/2001/02/19/296863/index.htm.
Thomas, A. 2007. *Gemstones: Properties, Identification and Use*. London: New Holland Publishers. (1st ed. 1922).
Thomas, W., and K. Pavitt. 1997. *The Book of Talismans, Amulets and Zodiacal Gems*. 3rd rev. ed. Whitefish, MT: Kessinger. 1st ed., 1922.
Wallis, K. 2007. *Gemstones: Understanding, Identifying, Buying*. Suffolk, UK: Antique Collectors' Club Ltd.
Webster, R. 1962. *Gems: Their Sources, Descriptions and Identification*. 1st ed. Vols. 1 & 2. London: Butterworth & Co.
Williams, G. F. 1902. *The Diamond Mines of South Africa*. New York: Macmillan.
Wilson, W. E., J. A. Bartsch, and M. Mauthner. 2004. *Masterpieces of the Mineral World: Treasures from the Houston Museum of Natural Science*. New York: Abrams.
Zancanella, V. 2004. *Tanzanite: The True Story*. Cavalese, Italy: Naturalis Historia.
Zoellner, T. 2006. *The Heartless Stone: A Journey through the World of Diamonds, Deceit, and Desire*. New York: St. Martin's.
Zucker, B. 2003. *Gems and Jewels: A Connoisseur's Guide*. New York: Overlook Press.

GENERAL WEB SITES

Chatham Jewelery. www.chatham.com.
"Conflict Diamonds: Sanctions and War; General Assembly Adopts Resolution on 'Conflict Diamonds.'" www.un.org/peace/africa/Diamond.html.
De Beers. www.DeBeers.com.
Diamond Facts. www.diamondfacts.org.
Gemological Institute of America (GIA), http://www.gia.edu.
International Colored Gemstone Association, http://gemstone.org.
Kimberley Process. www.kimberleyprocess.com.
Tiffany & Co. www.tiffany.com.

INDEX OF GEM, GEMSTONE, AND OTHER MINERAL NAMES }

SUBJECT INDEX

DESIGN AND COMPOSITION Jill Shimabukuro

COLOR SEPARATIONS Prographics, Inc., Rockford, Illinois

PRINTING AND BINDING R.R. Donnelley, Willard, Ohio

TYPEFACES Hoefler & Frere-Jones's Archer, with
Monotype Rockwell Extra Bold and Martin Majoor's Scala Sans

TEXT STOCK 80# Somerset Matte

BINDING Smythe sewn and bound in ICG Kennett Linen cloth